The Theory of Critical Distances

The Theory of Critical Distances

A New Perspective in Fracture Mechanics

David Taylor

ELSEVIER

Amsterdam • Boston • Heidelberg • London • New york • Oxford
Paris • San Diego • San Francisco • Singapore • Sydney • Tokyo

Elsevier
The Boulevard, Langford Lane, Kidlington, Oxford, OX5 1GB, UK
Radarweg 29, PO Box 211, 1000 AE Amsterdam, The Netherlands

First edition 2007
Reprinted 2008

British Library Cataloguing in Publication Data
Taylor, David
The theory of critical distances : a new perspective in fracture mechanics
1. Fracture mechanics 2. Fracture mechanics-Mathematical models
I. Title
620.1'126

Library of Congress Catalog Number: 2006940584

ISBN: 978-0-08-044478-9

Printed and bound in the United Kingdom
Transferred to Digital Print 2010

'I have made a bouquet from other men's flowers, and the only thing that I have added of my own is the cord that binds them.'

Michel de Montaigne

'Thus, the task is not so much to see what no one has yet seen but to think what nobody has yet thought, about that which everybody sees.'

Erwin Schrodinger

Contents

Preface

In 1998, I had a moment of inspiration. I was thinking about the problem of predicting fatigue limits for specimens containing notches and short cracks. It was already known that some notches – the relatively sharp ones – behaved much the same as cracks, whilst other, blunter notches behaved quite differently. I realised that I could make accurate predictions for both types of notches if, instead of looking at the stresses at the notch root, I shifted the focus of my attention slightly, to a point nearby. By choosing a suitable distance away from the notch, and using the stresses at that point, I found that the behaviour of both blunt and sharp notches could be predicted. Not only that, but I shortly realised that the same approach, when applied to sharp cracks, would allow predictions to be made of the well-known 'short crack effect', whereby fatigue cracks of small size are found to grow much faster than predicted by linear elastic fracture mechanics (LEFM). Combining my original ideas with some LEFM concepts, I was able to predict a value for this critical distance from first principles, allowing predictions to be made *a priori*.

It is my custom, if I have a good idea, to jump in the air and click my heels together; that day I jumped so high I hit my head on the ceiling. I soon found, however, that I was not the first person to walk down this particular road. Indeed the same basic idea – using stress values within a material-dependant critical distance from the notch – had been proposed as early as the 1930s and was the basis for the notch sensitivity rules devised by Neuber, Peterson and others. Although I was aware of these rules – they are widely used in industry – I had not appreciated that they were based on critical distance ideas because the form in which they are normally presented obscures their origins. I also found that the link between critical distance theories of notch behaviour and fracture mechanics concepts for crack behaviour had previously been made by researchers as far back as the 1970s. Further research on my part showed that the same basic idea had been invented, several different times and quite independently, by workers not only in the area of fatigue but also in the field of brittle fracture in polymers and composites. This encouraged me to investigate the use of the method in areas which had not previously received much attention, especially brittle fracture in metals and ceramics, fretting fatigue and fatigue of polymers, and also to apply the method to problems in the design and failure analysis of engineering components. In the process, I came across several other theories which, whilst not exactly the same as 'my' critical distance theory, nevertheless contained some of the same elements.

Here then was a theory of material behaviour, a theory capable of predicting a range of different types of failure caused by cracking, arising in the stress fields created by notches, cracks and other stress-concentration features, a theory which had been invented and reinvented by different workers studying different problems. But it was a theory which had no name, a quiet, shy theory whose proponents were largely unaware of each others work. I decided to give this theory a name – the Theory of Critical Distances (TCD) – and to do what I could to develop and enhance its use and to make others aware of its existence. Hence this book, which is the first, but I hope not the last, to treat this topic.

It is interesting to contrast the development, or rather lack of development, of the TCD with that of another science which started at about the same time, and which we now call 'Linear Elastic Fracture Mechanics'. LEFM was born in the work of Griffith, beginning in the 1920s. It faced many difficulties and setbacks, for example its application to fracture and fatigue in metals was resisted on the grounds that it could not take account of plastic deformation at the crack tip. But over the decades LEFM developed into a large undertaking, used extensively in industry and the subject of many books and university courses. It developed because it was able to make predictions of experimental phenomena which people needed to know about, especially the growth rates of fatigue cracks, and to define quantitatively the important mechanical property of toughness. These successes provided the stimulus to develop a theoretical understanding which acted, retrospectively, as a justification for the theory.

The TCD, on the other hand, though it began almost as long ago, did not develop into a coherent science in the same way. Although it is used industrially in the form of certain empirical equations, and more explicitly by a few individuals, it has not received the same attention as LEFM and consequently has not developed the all-important theoretical foundations that would inspire confidence.

This book is a first attempt to redress the balance, to bring together in one volume everything that we know about the TCD. Here I will be advocating a particular approach, firmly grounded in continuum mechanics, which emphasises the links between the TCD and LEFM, and allows me to develop a justification for the TCD on theoretical grounds. This is by no means the only way to use critical distance concepts, and I will be discussing and comparing a variety of approaches advocated by other workers: indeed the current trend seems to be for the inclusion of some form of length constant in almost all theories of fracture.

The structure of this book is as follows. The first four chapters form an introduction to the TCD and to other theories used to predict material failure. The next six chapters examine different aspects in detail, covering brittle fracture in ceramics, polymers, metals and composites and also covering failures due to fatigue and contact problems such as fretting. In each of these chapters the basic idea is the same, to first demonstrate how the TCD can be used to predict experimental data and, having established its success and noted any shortcomings, to discuss these results in the light of the known mechanisms and theories of failure. Chapters 11 and 12 consider the complications that arise in multiaxial stress fields and in real engineering components, providing a number of case studies. Finally, in Chapter 13, findings from the previous chapters are brought together to consider the theoretical basis of the approach.

When writing a book it is useful to imagine who may read it. I have considered two different types of reader; the first is a researcher working in a university or large company, who is interested in understanding material failure at a fundamental level. The second is an engineering designer who requires a practical tool for predicting failure in real components and structures. These two readers will approach the book in different ways, focusing on different chapters, but I hope that both will find something useful.

Many people helped with this work. I would like especially to mention three individuals who made pivotal contributions: Luca Susmel, Pietro Cornetti and Danny Bellett – it has been a particular pleasure to work with you guys. Many others contributed to the work of my group through their research theses on critical distance concepts, including Wang Ge, Niall Barrett, Susanne Wiersma, Giuseppe Crupi and Saeid Kasiri. However, at the end of the day it was I who wrote the book and who therefore must be responsible for the errors and omissions which it surely contains.

I would like to finish with a big thank you to my wife, Niamh, without whom I could never do anything at all.

David Taylor
Dublin
April 2006

Nomenclature

What follows is the standard nomenclature which I have used throughout the book. Symbols different from these appear from time to time in cases where I have quoted from other authors, using their nomenclature; in these cases the symbols are explained as they appear.

a; Δa	Crack length; crack growth increment used in FFM
a_h	Hole radius
a_o	ElHaddad's constant used in the ICM and FFM
b	Exponent in the Weibull equation
B	Specimen thickness
CMM	The crack modelling method
d	Grain size
D	Notch depth
E	Young's modulus
EPFM	Elastic plastic fracture mechanics
f_c, f_n, f_p	Correction factors from tension to torsion for cracked, notched and plain specimens, respectively
F	Geometry constant used in the equation for K
G, G_c	Strain energy release rate during crack growth; its critical value for brittle fracture
HCF	High-cycle fatigue
FFM	Finite fracture mechanics
ICM	The imaginary crack method
J	The J integral used in fracture mechanics
K	Stress intensity factor
K_c	Critical stress intensity for brittle fracture (the fracture toughness); note that this symbol is used for all toughness values, irrespective of the degree of constraint, i.e. we do not use the convention whereby the plain-strain toughness is denoted K_{IC}
K_{cm}	Measured fracture toughness using a notch of finite root radius instead of a crack
K_f	Fatigue strength reduction factor for a notch
K_t	Stress concentration factor of a notch
L	The critical distance
LCF	Low-cycle fatigue

LEFM	Linear elastic fracture mechanics
LM	The line method
N	Number of cycles
N_f	Number of cycles to failure
NSIF	The notch stress intensity factor method
P_f	Probability of failure
PM	The point method
r	Distance measured from the point of maximum stress for any stress concentration feature
R	Stress ratio in cyclic loading (the ratio of minimum stress to maximum stress in the cycle).
RKR	The Ritchie Knott and Rice model
UTS	Ultimate tensile strength
W	Specimen width
W	Strain energy
α	Notch angle
Δa	Crack growth increment used in FFM
$\Delta\sigma$	Range of cyclic stress
$\Delta\sigma_o$	Fatigue strength of a plain specimen
$\Delta\sigma_{on}$	Fatigue strength of a notched specimen (nominal stress)
ΔK	Range of cyclic stress intensity
ΔK_{th}	Fatigue crack propagation threshold
λ	Exponent used in describing the stress field of a sharp V-shaped notch
θ	Angle defining the path along which r is measured
ρ	Root radius of a notch or other stress concentration feature
σ	Stress; unless otherwise specified this is the nominal stress when applied to a notched or cracked specimen
σ_f	Fracture stress: the nominal stress to cause brittle fracture in a cracked body
σ_o	Characteristic stress used with the PM and LM
σ^*	Material constant used in the Weibull equation
σ_u	Tensile strength
σ_y	Yield strength
$\sigma(r)$	Stress as a function of distance r
τ	Shear stress
Ψ	Notch stress intensity factor (NSIF)

CHAPTER 1

Introduction

Materials Under Stress

Fig. 1.1. Examples of ductile fracture (left) and brittle fracture (right) in bolts (Wulpi, 1985).

It is assumed that the reader is familiar with some basic theory regarding the mechanical properties of materials, as can be found in textbooks such as Ashby and Jones' *Engineering Materials* (2005) or Hertzberg's *Deformation and Fracture Mechanics of Engineering Materials* (1995), and also with the fundamentals of solid mechanics and fracture mechanics, for which many useful textbooks also exist (Broberg, 1999; Janssen et al., 2002; Knott, 1973). Nevertheless, in this chapter we will briefly review the background material and introduce symbols and terminology, which will be used in the rest of the book. We will be concerned, in general, with the deformation and failure of materials under stress, but emphasis will be placed on those types of failure which will be the main subjects of the book, especially brittle fracture and fatigue, but also including ductile fracture and certain tribological failure modes such as fretting fatigue. Of special interest from a mechanics point of view will be cracks, notches and other combinations of geometry and loading, which give rise to stress concentrations and stress gradients. In this respect, the use of computer-based methods such as finite element analysis (FEA) will also be discussed. We will finish with critical appraisal of the use of traditional fracture mechanics and solid mechanics in failure prediction, setting the scene for the developments to be described in the rest of this book.

1.1 Stress–Strain Curves

A fundamental way to obtain information about the mechanical properties of a material is to record its stress–strain curve, usually by applying a gradually increasing tensile strain to a specimen of constant cross section. Figure 1.2 shows, in schematic form, some typical results; note that here we are plotting the true stress (σ) and true strain (ε), thus taking account of changes in specimen cross section and length during the test. Most materials display a region of linear, elastic behaviour at low strains, and in some cases (line 1) this continues all the way to failure. This is the behaviour of classic brittle materials such as glass and certain engineering ceramics. More commonly, some deviation from linearity occurs before final failure (line 2). This non-linearity has three different sources: (i) non-linear elasticity, which is common in polymers; (ii) plasticity, that is the creation of permanent deformation, which occurs principally in metals and; (iii) damage, which is important in ceramics and composite materials. We will define the stress at failure in all cases as the maximum point in the curve, and refer to it as σ_u or the Ultimate Tensile Strength (UTS). In some cases (line 3) complete separation does not occur at σ_u, rather some reduced load-bearing capacity is maintained. This happens when damage such as splitting and cracking becomes widespread, for example in fibre composites. Finally, some stress–strain curves display other features (line 4) such as a drop in stress after yielding (in some metals and polymers) and a long post-yield plateau terminating in a rapid upturn in stress just before failure: this occurs in polymers which display plastic stability due to molecular rearrangements.

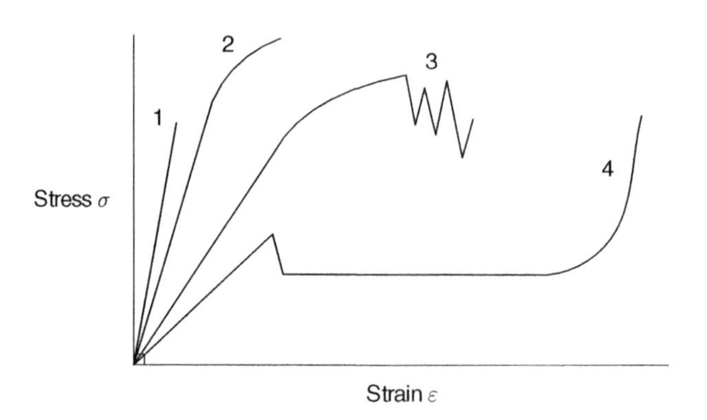

Fig. 1.2. Some typical stress–strain curves.

1.2 Failure Mechanisms

1.2.1 Failure at the atomic level

The study of failure mechanisms in materials has a tendency to get complicated, so it is worth remembering that, at the smallest scale, there are only two mechanisms by which materials can break, which I will call *cleavage* and *tearing*. Cleavage involves the fracture of atomic bonds; a crack can form by breaking the bonds linking two parallel planes of atoms, and this crack can then grow by the fracture of successive bonds near the crack tip, essentially unzipping the material in directions corresponding to atomic lattice planes. The fracture surface consists of a series of flat facets corresponding to the grains of the material. Tearing, on the other hand, occurs when material separates due to plastic deformation: atoms move around to create high levels of strain so that the material literally tears itself apart. This can manifest itself in various different ways, from macroscopic thinning (necking) or sliding (shearing) of material to microscopic void formation and growth. These two atomic failure mechanisms are often referred to as 'brittle' and 'ductile'; however, I have avoided using these terms because they are also used with different meanings to describe failure modes at the macroscopic scale as discussed below.

1.2.2 Failure modes in engineering components

The failures of engineering components and structures occur by one of seven different modes: elastic, ductile, brittle, fatigue, stress-corrosion, creep, and wear.

Elastic failures are those failures which occur as a result of a low value of Young's modulus, E. Two types of elastic failure can be mentioned. The first is excessive deflection, which may prevent the correct functioning of a structure – examples include bridges and vehicle suspensions. The second is buckling, by which, at a certain critical combination of load and elastic modulus, the deflections of a structure become unstable so that small deviations become magnified. A classic example is the collapse of a thin column loaded in compression.

Ductile fracture is the term used to describe failure occurring due to macroscopic plastic deformation; the material's yield strength is exceeded over a large region so that plastic strain can occur throughout the load-bearing section, causing either fracture or a major change in shape so that the component can no longer function. In principle the prediction of this type of failure is simple, since the only consideration is that the stress in the part should exceed the yield strength. In practice, however, the spread of plasticity and the resulting redistribution of stresses and strains makes the prediction of plastic collapse loads a difficult analytical problem. For complex engineering structures, solutions are usually obtained using FEA and other computer simulations.

Brittle fracture refers to failures which occur as a result of rapid crack propagation. The crack in question may already exist (for example, in the form of a manufacturing defect or slowly growing fatigue crack) or it may form as a result of locally high stresses, for example near a notch. Once formed, the crack is able to grow, if the applied loads are high enough, by fracture of material near its tip. This material may fail by either cleavage or tearing. In classic brittle fracture, the process of crack growth is unstable, leading to almost instantaneous failure of the component. In such situations any plastic deformation is confined to the immediate vicinity of the crack, so there may be little sign of macroscopic plasticity. Figure 1.1 illustrates the difference between ductile and brittle failures in bolts tested in tension. This simple distinction between brittle and ductile fracture is complicated by the fact that intermediate situations can often arise: crack growth can occur more slowly and gradually, requiring a monotonically increasing load, if there is a significant amount of plasticity or damage near the crack tip. The study of crack propagation has created the science of Fracture Mechanics, which will be discussed in more detail below.

Fatigue is a process of crack initiation and growth, which occurs as a result of cyclic loading. A regular cycle of stress, such as a sine wave (Fig. 1.3), can be described using two parameters: the stress range $\Delta\sigma$ and the mean stress σ_{mean}. Another common descriptor is the load ratio R, defined as the ratio of the minimum and maximum stresses in the cycle:

$$R = \frac{\sigma_{min}}{\sigma_{max}} \tag{1.1}$$

The most common type of fatigue test involves applying a cyclic stress to a test specimen and counting the number of cycles to failure N_f. Separation will occur when a crack has grown to a sufficient length to cause a ductile or brittle fracture of the remaining cross section: some workers prefer to define failure as the creation of a crack of a specified size, usually a few millimetres. Figure 1.4 shows typical stress-life curves, describing the dependence of N_f on $\Delta\sigma$ and σ_{mean}. In some materials the curve becomes effectively horizontal for N_f values in the range 10^6–10^7 cycles, allowing one to define a *fatigue limit*, $\Delta\sigma_o$; often, however, there is no clear asymptote in which case the fatigue limit is defined at a specified number of cycles, when it is often called the *fatigue strength*. Recent work, which will be discussed further in Chapter 9, has shown that this asymptote can be somewhat illusory: in some materials failures can occur at very large numbers of cycles, in excess of 10^9, at low values of stress range. Changing the mean stress or R ratio will shift the entire curve. If the applied stress is high enough to cause large-scale plastic deformation on every cycle, then a non-linear stress–strain relationship will occur, the nature of which may change during cycling as the material hardens or softens. In such cases it is common practice to use the strain range $\Delta\varepsilon$ as the characterising parameter, instead of $\Delta\sigma$. In this situation the number of cycles to failure is generally

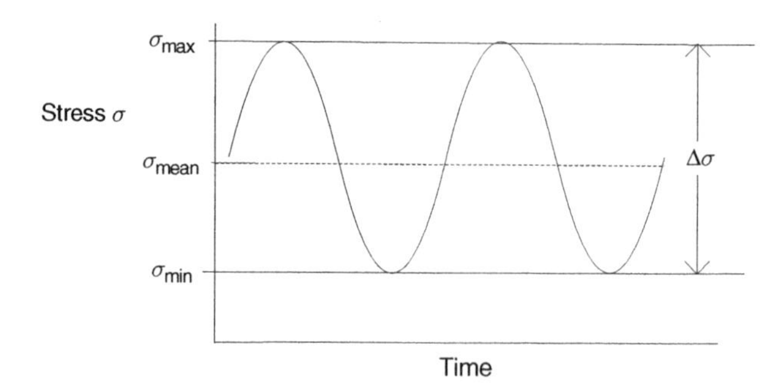

Fig. 1.3. Definition of parameters for cyclic loading.

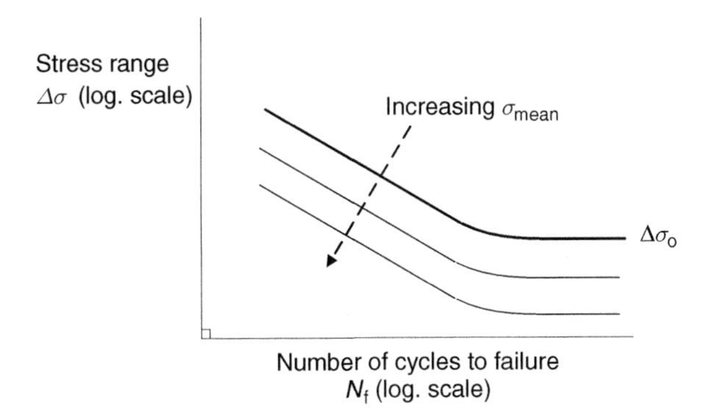

Fig. 1.4. Typical fatigue stress-life curves.

low: this type of fatigue is referred to as low-cycle fatigue (LCF) to distinguish it from high-cycle fatigue (HCF) which occurs under nominally elastic conditions.

In real engineering components a crack may already exist, in which case one is interested in how fast it is propagating. The crack growth rate is usually expressed in terms of number of cycles, da/dN, rather than time, because normally the amount of crack growth per cycle is rather insensitive to the cycling frequency; there are, however, important exceptions to this rule, especially among polymers. The crack growth rate has been found to be a function of ΔK, the range of stress intensity, K, which will be defined below in Section 1.4.

Figure 1.5 shows the typical dependence of crack growth rate on stress intensity range, which displays two asymptotes: a growth threshold, ΔK_{th}, below which crack growth effectively ceases, and an upper limit where the conditions for rapid, brittle fracture are approached. Changing the mean K, or R ratio, shifts the curve as shown.

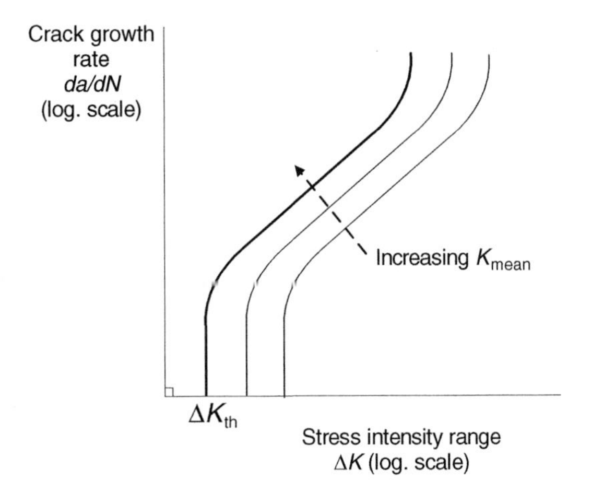

Fig. 1.5. Typical fatigue crack growth rate curves.

Stress-corrosion cracking (SCC) is a form of gradual failure which is rather like fatigue in that it proceeds by crack initiation and gradual propagation. However, in this case the crucial factor is not a cyclic stress but the existence of a corrosive chemical environment. The mechanisms of SCC are many and varied but usually involve some kind of synergistic action between the chemical process and the applied stress. This type of failure will not be discussed in any detail in this book; it is quite likely that the Theory of Critical Distances (TCD) could be used to predict failures that occur by SCC, but to date this has not been investigated.

Creep is a process of plastic deformation that occurs gradually. In fact all plastic deformation processes are thermally activated, proceeding more easily as the temperature is increased towards the material's melting point. Creep failures can also involve the creation and growth of crack-like damage. Critical distance methods have been used to study creep (see Section 9.7) but not in enough detail to merit discussion in this book.

Wear is the general name given to tribological failures, that is to failures which occur due to the rubbing action between two surfaces. If compressive stress and a sliding (shear motion) occur across a material interface, then very high local stresses will arise due to small surface irregularities, creating conditions in which material can be removed from one or both surfaces. There are various mechanisms of wear: the one that will be of most interest to us is known as *contact fatigue* and involves the creation of cracks at or near the point of contact. These cracks can grow to cause removal of surface material by spalling, for example in gear teeth. If there are also cyclic body forces in the component, then cracks which are initiated by contact fatigue can subsequently grow into the component by conventional fatigue processes: this type of failure is known as *fretting fatigue*. The prediction of tribologically induced failures such as these is difficult owing to the problems involved in estimating local stresses, which are affected by surface roughness, surface deformation and lubrication.

1.3 Stress Concentrations

It is almost inevitable that, in any engineering component, stresses will vary from place to place, and that failure will occur in locations where stresses are relatively high. One can think of a few exceptions to this rule – wires and tie-bars under pure tensile loading, for example – but apart from these we can say that the phenomenon of stress concentration is responsible for all mechanical failures in practice. Stress concentration has two causes: loading and geometry. Loading modes which cause stress gradients include bending and torsion, both of which tend to concentrate stresses at the surface. But, as we shall see, this type of stress concentration is generally very mild in comparison to the effect of geometric features such as holes, corners, bends and grooves.

To illustrate the magnitude of stress concentrations, it is useful to consider a specific example. Consider a rectangular bar of material of width 30 mm – the length and thickness of the bar are not important. If loaded in tension with a stress of 100 MPa the stress at any point in the bar will, of course, be the same. If we introduce a central hole, of radius 3 mm, then the stress will become much higher at two points (the 'hot spots') on the circumference of the hole. In fact the local stress will be approximately

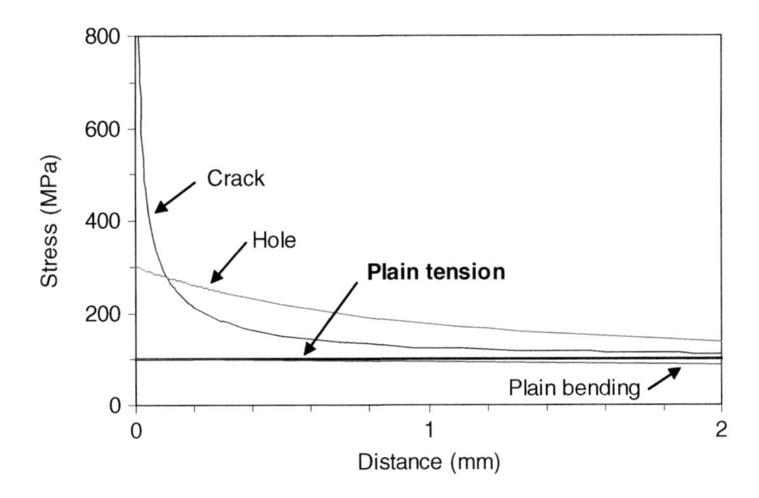

Fig. 1.6. Examples of stress–distance curves showing the effect of geometric features (a crack and a circular hole of the same size) and of bending loads. In all cases the nominal stress is 100 MPa; stresses are plotted as a function of distance from the point of maximum stress.

300 MPa, because the stress concentration factor K_t for a hole is 3 (actually this is the K_t factor for a hole in a body of infinite width, but it is close enough for our present purposes). Figure 1.6 shows how the stress decreases with distance from the hole, along a line drawn perpendicular to the hole surface (and therefore also perpendicular to the axis of principal stress). Also shown is the stress distribution that would arise if we were to replace the hole with a crack of the same size (i.e. one whose half-length a is equal to the radius of the hole). It is important to note that the stress being plotted here is the elastic stress, that is we have calculated the stress assuming that there will be no yielding or other non-linear deformation behaviour in the material. This is an assumption that we will return to later on. Finally, the graph also shows the stress distribution that will occur if, instead of introducing a hole or crack, we subject the bar to pure bending with the same nominal stress; in this case the graph shows stress as a function of distance from the surface of the bar at which the maximum tensile stress occurs.

It is clear that the stress concentration effect due to bending is much smaller than that caused by either of the geometric features. Of course, the gradient of stress in bending will increase if we decrease the width of the bar, but we would have to make the bar very narrow indeed to create the stress gradients caused by the hole and the crack. The maximum stress due to the crack is theoretically infinite because, having zero radius at its tip, it creates a singularity in the stress field. It is interesting to note that, whilst the crack causes much higher stresses in its immediate vicinity, the stresses due to the hole are actually higher than those for the crack at larger distances. We shall see later that this observation turns out to be very significant. In order to understand the effect of geometric features on mechanical failure, it is necessary to consider not only the maximum stresses which they create, but also how these stresses change with distance. Indeed the majority of the analyses conducted in this book will make use of these elastic stress–distance curves.

1.4 Elastic Stress Fields for Notches and Cracks

The study of stress concentration effects is mostly carried out using notches. As Fig. 1.7 shows, a notch can be defined by three parameters: its depth D, root radius ρ and opening angle α. To be precise one should add a fourth feature, the notch shape, to include the fact that the sides of the notch can have different amounts of curvature. However, in practice the two features which mostly control stress concentration are D and ρ, with notch angle having a secondary effect which becomes significant at large values ($\alpha > 90°$). In considering notch stress fields we will normally use coordinates centred on the point of maximum stress, at the notch root: Fig. 1.7 shows a polar coordinate system (r, θ).

The reason that researchers use notches to study stress concentration effects is because they are relatively simple to make and test experimentally and to analyse theoretically. Much of the work described in this book (especially in Chapters 5–9 and 11) will be concerned with the effect of notches. However, it should not be forgotten that our real purpose in doing all this is to predict the behaviour of stress concentration features in engineering structures and components, which can be geometrically much more complex. For this reason we will consider some components in Chapter 12, along with features such as corners and joints, and in Chapter 10 we will consider stress concentrations which arise due to contact, causing fretting fatigue and other tribological failures. Defects such as porosity and inclusions also fall into the general category of stress concentrations.

To return to notches, some simple analytical solutions exist in certain cases. For example, the stress field created by a circular hole in a body of infinite size can be described as a function of applied nominal stress σ and hole radius a_h. For the case of $\theta = 0$ the result is

$$\begin{aligned} \sigma_{\theta\theta} &= \sigma\left(1+\frac{1}{2}\left(\frac{a_h}{r+a_h}\right)^2+\frac{3}{2}\left(\frac{a_h}{r+a_h}\right)^4\right) \\ \sigma_{rr} &= \frac{3\sigma}{2}\left(\left(\frac{a_h}{r+a_h}\right)^2+\left(\frac{a_h}{r+a_h}\right)^4\right) \end{aligned} \tag{1.2}$$

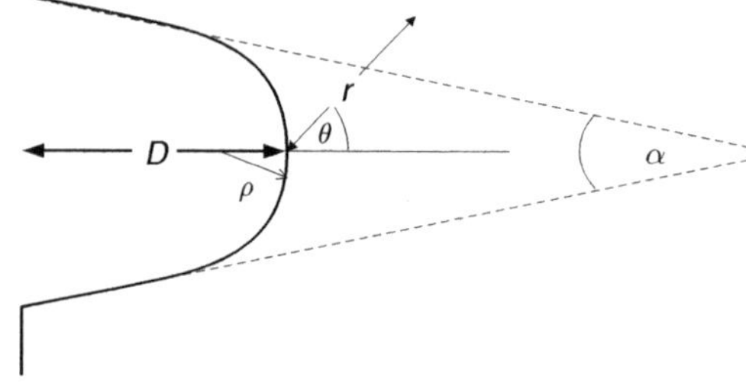

Fig. 1.7. Definition of parameters for notch geometry: length D, root radius ρ and angle α. Stresses are normally defined using cylindrical coordinates (r, θ) centred on the point of maximum stress at the notch root.

Here $\sigma_{\theta\theta}$ is the tensile stress in the circumferential direction and σ_{rr} the tensile stress in the radial direction. This is a two-dimensional (2D) analysis – stresses can also arise in the thickness (z) direction, depending on the degree of constraint: this will be considered below in the section on Fracture Mechanics. Note that for this hole, as for any notch, stresses arise not only in the direction of the applied tension, but also in other directions, creating in general a complex three-dimensional (3D) stress field. The effects of these other stresses can be important, and will be considered in detail in Chapter 11 on multiaxial loading.

The maximum value of $\sigma_{\theta\theta}$ (occurring at $r = 0$) is 3σ, giving a stress concentration factor of $K_t = 3$, as mentioned above. Stress concentration factors – determined by analytical solution, computer simulation or experimental stress analysis – have been recorded for many different types of notches and other features (see, for example, Peterson, 1974). One useful result, which is strictly valid only for elliptical holes but is reasonably accurate for most notches, gives K_t as a function of notch depth D (equal to the length of the semi-major axis of the ellipse) and root radius ρ (defined at the point of minimum radius) as:

$$K_t = 1 + 2\sqrt{\frac{D}{\rho}} \tag{1.3}$$

Creager and Paris developed a simple equation to describe the stress–distance curve ahead of a narrow slot, that is a notch in which $\rho << D$ (Creager and Paris, 1967), which we will make considerable use of in later chapters (see, for example, Eq. 5.10 in Chapter 5). In the limit where $\rho = 0$ we have a crack, and in this case also the stress field can be predicted analytically. The K_t factor becomes infinity; the result for stress $\sigma(r)$ as a function of distance r from the crack tip, for a through-thickness crack of half-length a in an infinite body subjected to tensile stress σ, is (Westergaard, 1939)

$$\sigma(r) = \frac{\sigma}{\left[1 - \left(\frac{a}{a+r}\right)^2\right]^{1/2}} \tag{1.4}$$

Here $\sigma(r)$ is the tensile stress in the same direction as the applied stress: this is also the direction perpendicular to the applied crack faces, so this stress is often referred to as the crack-opening stress and is usually the most important stress controlling crack propagation. For points close to the crack tip (i.e. $r << a$), Eq. (1.4) reduces to a simpler form, thus:

$$\sigma(r) = \sigma\sqrt{\frac{a}{2r}} \tag{1.5}$$

Combining the stress and the crack length we can define the stress intensity K as:

$$K = \sigma\sqrt{\pi a} \tag{1.6}$$

The convenience of this definition, and the reason for the insertion of the constant π will be explained in the following section. Now the stress field depends only on K and a: this result is precisely true only for the particular geometry of an infinite body containing a straight, through-thickness crack; however, it turns out that, for many other cases, Eq. (1.6) retains its same general form with the inclusion of a constant F giving:

$$K = F\sigma\sqrt{\pi a} \tag{1.7}$$

Here F is a function of various parameters including crack shape and location and the type of loading. Values for F have been calibrated for many cases of interest (e.g. Murakami, 1987).

For notches in which $\rho = 0$ but $\theta > 0$ (sharp, V-shaped notches), Eq. (1.5) retains the same general form, but the dependence on r changes, thus:

$$\sigma(r) = \Psi r^{-\lambda} \tag{1.8}$$

Here ψ has the same meaning as K except that the square root is replaced by the exponent $(-\lambda)$, whose value is a function of θ (Williams, 1952). As a general problem, the full analytical description of stress fields for notches, especially in bodies of finite width, presents significant challenges: useful solutions have been obtained for various cases (Atzori et al., 2001; Filippi and Lazzarin, 2004). In practice, closed-form solutions cannot be determined for most of the stress concentration features which exist in components, but fortunately this information can now be obtained using computer simulations such as FEA.

1.4.1 Stress fields at the microstructural level

It is important to remember another assumption of the above analyses, which is that the material behaves as a homogeneous continuum. In practice, of course, materials are not continuous, a fact which had been suspected since the time of the Greek philosopher Leucippus (fifth century BC), who first proposed that material is made up of atoms. Atomic structure is of course important, but for most materials, properties such as strength and toughness are strongly affected by behaviour at the microstructural level, where features such as grains, precipitates and inclusions exert both positive and negative effects. A fact which is often overlooked is that if we examine stress and strain fields at this small scale, we find that they are strongly inhomogeneous, affected by microstructural parameters such as local grain orientation, disparities in the elastic stiffness of different phases, and the properties of grain boundaries and other interfaces. Experimental measurements (Delaire et al., 2000) and computer models (Bruckner-Foit et al., 2004) have revealed the large extent of these local variations in stress and strain, which can be as high as a factor of 10.

These effects may be of relatively little importance if the scale of the fracture process is large – for example, if the size of the plastic zone (see Section 1.5.2) is much larger than any microstructural feature, in which case it may be satisfactory to think of the stresses calculated by continuum analysis as average quantities, ignoring their local variations. However, the fact is that many failure processes happen on the microstructural scale. For example, the sizes of zones of plasticity and damage during the fracture of brittle materials and the HCF of metals are generally the same as the sizes of grains and other components of the microstructure. Under these circumstances it is rather remarkable that we can make meaningful predictions of failure using continuum mechanics theory. This implies that, at least under some circumstances, we will need to modify continuum mechanics to take account of crucial length scales in a material; this is the main subject of this book.

1.5 Fracture Mechanics

Fracture mechanics – the science which describes the behaviour of bodies containing cracks – is one of the most important developments in the entire field of mechanics. The great success of fracture mechanics has been to show that, under certain well-defined conditions, the propagation of the crack can be predicted using some very simple linear elastic analysis. When these conditions prevail, we are in the realm of Linear Elastic Fracture Mechanics (LEFM). We will first describe the basic theory of LEFM, leaving discussion of its limitations and assumptions for later. What follows is necessarily only a brief outline: for more detailed treatment the reader is referred to some of the excellent books which have been written on this subject (Broberg, 1999; Janssen et al., 2002; Knott, 1973).

We can predict the conditions necessary for brittle fracture, and also for slow crack growth by fatigue and stress-corrosion cracking, assuming that a crack already exists. This is much simpler in the case of a crack than in the more general case of a notch, because all these fracture modes involve a cracking process: if a crack is not present then it will have to be created during the failure. If the crack is already there, on the other hand, we merely have to consider its propagation. Propagation can be defined as any increase in crack length, δa; if we consider the limit in which δa is vanishingly small, then we can assume no significant change in the stress conditions near the crack tip during propagation. We say that the crack extends under *steady state* conditions. A further simplifying assumption is that crack growth is under *local control*, by which we mean that the criteria for propagation can be entirely determined by stress conditions in the immediate vicinity of the crack tip. The opposite of local control is *global control*, which implies that other aspects, such as for example the type of remote loading being applied, influence crack behaviour.

Within these limitations, the behaviour of the crack can be described using the parameter K, the stress intensity, which was defined in the previous section where we saw that it uniquely determines the magnitude of the stress field in the vicinity of a crack. The argument goes that two different cracks (e.g. cracks of different length and shape in different bodies) will have the same stress fields if K is the same for both; therefore if one crack can propagate, then so can the other. As pointed out in the previous section, this only applies to the stress fields close to the crack tip ($r << a$), hence the assumption of local control.

An alternative, and rather more persuasive, argument for the uniqueness of the K parameter is a thermodynamic one first formulated by Griffith and further developed by Irwin (1964). This is a virtual work argument, in which we imagine a small amount of crack extension and compute the energy changes which occur. The problem can be simplified by assuming a so-called 'fixed grips' type of loading, in which the cracked specimen is held tightly between two loading grips which do not move during the experiment, so that there is no external work done on the specimen. Griffith proposed that the energy necessary for crack extension was equal to the energy needed to create the new surfaces, thus (for a through-thickness crack in a specimen of unit thickness) this is simply equal to $2\gamma(\delta a)$, where γ is the surface energy and the factor 2 arises because two surfaces are being created. In fact, though this is an accurate estimate in the case of certain very brittle materials such as glass, crack propagation in most other materials requires more energy, due to various toughening mechanisms which operate in front of or behind the crack tip (see Section 1.6); we can lump these together to define a general crack-propagation term G_c, so that the energy for crack extension becomes $G_c(\delta a)$.

The energy which is available to drive crack propagation, in the absence of any external work, is the elastic energy released when the crack grows. This can be visualised as the energy released when atomic bonds near the crack tip are broken and, more importantly, when the strains in the surrounding atomic bonds are reduced. The decrease in elastic energy, δW (per unit thickness), accompanying crack extension δa, can be shown to be:

$$\delta W = \frac{\sigma^2}{E} \pi a \, \delta a \tag{1.9}$$

Equating this to $G_c(\delta a)$, we can find the stress needed for brittle fracture, that is the stress at which there will be just enough elastic energy stored in the body to drive crack propagation. This is the brittle fracture strength, σ_f; the result is

$$\sigma_f = \sqrt{\frac{G_c E}{\pi a}} \tag{1.10}$$

We note that fracture strength depends only on crack length and two material parameters, G_c and E; combining these we can rewrite the equation as:

$$\sigma_f = \frac{K_c}{\sqrt{\pi a}} \tag{1.11}$$

where K_c is defined as:

$$K_c = \sqrt{G_c E} \tag{1.12}$$

Equation (1.11) is exactly the same as Eq. (1.6) except that Eq. (1.6) defines the general parameters σ and K whilst Eq. (1.11) defines their critical values σ_f and K_c, the latter being a material constant known as the *fracture toughness*. Just as Eq. (1.6) can be generalised into Eq. (1.7), so can Eq. (1.11) be converted for use in any arbitrary geometry of cracked body, provided the appropriate F factor has been determined.

We saw above that this parameter K can also be used to describe crack growth in fatigue. Here we use the range of stress intensity, ΔK, defined as:

$$\Delta K = F \Delta\sigma \sqrt{\pi a} \tag{1.13}$$

As Fig. 1.5 showed, the crack growth rate (for a given R) is a function of ΔK and R. At values of K in the mid-range, the following equation (Paris, 1964) applies:

$$\frac{da}{dN} = A(\Delta K)^n \tag{1.14}$$

Here A and n are material constants for a given R. At low values, the line curves down to a threshold ΔK_{th} below which crack growth is negligible. Similar dependencies can also be defined for stress-corrosion cracking.

The above calculations all assumed that the crack was being loaded by a tensile stress applied perpendicular to its faces. This is certainly the most important case: compressive stresses, or tensile stresses applied in orthogonal directions (parallel to the crack faces or in the through-thickness direction) do not generally have any effect because they do not cause stress concentration, though exceptions can occur in anisotropic materials.

However, local stress fields (and therefore, potentially, crack propagation) can occur due to shear loadings, applied parallel to the crack faces, in one of two orthogonal directions. Figure 1.8 illustrates the three important types of loading: simple tension (which is referred to as mode I), in-plane shear (mode II) and out-of-plane shear (mode III). Multiaxial loading, mixtures of these three modes, will be considered specifically in Chapter 11 and will arise in the context of contact problems (Chapter 10) and component failure (Chapter 12).

1.5.1 The effect of constraint on fracture toughness

The above analysis assumed a body of constant thickness B but did not consider any particular values for that thickness. If B is small, plane stress conditions will occur, in which the through-thickness stress σ_z is zero. In thicker specimens, however, material near the crack tip in the centre of the specimen will experience plane-strain conditions, in which σ_z is finite and varies with r. The net effect of this, especially for metals

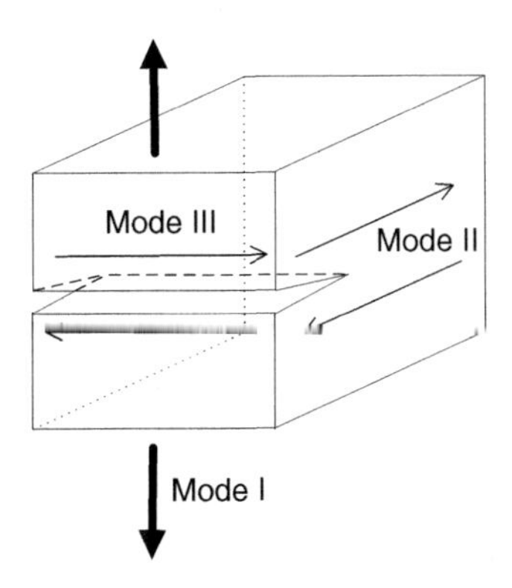

Fig. 1.8. Definition of loading modes applied to a crack: mode I (tension); mode II (in-plane shear) and mode III (out-of-plane shear).

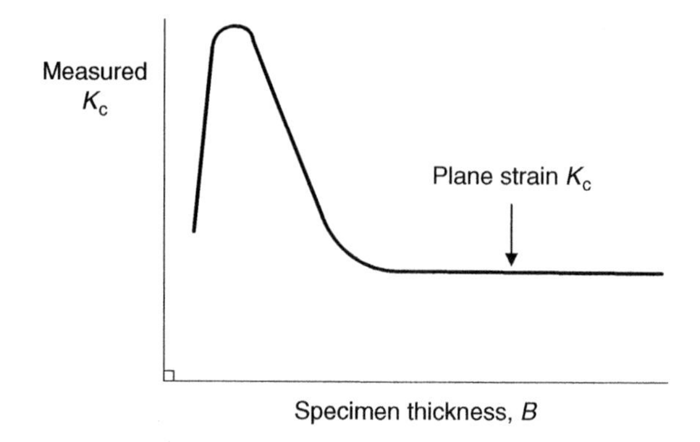

Fig. 1.9. Typical variation of measured K_c with specimen thickness B, illustrating the effect of out-of-plane constraint.

and other materials which develop plastic zones (see Section 1.5.4 below) is that crack propagation is easier, and therefore K_c is lower, when plane strain is present. Figure 1.9 shows the typical variation of measured K_c with thickness: the plane-strain toughness can be measured, provided B is sufficiently large. For thin specimens, the picture is more complicated: a reliable value of K_c is difficult to measure due to out-of-plane forces, and since it is certainly larger, one is more likely to encounter other limitations arising from the size of the plastic zone (see below). For these reasons the plane-strain fracture toughness is the value which is generally measured and quoted: there is a convention by which this is distinguished using the symbol K_{IC}, however this convention will not be used in this book; we will refer to the fracture toughness, however measured, as K_c.

However, the problem just described is actually only one, relatively simple, case of the more general problem of constraint. The change in stress pattern in thick specimens due to finite values of σ_z is known as 'out-of-plane constraint'; in fact, stresses can also arise in the in-plane directions, which we can refer to as σ_x and σ_y, these being directions parallel and perpendicular to the crack direction, respectively. We noted above that the stress field described by Eq. (1.6), which is a necessary form for the definition of K, is a simplification of the true, 3D stress field. It turns out to be sufficiently accurate, provided failure occurs at relatively low applied stresses, but otherwise it ignores stress terms which become significant and which, like σ_z, have real effects on the strength of the material. The problem of constraint will be returned to later, in Chapters 6 and 7.

1.5.2 Non-linear behaviour: Plasticity and damage zones

The stress fields predicted in Section 1.4 often do not occur in practice because when stresses become locally very high, a number of other factors come into play. There are essentially three mechanisms, which modify stresses near the tip of a crack or notch. First, elastic behaviour may become non-linear: this is generally not taken into account, though it may have significant modifying effects, especially in certain polymers and ceramics. Secondly, yielding may occur, creating a plastic zone. Thirdly, the high stresses may cause damage, for example in the form of microscopic cracks or delaminations, creating a damage zone. The term 'process zone' is sometimes used as a general term to describe the region near the stress concentration feature in which any of these non-linear processes are occurring (though some workers use this term to mean only the zone in which fracture is occurring).

The effect of these non-linear processes is to reduce peak stress in any situation where there is a stress gradient, including plain beams in bending or torsion as well as stress concentration features. The details of the stress field inside the plastic zone or damage zone are difficult to estimate, since they depend on the precise mechanisms which are operating and how these mechanisms are affected by the 3D stress field. For example, stresses rise considerably higher in a plastic zone which is subjected to high constraint, due to suppression of yielding, because yielding is controlled by shear stress and thus by differences between the three principal stresses. This phenomenon is the basis for the effect of constraint on K_c that was mentioned above.

Failure, when it occurs, is invariably initiated within the zone of plasticity or damage. An existing crack may extend, or a crack may form at the root of an existing notch; alternatively cracks may form elsewhere in the process zone and link back to the main crack or notch. In many cases the detailed processes of failure at the microscopic level are still poorly understood. What is clear, however, is that materials which have high toughness invariably form large zones of plasticity or damage before failure. But if the failure process always involves these highly non-linear mechanisms, how is it that a simple linear-elastic theory such as LEFM can be used? This is a question that theoreticians have struggled with for some time. The justification for using LEFM is generally explained as follows: provided the non-linear zone is small compared to the dimensions of the body – that is provided the surrounding linear-elastic zone is much larger than the process zone – then conditions of stress and strain inside the non-linear zone, though they may be poorly known, are nevertheless uniquely characterised by conditions within the linear zone. This statement is much easier to make than it is to prove – for the interested reader I feel that Broberg, in his recent book, probably comes closest to a theoretical proof (Broberg, 1999). Most readers will be more convinced by the experimental evidence which shows overwhelmingly that, provided this so-called 'small-scale yielding' criterion is obeyed, the brittle fracture strength and HCF strength of specimens containing cracks can be accurately predicted using the stress-intensity parameter, K. There are, in addition, some other limitations to the successful use of LEFM, for example the crack length must also be large compared to the plastic zone size. These issues are covered in detailed testing standards which have been developed by various national and international bodies.

In addition to the processes which occur in front of the crack tip, some mechanisms operate behind the crack tip, in the region which is referred to as the crack wake. Here we find the remains of the crack-tip plastic zone, in which there are often significant residual stresses. These residual stresses can affect subsequent crack propagation, especially in fatigue where they alter the level of crack closure (see Section 9.1.2). In materials which do not display much plasticity there are a variety of crack-wake mechanisms which may act to improve toughness, such as bridging of the crack faces by fibres or unbroken ligaments of material. This is one reason why short cracks – in which these mechanisms have not had space to develop – may show different behaviour from long cracks.

1.5.3 Elastic–plastic fracture mechanics

The most unfortunate thing about LEFM is that it cannot be applied to many of the practical situations for which we would really like to use it, namely to predict fracture in components made from tough materials such as metals and composites. Most components made from these materials sustain large zones of plasticity or damage before failure, thus violating the small-scale yielding criterion. Notable exceptions are ferritic steels at low temperatures which fail by cleavage at low stress intensity, and some structures in which exceptionally long cracks may occur, such as pressure vessels or pipelines.

This problem has been addressed by the development of modified forms of fracture mechanics. These innovations have occurred particularly in relation to metallic materials, where they are known by the general heading of elastic–plastic fracture mechanics (EPFM). A number of parameters have been developed to replace K, notably the crack-opening displacement δ and the J integral. A particular problem here is that, when the conditions for LEFM are violated, this is often accompanied by a change in fracture mechanism. For example, brittle fracture, which classically involves sudden, unstable crack propagation, may, under conditions of increased plasticity, change into a process of gradual, stable crack extension, the amount of crack growth gradually increasing with applied load. This stable crack growth may continue indefinitely, or may become unstable at some critical load. In some cases the location of cracking may shift from the main crack to the centre of the specimen, where higher levels of constraint occur. Regarding the mechanics of the situation, the presence of large-scale yielding usually implies a loss of local control of the fracture process, so that the nature of the external loading (for example, whether the body is under load control or displacement control) will now have an effect.

Materials such as ceramics, which achieve toughness by the development of damage zones and crack-wake mechanisms rather than by plasticity, may not display such dramatic changes in failure mechanism, but their mechanical properties may be significantly different in cases where the scale of damage becomes large in comparison to the size of the specimen. A good example is concrete, whose measured strength and toughness are strongly affected by specimen size. Such materials are referred to as *quasi-brittle* to indicate that, whilst the mechanism of failure may be brittle, LEFM conditions may still not occur.

A detailed treatment of EPFM is beyond the scope of this book: a very readable introduction to this subject is provided by Janssen et al. (2002). The current situation is that parameters such as δ and J, whilst being useful measurements of a material's toughness and thus allowing materials to be ranked and compared with each other, are of very limited use when it comes to predicting failure in an industrial component.

1.6 The Failure of Notched Specimens

The introduction of a notch into a specimen creates conditions which are intermediate between those of a plain specimen and one containing a sharp crack. The behaviour of these specimens will be a major preoccupation in this book. To summarise the situation very briefly, we may say that some notched specimens behave in a manner similar to that of plain specimens, once the stress-concentration factor is taken into account. In these cases, failure occurs (either under monotonic loading or cyclic loading) when the local notch-root stress reaches the plain-specimen strength (σ_u or $\Delta\sigma_o$). At the other extreme, some notches behave exactly like cracks of the same length: provided the notch-root radius ρ is sufficiently small we may expect failure at $K = K_c$ (or, in cyclic loading, $\Delta K = \Delta K_{th}$).

Unfortunately, many notches do not conform to either of these extreme cases: at failure the notch-root stress is often greater than σ_u, and K is often greater than K_c; thus the notch is stronger than would be expected, sometimes to such a large extent that these calculations cannot be used even as conservative estimates. Notches also display complex size effects (related to both the size of the notch and the size of the specimen containing it), so that small notches (and small cracks) can fail with a local stress greater than σ_u but a stress intensity less than K_c. Similar problems arise in the prediction of fatigue failure.

Various methods have been devised for dealing with this problem: one of the few which is in common use in engineering design is the method of Neuber, by which strain is used as the characterising parameter instead of stress (Neuber, 1958). This approach is often useful, but has some important limitations: it tends to break down at high K_t and cannot predict the size effect.

It would be particularly desirable to have a theory which is generally applicable, that is one which is valid for all kinds of stress concentration, including the extreme cases of a sharp crack and a plain test specimen and also including stress-concentration features of non-standard shape. The main aim of this book is to describe theories which fulfil these requirements.

1.7 Finite Element Analysis

The last few decades have seen an enormous rise in computing power and, with it, methods of numerical analysis which allow us to simulate complex systems. This has had a profound effect on engineering design: today, techniques to estimate the forces and stresses in components such as multi-body analysis and FEA are available to designers even in relatively small engineering companies. This is bringing about a qualitative change in the way in which components are being designed, as we move away from simplified analytical calculations and empirical rules towards computer simulations.

The same changes are being witnessed in many other fields of science and engineering. A good example is weather forecasting, where systems which are so complex that analytical solutions will never be possible can now be tackled using large computer models. These developments have naturally brought about corresponding changes in the way in which research is being conducted. It now becomes more relevant to study those kinds of theoretical approaches which can be incorporated into computer models, rather than approaches based on the solution of analytical expressions, though the latter will always be of value at a scientific level.

A computer model will only ever be as accurate as our knowledge of its boundary conditions, such as the applied loads and restraints, and FEA still has some important limitations with regard to the size and complexity of components that can be modelled, especially when using accurate material descriptions incorporating non-linear and anisotropic behaviour. However, the critical distance methods described in this book require only linear-elastic stress analyses. The necessary stress–distance data can already be obtained for many engineering components using the kinds of FE model already employed routinely in engineering companies.

1.8 Concluding Remarks: Limitations and Challenges in Failure Prediction

In this chapter we have described, in summary form, the state of the art in the prediction of material failure as articulated in national standards and specifications and as used in practice in engineering companies. We have not discussed here many of the more advanced techniques, which will be described in subsequent chapters; however, these techniques are, for the most part, used only in academic research and not in engineering practice. The current position is unsatisfactory, containing limitations which ultimately affect our ability to design load-bearing structures with confidence.

We can predict material failure with precision only in two rather special cases. The first is simple tension, as described by the stress–strain curve, and the second is the propagation of pre-existing cracks as described by LEFM. The tensile test is of limited practical value because conditions of pure tension arise only rarely in real components. In fact, the strength of the material as measured in a tensile test (σ_u) can often be misleading. Ductile materials fail in a tensile test by a process of plastic instability (necking) which does not occur in other types of loading such as bending or tension, and the tensile strength of brittle materials is usually determined by small pre-existing flaws, the size of which will depend on processing parameters and specimen size.

The LEFM, as we have seen, is a wonderful tool in those cases where it is applicable, but more often than not, when we want to use it, we find that it is not applicable. As regards brittle fracture occurring under constant or monotonically increasing loads, LEFM can only be used for components which contain pre-existing cracks of sufficient length, in components which are sufficiently large to maintain the small-scale yielding criterion. This effectively rules out many components of moderate size, made from relatively tough materials. As regards cyclic loading, LEFM finds an important application – probably its most important practical use – in the assessment of fatigue cracks in critical structures such as aircraft, offshore structures and chemical plant. In this respect, its applications are limited to those components which can sustain relatively large cracks before failure (usually of the order of centimetres) and in which regular inspection procedures can be used to monitor the growth of the cracks over long periods of time. For this reason, LEFM is of very limited value in, for example, car components or other mass-produced consumer products.

In between the two extremes of plain, tensile specimens and bodies containing long, sharp cracks lie all the other stress-concentration features which we may find on components: geometric irregularities such as notches, defects such as inclusions, joints such as welds and contact features such as bearings. The challenge, which will be addressed in the remainder of this book, is to predict failure in all these situations, in a manner which can be incorporated into modern, computer-aided design procedures.

References

Ashby, M.F. and Jones, D.R.H. (2005) *Engineering materials 1*. Elsevier, Oxford UK.

Atzori, B., Lazzarin, P., and Filippi, S. (2001) Cracks and notches: Analogies and differences of the relevant stress distributions and practical consequences in fatigue limit predictions. *International Journal of Fatigue* **23**, 355–362.

Broberg, K.B. (1999) *Cracks and fracture*. Academic Press, London UK.

Bruckner-Foit, A., Huang, X., and Motoyashiki, Y. (2004) Mesoscopic simulations of damage accumulation under fatigue loading. In *Proceedings of the 15th European Conference of Fracture* (Edited by Nilsson, F.) pp. 3–12. KTH, Stockholm, Sweden.

Creager, M. and Paris, P.C. (1967) Elastic field equations for blunt cracks with reference to stress corrosion cracking. *International Journal of Fracture Mechanics* **3**, 247–252.

Delaire, F., Raphanel, J.L., and Rey, C. (2000) Plastic heterogeneities of a copper multicrystal deformed in uniaxial tension: Experimental study and finite element simulations. *Acta Mater.* **48**, 1075–1087.

Filippi, S. and Lazzarin, P. (2004) Distributions of the elastic principal stress due to notches in finite size plates and rounded bars uniaxially loaded. *International Journal of Fatigue* **26**, 377–391.

Hertzberg, R.W. (1995) *Deformation and fracture mechanics of engineering materials*. Wiley, New York USA.

Irwin, G.R. (1964) Structural aspects of brittle fracture. *Applied Materials Research* **3**, 65–81.

Janssen, M., Zuidema, J., and Wanhill, R. (2002) *Fracture mechanics*. Spon, London UK.

Knott, J.F. (1973) *Fundamentals of fracture mechanics*. Butterworths, London.

Murakami, Y. (1987) *Stress intensity factors handbook*. Pergamon, Oxford UK.

Neuber, H. (1958) *Theory of notch stresses: Principles for exact calculation of strength with reference to structural form and material*. Springer Verlag, Berlin.

Paris, P.C. (1964) Fatigue – An interdisciplinary approach. In *Proc. 10th Sagamore Conference* pp. 107–117. Syracuse University Press, Syracuse, New York USA.

Peterson, R.E. (1974) *Stress concentration factors*. Wiley, New York USA.

Westergaard, H.M. (1939) Bearing pressures and cracks. *Journal of Applied Mechanics A* 49–53.

Williams, M.L. (1952) Stress singularities resulting from various boundary conditions in angular corners of plates in extension. *Journal of Applied Mechanics* **19**, 526–528.

Wulpi, D.J. (1985) *Understanding how components fail*. ASM, Ohio USA.

CHAPTER 2

The Theory of Critical Distances: Basics

An Introduction to the Basic Methodology of the TCD

2.1 Introduction

This chapter will introduce the basic methodology of the TCD, showing how it can be used in its simplest forms. In fact, the TCD is not one method but a group of methods which have certain features in common – principally the use of a characteristic material length parameter, the critical distance L. In this chapter we will start with the simplest method of analysis, which we call the Point Method (PM) and proceed to some slightly more complex methods: the Line Method (LM), Area Method (AM) and Volume Method (VM). The aim here will be to show how predictions of brittle fracture and fatigue can be made very easily, for situations where the elastic stress field around the stress concentration feature is known, for example from FEA. In Chapters 3 and 4 we will look at the TCD in more detail, charting its history and discussing it in the context of other methods of failure prediction, especially those which use some form of material length parameter. At that stage we will show that some other methods – essentially modifications of LEFM – can also be considered to be TCD methods.

In describing how to implement the TCD, it is convenient to use a series of specific examples. The first example will be the prediction of brittle fracture in a test specimen containing a notch: we will use this example to introduce the PM. In the second example, we will consider fatigue failure in an engineering component, again using the PM. After developing some simple theory to make a link between the PM and LEFM and then introducing the other related methods (the LM, AM and VM), we will consider a final example which looks at the prediction of size effects for notches.

2.2 Example 1: Brittle Fracture in a Notched Specimen

Consider a simple notched tensile test specimen, in this case containing a pair of symmetrical edge notches of depth $D = 5\,\text{mm}$ and root radius $\rho = 2\,\text{mm}$, as shown in Fig. 2.1. The width of the specimen is 20 mm, reducing to 10 mm between the notches.

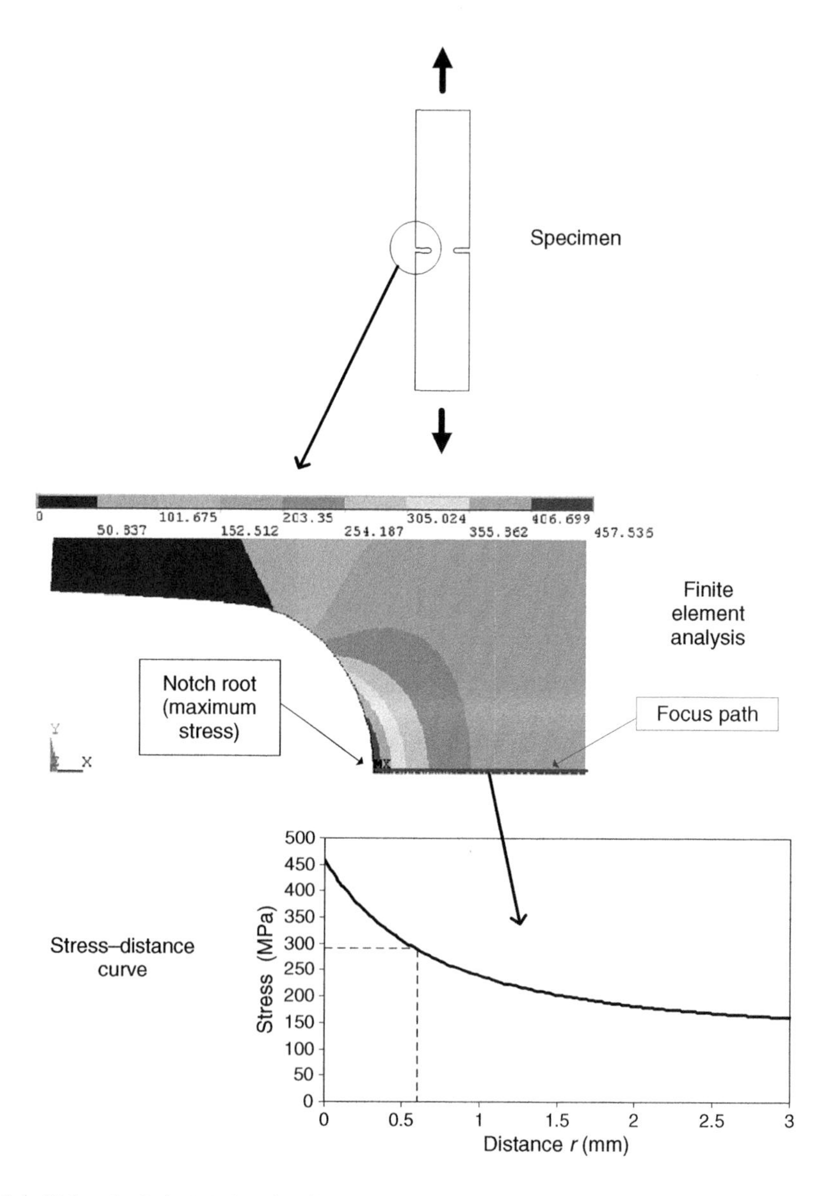

Fig. 2.1. Using the Point Method (PM) for prediction of brittle fracture in a notched test specimen.

A uniform axial tensile load is applied, giving a nominal cross-section stress (the stress at a cross section remote from the notches) of σ. The problem is to find the value of σ at which failure will occur by brittle fracture: we will call this the fracture stress, σ_f. In this case, we will assume that brittle fracture is the operative mode of failure; this will be the case, for example, in ceramic materials and many polymers such as Perspex, and also in metals under some circumstances (e.g. steels at very low temperatures). We will also assume that the material is not subject to time-dependant effects: that is when we apply a monotonically increasing load, the specimen will fail at a unique value of

σ_f, which is independent of the rate of application of load. To make the prediction we are going to use the simplest form of the TCD, which we call the PM.

2.2.1 Necessary information: The stress–distance curve and material parameters

In order to make predictions we require a stress analysis of the specimen, in particular of the region near the notches. We will use an elastic analysis, that is to say one which estimates the elastic stresses and strains in the material assuming that no yielding or damage occurs to cause permanent strains or non-linear stress–strain behaviour. In reality, of course, this will often be an unrealistic analysis, especially if the notch is quite sharp, because almost all materials will display some deviation from simple elastic behaviour if the stresses rise to high enough values. It is an important aspect of the TCD that we can use an elastic analysis even in these circumstances: the theoretical explanation for this will be left until a later stage in this book; suffice it to say that all the predictions which we will carry out using the TCD will be done using elastic stress information only. For our purposes it does not really matter how this stress analysis is carried out; in practice it will usually be done using FEA or some other numerical method, because analytical solutions are available only for a small number of cases, in which the geometry is very simple. Figure 2.1 shows the results of FEA applied to our test specimen, illustrating part of the stress field in the vicinity of one of the notches; the stress shown is the maximum principal stress. Of course, both notches are identical so we only need to consider one, and we can take advantage of symmetry by only modelling half of the notch. In the FEA, the nominal stress applied to the specimen was $\sigma = 100\,\text{MPa}$; this gave rise to a maximum stress (at the root of the notch) of 457.5 MPa, thus the stress concentration factor of this notch (relating the maximum stress to the nominal stress on the cross section) is approximately 4.6.

Figure 2.1 also shows a graph obtained from the FEA, which plots the stress as a function of distance from the notch root, taken along a line drawn horizontally through the specimen. We shall make much use of this type of plot, which we call the 'stress–distance curve'. The line is called the focus path: in the present example it seems fairly obvious that, if we wish to get a feel for the stress field in the vicinity of the notch, then it makes sense to use the maximum principal stress (which in any case will be equal to the tensile stress in the axial direction) and to draw the focus path starting at the notch root and running across the specimen, perpendicular to the loading axis. In other problems which will be considered below, these choices will be less obvious and will need some discussion.

When making predictions using the TCD, we need two material parameters: a critical stress σ_o and a critical distance L. For the time being we will not consider how these parameters are obtained – this will be dealt with below. Suppose, for the sake of argument, that the material in our test specimen has the following values for these material constants: $\sigma_o = 420\,\text{MPa}$ and $L = 1.2\,\text{mm}$.

2.2.2 *The point method*

The PM – the simplest form of the TCD – uses a failure criterion which can be stated as follows: 'Failure will occur when the stress at a distance $L/2$ from the notch root is equal to σ_o.' Putting this in mathematical form, if we denote distance on the stress–distance curve by r and stress on this curve by $\sigma(r)$, then the PM prediction can be written as:

$$\sigma(L/2) = \sigma_o \tag{2.1}$$

As the figure shows, the stress at a distance of $L/2$ (0.6 mm in this case) is 289.6 MPa, so since this is less than σ_o we predict that no failure will occur under these loading conditions, that is with a nominal stress of 100 MPa. To predict failure we need to find the nominal stress for which Eq. (2.1) is satisfied. In this particular case (and indeed in many cases in practice) this is very easy to do because the FEA is not only elastic but also linear: the stress at every point is directly proportional to the applied load. This is not always true: there are problems in which, though the analysis is elastic, it may not be linear. Non-linearities can occur due to material properties, and also due to geometric effects; the latter will occur if the geometry of the body changes during loading in a way which can affect the local stresses – for example, an applied bending moment may change due to deflection. In such cases it may be necessary to conduct a series of FEAs to find the stress–distance curves corresponding to different applied loads.

Assuming that these non-linearities do not occur, we can find σ_f without the need to make further FEAs, simply by scaling the stress–distance curve. So in this case the value of σ_f will be $100 \times (420/289.6) = 145.0$ MPa. Figure 2.2 shows the stress–distance curve corresponding to this applied stress, confirming that the criterion stated in the PM is indeed fulfilled: the stress at a distance of $L/2$ is equal to σ_o.

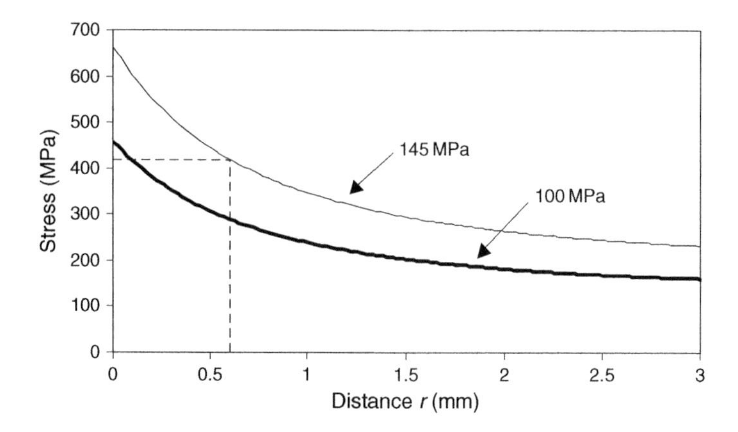

Fig. 2.2. Stress–distance curves for the notched specimen of Fig. 2.1, at applied stresses of 100 MPa and 145 MPa, the latter being the predicted fracture stress, σ_f. The dashed lines confirm that σ_o (=420 MPa) occurs at $r = L/2$ (=0.6 mm).

2.3 Example 2: Fatigue Failure in an Engineering Component

The same approach can also be used to predict fatigue. Figure 2.3 illustrates an example: the crankshaft of a car engine, which is prone to fatigue cracking at the right-angle corners which occur near the bearings. The figure shows the stress field obtained from FEA, and the stress–distance curve. Note that in this case we have used, for the focus path, a line drawn starting at the point of maximum stress (on the surface at the corner), the line being perpendicular to the surface at that point. The stress parameter used is again the maximum principal stress, so the focus path is also perpendicular to the direction of maximum principal stress at the point of maximum stress, a point which

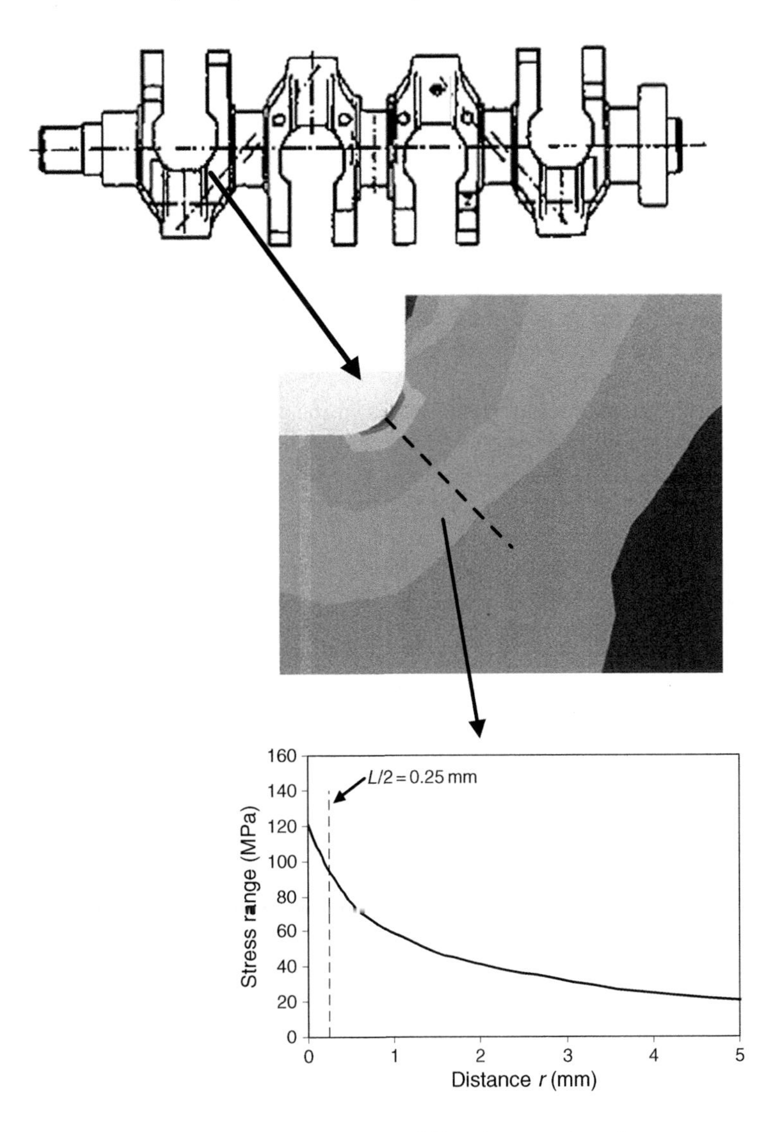

Fig. 2.3. Using the PM to predict fatigue failure in an engineering component.

is often referred to as the 'hot spot'. In this case, these choices – the use of maximum principal stress and the direction of the focus path – are not as intuitively obvious as they were in the first example. This matter will be considered in detail at a later stage in this book, when we consider multiaxial stress fields (Chapter 11) and complex geometrical features (Chapter 12).

In the present example, our problem is to predict whether or not fatigue failure will occur from this corner in the crankshaft, when it is subjected to the loadings which occur during normal operation. These stresses will be cyclic, which is why the stress–distance curve shown in this figure uses the stress range, $\Delta\sigma(r)$. We assume that this stress range has been accurately estimated by applying the appropriate cyclic loads to the FE model and, for the sake of simplicity, we will assume that the resulting cyclic stress at any point near the corner has the form of a sine wave of constant amplitude and constant R ratio.

Again some material constants are needed: a critical distance L which we will assume has a value of 0.5 mm, and a critical stress (in this case a cyclic stress range) of $\Delta\sigma_o = 590\,\text{MPa}$. These values will not be solely material constants because they will also depend on the R ratio of the cycle and on the number of cycles to failure – here we wish to know whether fatigue will occur after a large number of cycles, that is to say we are interested in the fatigue limit of the material. Again we assume that these two parameters are known, without going into details as to how they have been obtained.

The PM is used in exactly the same way as in the first example, except we now use cyclic range values for the stresses, so the condition for fatigue failure can be written as:

$$\Delta\sigma(L/2) = \Delta\sigma_o \tag{2.2}$$

Looking at the stress–distance curve in Fig. 2.3, we can predict that fatigue will not occur in this case. The cyclic stress at $r = L/2$ is 96 MPa, much lower than $\Delta\sigma_o$. We can define a safety factor – always a useful quantity in engineering design – as the ratio between $\Delta\sigma_o$ and $\Delta\sigma(L/2)$; in this case the result is 6.1 which would probably be considered sufficient to ensure safety in this kind of component.

2.4 Relating the TCD to LEFM

Having explained the TCD – at least in its simplest form as the PM – it is useful to show how it is possible to make a theoretical link between the TCD and traditional LEFM. Consider the case of a notch in which $\rho = 0$, that is a sharp crack. In this unique case, we can make predictions using both methods: the TCD and LEFM. Brittle fracture will occur when the stress intensity, K, is equal to the fracture toughness, K_c. Since K_c is a material constant, it follows that there must be some relationship between K_c and the constants used in the PM. This relationship can easily be deduced as follows. Recall from Chapter 1 that K_c is related to σ_f and the crack length, a, by:

$$\sigma_f = \frac{K_c}{\sqrt{\pi a}} \tag{2.3}$$

Recall also that the stress–distance curve for the case of a crack can be expressed analytically as follows:

$$\sigma(r) = \sigma\sqrt{\frac{a}{2r}} \tag{2.4}$$

This equation is only valid for $r << a$, so it will be sufficient for our purposes, provided we only wish to examine stresses close to the crack tip. Effectively this means that the crack length a must be much larger than the critical distance, L. If we combine Eqs (2.3) and (2.4) with the criterion for the PM (Eq. 2.1) the result is

$$L = \frac{1}{\pi}\left(\frac{K_c}{\sigma_o}\right)^2 \tag{2.5}$$

This equation gives a relationship between the fracture toughness and the two material constants of the TCD. This is a very important relationship which we will make considerable use of throughout this book. Note that Eqs (2.3) and (2.4) are strictly valid only for the particular case of a central through crack in a plate of infinite dimensions. For other cracks we need to introduce the geometry factor F (see Eq. 1.7) but this does not affect the generality of Eq. (2.5) because a given value of K (and therefore of K_c) is associated with a unique stress–distance curve near the crack tip. Exactly the same type of equation can be deduced for fatigue, simply replacing the static parameters with cyclic ones. The cyclic equivalent of K_c (at the fatigue limit) is the fatigue crack propagation threshold ΔK_{th}, thus the appropriate critical distance for fatigue limit predictions will be

$$L = \frac{1}{\pi}\left(\frac{\Delta K_{th}}{\Delta\sigma_o}\right)^2 \tag{2.6}$$

2.5 Finding Values for the Material Constants

Up to now we have assumed that the two material constants L and σ_o are known. How can we obtain values for them for a particular material? In principle, since there are two parameters, we can deduce their values from experimental data obtained from tests on specimens containing any two different stress concentration features. For example, we could use two different notches (notches with different values of D and/or ρ). The choice is somewhat arbitrary, but from an experimental point of view, the accuracy of our determined values will be increased if we use two very different notches. So far we have implicitly assumed that this method of prediction can be applied to any geometry of notch or stress concentration feature. The two extreme cases we can imagine are a sharp crack and a plain, unnotched specimen, so it would make sense to choose these two specimen types when determining the material constants.

The case of a plain specimen is trivial: at failure in a tensile test the stress is equal to the ultimate tensile strength, σ_u, at all points in the specimen, so this must correspond to our value of σ_o. For the case of a long, sharp crack, failure will occur when $K = K_c$

and we have already deduced a relationship linking this to the other two parameters. Rewriting Eq. (2.5) with $\sigma_o = \sigma_u$ gives

$$L = \frac{1}{\pi}\left(\frac{K_c}{\sigma_u}\right)^2 \tag{2.7}$$

Thus we can obtain the material constants we need by using two parameters which are commonly available: the ultimate tensile strength and the fracture toughness. The same also applies for fatigue: Eq. (2.6) can be used, with $\Delta\sigma_o$ understood to be the plain-specimen fatigue limit.

This method of deriving the material constants will be valid, provided our assumption holds true that the TCD can be used for all kinds of stress concentration feature, even including the two extreme cases which correspond to an infinite stress concentration and no stress concentration. In practice, we will find that this assumption does indeed hold true in quite a lot of cases, such as the brittle fracture of ceramic materials and fibre composites, and the fatigue of metals. In certain other cases, however, a comparison with experimental data reveals that the TCD is not valid for plain specimens. Examples in which this problem arises include the brittle fracture of polymers and metals. In these cases a different approach has to be used to find the value of σ_o and there is, as a consequence, a somewhat smaller range to the validity of the TCD. These matters will be explored comprehensively in the chapters which deal with different types of materials and different failure mechanisms (Chapters 5–11).

2.6 Some Other TCD Methods: The LM, AM and VM

Three other methods can be identified, alternatives to the PM, which also use the elastic stress field in the vicinity of the notch. In these methods the appropriate stress parameter, rather than being the stress at a particular point, is defined as the average stress over some region of the stress field.

2.6.1 The line method

In the LM, we use the same line – the focus path – as defined previously for the PM. However, in this case the stress parameter used is the average stress over some distance starting at $r = 0$. Let the distance be d, in which case we can write the LM criterion for brittle fracture as:

$$\frac{1}{d}\int_o^d \sigma(r)\,dr = \sigma_o \tag{2.8}$$

This is illustrated schematically in Fig. 2.4. We can again make use of the link with fracture mechanics to find the distance over which this average should be obtained. Taking the case of a long, sharp crack and therefore using Eq. (2.4) to describe the stress–distance curve, Eq. (2.8) leads to:

$$d = \frac{2}{\pi}\left(\frac{K_c}{\sigma_o}\right)^2 \tag{2.9}$$

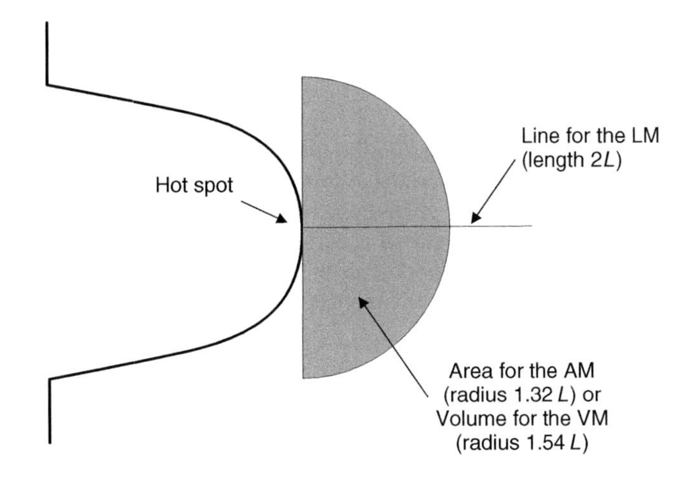

Fig. 2.4. The Line, Area and Volume Methods applied to an edge notch.

This is equal to $2L$, defining L as previously (Eq. 2.5); this shows that there is a simple relationship between the distances used for the PM and the LM: the PM uses a distance of $L/2$ whilst the LM uses $2L$. Thus the LM criterion can be written, using the same definition of L as above, as:

$$\frac{1}{2L}\int_{o}^{2L}\sigma(r)\,dr = \sigma_o \tag{2.10}$$

This ensures that the predictions of the PM and LM will be identical for long cracks; trivially, the two predictions will also be identical for plain tensile specimens. However, there is no guarantee that the predictions will be identical for any other problem and, indeed, they are not. However, as will be shown in subsequent chapters, the differences between the PM and LM predictions are almost always small. We will see that the PM is more accurate in some cases and the LM is better in others, and there is some evidence that this may be related to the operative mechanisms of failure. However, in the great majority of cases, the difference between the PM and LM predictions is so small that both are quite adequate for describing experimental data that inevitably contains a certain amount of scatter.

2.6.2 The area and volume methods

The AM involves averaging the stresses over some area in the vicinity of the notch; the VM, likewise, makes use of a volume average. In both cases the same value is used for the critical stress σ_0. In these cases the analysis is somewhat more complicated, but even so it is not difficult to implement such averaging procedures as part of the post-processing of an FEA. The results will obviously depend on the shape of the area or volume chosen. We could choose, for example, a semicircular area (or hemispherical volume) centred on the point of maximum stress (Fig. 2.4). In that case, it is possible to show, by suitable integrations of the stress field ahead of a sharp crack (Bellett et al., 2005), that the radius of the semicircular area will be $1.32L$ and that of the hemispherical volume will be $1.54L$. Thus, we see that the same definition of L can be used for all four of these methods.

In what follows in the rest of this book, the great majority of analyses will use either the PM or the LM. This is because we have found from experience that, whilst the AM and VM are also capable of valid predictions, these methods are more difficult to use and do not seem to confer any increased accuracy when compared to the experimental data. We include them here partly for completeness and partly because they will be used again in later discussions relating to the theoretical basis of the TCD in general.

2.7 Example 3: Predicting Size Effects

Let us return to the example of the notched specimen used in Section 2.2, but consider now what will happen if we change its size. Our original specimen had a notch depth of $D = 5$ mm: Fig. 2.5 shows stress–distance curves for this specimen and also for a half-size specimen ($D = 2.5$ mm) and one double the size ($D = 10$ mm). All other dimensions have been changed in proportion. In fact, it is very easy to draw these curves because the stress analysis is linear and elastic, so we can use the original curve and simply change the length scale by a factor of 2 or ½, respectively.

Using the same value for $L/2$ (0.6 mm) we can see that the stress at this distance increases as the size of the notch increases. Therefore, if we make predictions using the PM we will conclude that the fracture strength σ_f will decrease with increasing size. In this example, the predicted strengths of the 2.5, 5 and 10 mm notches are, respectively, 189 MPa, 145 MPa and 119 MPa, quite significantly different.

Further examination of the curves in Fig. 2.5 shows that the differences between them are not constant. They coincide at $r = 0$ (the notch root) showing that the K_t factor for these three notches is identical: this is a necessary condition since we have not changed

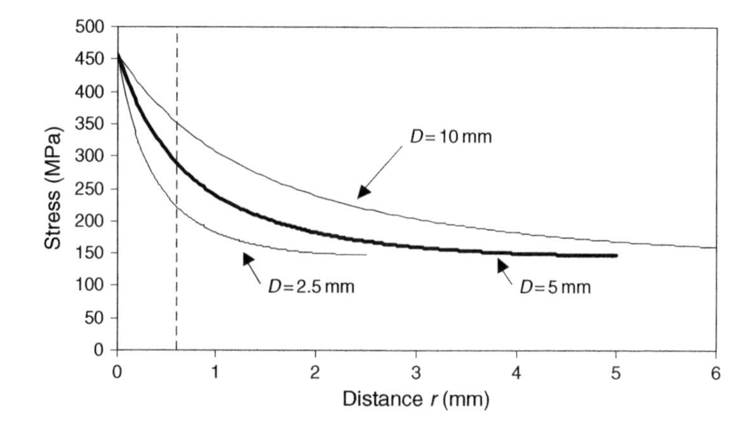

Fig. 2.5. Stress–distance curves for the specimen shown in Fig. 2.1 ($D = 5$ mm) and also for specimens of double and half size ($D = 10$ mm and $D = 2.5$ mm).

the geometry of the problem, only the scale. So we can see that K_t is a parameter which is insensitive to scale and cannot be used to predict size changes. The three stress–distance curves also tend to similar values at large distances, remote from the notch. These characteristics of the stress–distance curve suggest that the magnitude of the scaling effect, that is the relative differences between the strengths of these specimens, will be affected by the value of L. If L is very small, or very large, then the critical distance will occur at a point on the graph where there is little difference between the three curves, and therefore very little difference between the predicted strengths. This hints at the fact that scaling laws for material strength are complex and depend on a number of factors, some of which are geometric and some material based. Many examples of size effects will be shown in the subsequent chapters of this book, and the general theoretical problem of scaling will be addressed in Chapter 12. At this stage, it is sufficient for us to note that the methods of the TCD are capable of predicting the existence of size effects.

2.8 Concluding Remarks

In this chapter, we have introduced the basic methods of the TCD which will be used extensively throughout this book – the analysis of stress–distance curves along the focus path, using elastic stress fields obtained from FEA or other techniques. We have seen that only two material parameters are needed, σ_0 and L, and that the analyses can be performed very simply and quickly, especially if we can take advantage of linear scaling laws to find the effect of changing the applied loads or dimensions.

References

Bellett, D., Taylor, D., Marco, S., Mazzeo, E., and Pircher, T. (2005) The fatigue behaviour of three-dimensional stress concentrations. *International Journal of Fatigue* **27**, 207–221.

CHAPTER 3

The Theory of Critical Distances in Detail

The History, Background and Precise Definition of the TCD

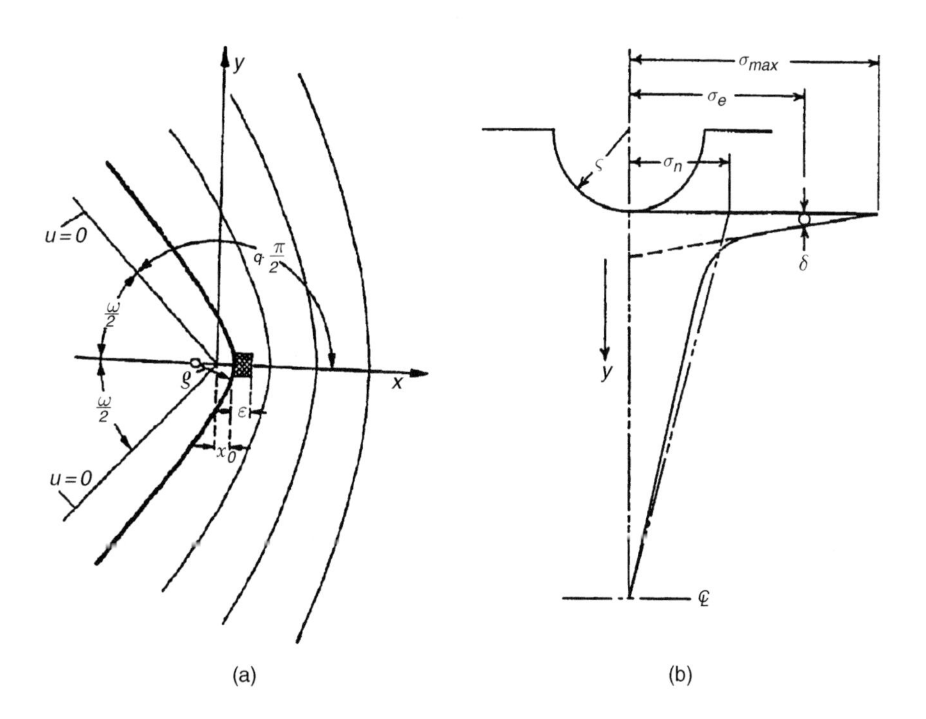

Fig. 3.1. Early diagrams illustrating critical distances: (a) Neuber's LM (Neuber, 1958), using the symbol ε, and; (b) Peterson's PM (Peterson, 1959) using the symbol δ.

3.1 Introduction

In this chapter, we will examine the TCD more carefully, starting with a brief history of the subject and then describing a number of theories which, whilst they differ in detail, can all be described as critical distance theories. This will lead us to a clear definition of what we mean by the TCD, which is not one single theory but a group of theories having certain features in common.

In Chapter 4, we will examine other methods which are used in fracture prediction: these methods aim to predict the same phenomena as the TCD and can loosely be thought of as the 'competitors' to the TCD. Some readers may prefer to skip these two chapters. Chapter 2 has already explained how the TCD can be used in practice, especially in its simplest forms of the PM and LM, so a reader whose intentions are purely practical – for example, an engineering designer who wishes to apply the theory to an FEA of a component – may prefer to move straight on to the chapters which explain how the theory is applied to different materials or different types of failure, finishing with the chapter on Case Studies and Practical Aspects (Chapter 12). Chapters 3, 4 and, later on, Chapter 13 were written for the reader who is more interested in the theoretical basis of the TCD and its relationship to other current theories, that is for the reader who asks 'Why does it work?'.

3.2 History

The history of critical distance methods is an interesting example of a phenomenon which occurs often in science: the repeated discovery of the same idea and its parallel development in different fields. We shall see that the TCD has been discovered not once but many times, by workers who were, for the most part, ignorant of each other's activities because they were studying different materials or different failure modes.

3.2.1 Early work

The story begins in the 1930s with Neuber in Germany and Peterson in the United States, both of whom were concerned with predicting fatigue failure in metallic components containing notches (Neuber, 1936; Peterson, 1938). Their ideas were more fully developed by the 1950s, and described in two important publications: Neuber's seminal work *Kerbspannungslehre* ('Theory of notch stresses', in its second edition by 1958) and Peterson's contribution to the book *Metal Fatigue* edited by Sines and Waisman. Figure 3.1 shows diagrams from these two publications (Neuber, 1958; Peterson, 1959) in which the critical distance principles are illustrated.

Neuber invented the method that we now call the Line Method (LM), in which the elastic stress is averaged over a critical distance from the notch root. It is interesting to note that, for Neuber, the initial motivation for this idea was not to predict fatigue failure but the more basic activity of stress analysis. Neuber believed that classical theories for predicting the elastic stresses in bodies were erroneous in situations of high surface curvature, and therefore high stress gradient. Describing the classical theory of elasticity, he wrote the following:

> As a hypothetical model, an infinitely small element with the edge dx dy dz is used; this is of fundamental importance. The tacit assumption of the arbitrary divisibility of the material, its lack of structure etc, obviously represents the criterion of applicability for the classical theory of elasticity . . . Conditions differ for strong surface curvature where stress variations occur over very small distances. The applicability of the classical theory of elasticity would now require that the material continue to be considered as non-structural within zones having the order of magnitude of the crystals. However the presence of the crystals themselves contradicts this . . . . Consequently the material will henceforth be conceived as composed of numerous small but finite particles.

Of course we have known for a very long time that materials are not truly continuous, but in many cases it is permissible to use continuum mechanics because the scale of the problem is larger than that of any material inhomogeneities. Neuber seems to have known little about the microstructure of the materials he was dealing with: he refers variously to 'crystals' and 'finite structural particles' but makes no particular attempt to link his ideas to the actual deformation behaviour at the microstructural level. Neuber's solution was to continue using continuum mechanics, but to modify it by introducing a parameter with the dimensions of length: instead of using infinitesimal calculus, he argued, one should move to the calculus of finite differences. But this caused a problem, in the words of Neuber:

> . . . the reader, who has perhaps at some time solved problems by means of the calculus of differences will surely regard the practical process as open to question, for such problems are known to be extremely tedious. In fact it would be practically impossible to get anywhere by this method.

It is very interesting to read these statements 50 years later, when finite difference and finite element methods are now used routinely in computer simulations. Indeed, sitting here at my personal computer, I could now take any of the problems in notch stress analysis found in Neuber's book – problems which he solved using such elegant mathematics – and obtain a solution within a few minutes using FEA. This indicates a fundamental change in the way that science and engineering are developing: of course, there is still an important place for analytical solutions, but increasingly we are finding that complex problems are much better solved using numerical simulations.

Returning to Neuber, his solution to the problem of material inhomogeneity was to calculate the stresses using classical theory and then average them over the length of the structural particle: this is the length which we now refer to as $2L$. In later work he went on to use this as the basis for predicting fatigue behaviour. Peterson was aware of Neuber's work, but chose a slightly different solution, using the stress at a single point. This is the method which we now refer to as the PM, with Peterson's critical distance corresponding to $L/2$.

These pioneers of the TCD faced two problems in using these methods. The first problem was what value to ascribe to the critical distance. Peterson speculated that it might be related to grain size, but this posed some measurement difficulties, so, like Neuber, he chose to determine the critical distance value empirically, fitting fatigue predictions to data. He did note, however, that for a particular class of materials (e.g. steels) the value of the critical distance seemed to be inversely related to the material's strength. The second problem faced by workers at that time was the accurate estimation of stresses in real components. Thanks to Neuber and others, elegant solutions existed for various standard notch geometries, but these would only ever be approximations to the features in real components. To avoid this problem, use was made of the fact that local stresses are largely determined by the root radius of the notch: knowing this, and the K_t factor, a reasonable approximate stress analysis can be achieved, and therefore used with the PM and LM. This lead to empirical equations involving K_t and the root radius (ρ), to predict the actual reduction in fatigue limit, the fatigue strength reduction factor K_f. Neuber's formula was as follows:

$$K_f = 1 + \frac{K_t - 1}{1 + \sqrt{\frac{\rho'}{\rho}}} \tag{3.1}$$

Here the critical distance parameter is denoted by ρ'. Peterson obtained a slightly different formula:

$$K_f = 1 + \frac{K_t - 1}{1 + \frac{\rho''}{\rho}} \tag{3.2}$$

Here ρ'' denotes the critical distance, though in Peterson's case this constant was found to be also a weak function of K_t. These formulae represented realistic attempts to use the PM and LM, given the technology of the time; however, they have several important limitations. Aside from the fact that they are based on approximate stress analysis, they require an estimate of K_t which, in most components, will not be a definable quantity, since to define K_t one must define a nominal stress – the stress that would occur if the notch were not present. Nominal stresses have no meaning for most components. Furthermore, these equations break down as ρ approaches zero, giving unreliable predictions for sharp notches.

The surprising thing is that these 50-year-old equations are still being used by many designers in engineering companies. Indeed many software packages used for fatigue analysis of components require the user to input a value of K_f for the feature under consideration. This is a rather silly situation, since an analysis using the PM or LM can now be carried out directly using the results of FEA, as we showed in the previous chapter. This situation has arisen because, though the equations of Neuber and Peterson have remained in use since their time, the underlying theory on which they were based has largely been forgotten.

3.2.2 Parallel developments

The work of Neuber and Peterson was applied quite extensively to problems in metal fatigue in the 1960s. In this decade also the PM and LM were suggested for the prediction of brittle fracture using the atomic spacing as the critical distance (McClintock and

Irwin, 1965; Novozhilov, 1969), an idea which has recently been revived in an attempt to predict the behaviour of very small material samples such as carbon nanotubes (see Section 5.5). An approach using averaged strain rather than stress was suggested as a failure criterion for conditions of extensive plasticity (McClintock, 1958).

A major step forward occurred in 1974, with the work of Whitney and Nuismer. These researchers were studying a different problem: monotonic failure of fibre composite materials. They developed theories identical to the PM and LM (which they called the Point Stress and Average Stress methods) to predict the effect of hole size and notch length on the static strength of long-fibre composite laminates (Whitney and Nuismer, 1974). They do not appear to have been aware of the earlier work of Neuber and Peterson. Whitney and Nuismer also took the crucial step of linking the PM and LM to LEFM, using the derivation given above in Chapter 2 (Section 2.4). This step is important because it allows the critical distance to be expressed as a function of the fracture toughness, K_c (Eq. 2.5), and also links the critical distances for the PM and LM ($L/2$ and $2L$, respectively). Whitney and Nuismer had the advantage over Neuber and Peterson that they were working at a time when LEFM had become well established in the field of brittle fracture. Though the theoretical derivation is identical, and equally valid, for HCF, this link was not made until a decade later (Tanaka, 1983). Tanaka presented the theoretical relationship (Eq. 2.6) but offered no experimental data for comparison. It seems that this paper was largely ignored, the idea being rediscovered and subjected to experimental validation sometime later (Lazzarin et al., 1997; Taylor, 1999; Taylor and Wang, 2000); being unaware of Tanaka's contribution, these workers developed the same theoretical derivation and went on to show that it was indeed possible to predict experimental fatigue limit data from specimens containing notches and cracks of varying sizes.

The work of Whitney and Nuismer was taken up by many other researchers in the field of composite materials: this work is described below in Chapter 8 – suffice it to say that the PM and LM are now established techniques for the prediction of failure in these materials. This differs radically from the field of metal fatigue, where, as we saw above, the theory itself was largely forgotten (though it remained in the form of some empirical equations) only to be rediscovered in recent decades.

The TCD can also be used to predict brittle fracture in polymers, and this fact was realised in the 1980s by Kinloch, Williams and co-workers (Kinloch and Williams, 1980; Kinloch et al., 1982). Again, these workers do not appear to have been familiar with the earlier work of Whitney and Nuismer, nor that of Neuber or Peterson. Their aim was rather different: the motivation for their work was to understand the effect of crack-tip blunting on fracture toughness. They developed a method which was essentially the same as the PM, but with the important difference that the critical stress parameter, σ_o, was not equal to the material's UTS. This modification turns out to be crucial to the use of the TCD in certain materials, as will be discussed in Chapters 6 and 7.

Surprisingly, this initial work on polymers does not seem to have been continued, either by these workers or others, so to this day the TCD is not being used to predict fracture in polymers, despite its extensive use in the closely related field of polymer-matrix composites. A small number of papers has appeared in the last decade, applying

TCD-like theories to the behaviour of very sharp V-shaped notches, and recently my own research group has examined a wide range of notches in polymethylmethacrylate (PMMA) (see Chapter 6).

The PM and LM can also be applied in three other fields: brittle failure in ceramic materials; brittle fracture in metals; and fatigue in polymers. I have demonstrated these applications using experimental data from the literature, and this work is described in later chapters of this book, and in a number of recent papers. To my knowledge, no other similar work has been carried out in these fields. That is not to say that other workers have not realised the importance of length scales, or the problems associated with material inhomogeneity, but these problems have been addressed in different ways, as will be described below in this chapter and the next.

3.3 Related Theories

In this section, we shall introduce some other theories that are used for fracture and fatigue prediction, theories which also involve a material length constant. It will be shown that these theories are closely related to each other, and to the PM and LM.

3.3.1 The imaginary radius

This approach was introduced by Neuber, who conceived it as a simple way to achieve the same result as the LM. Consider a notch, of depth D and root radius ρ. In order to predict the strength of this notch, Neuber suggested that one could imagine the notch to have a larger radius. The radius is increased by an amount, say ρ^*, which is assumed to be a material constant. The relevant stress parameter to use is then the maximum stress at the notch root, for this imaginary notch of depth D and radius $(\rho+\rho^*)$. Neuber attempted to show that this approach will give the same results as the LM: his proof is only approximate because it relies on an approximate solution for the notch stress field, a solution which is more accurate for some types of notches than others.

In Neuber's day this approach was useful, given the difficulty of determining the stress–distance curve. But today, with the wide availability of FEA, the method has little practical value. For the FE analyst, it is generally much easier to find the average stresses on the line of length $2L$ than it is to introduce the imaginary radius, because the latter activity involves modifying the geometry of the FE model. It goes without saying that the imaginary radius method has no physical meaning, it is simply a convenient way to reduce the notch-root stress for analytical purposes. One field in which the method is sometimes used is the analysis of welded joints (Sonsino et al., 1999). The actual root radius of features at the weld toe or in areas of incomplete penetration is rather variable in actual welds; this difficulty can be overcome by making the radius equal to ρ^* throughout, as illustrated in Fig. 3.2. Even here one can anticipate some difficulties – for example, the introduction of the radius will effectively change the area of the cross section (since some material must be removed) and so increase average stresses.

A possible modern equivalent to the imaginary radius model is the use of a FEA with a specified element size. Rather than refining the element size until convergence occurs,

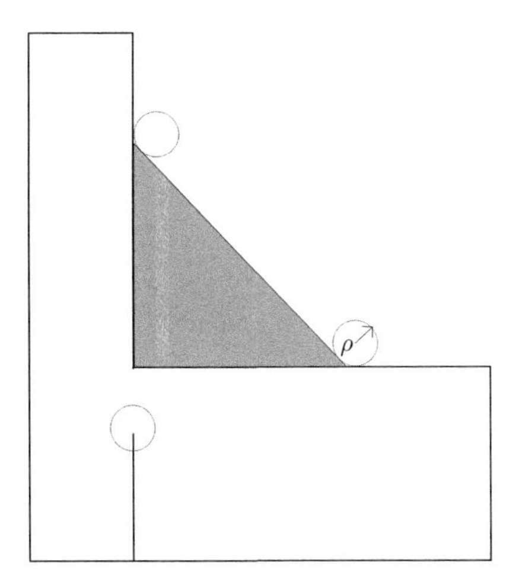

Fig. 3.2. An example of the use of the imaginary radius: a welded joint with radii introduced at the weld toes and at the end of the lack-of-penetration.

as would be the normal procedure in FEA, one can imagine using a mesh in which the element size in the vicinity of the notch is fixed.

The value of the maximum stress at the notch root will then be a function of element size, reflecting the fact that stresses are calculated by interpolating across the elements. In principle, then, the element size now becomes the critical distance. Some workers have suggested this approach but to date it has not been developed in any detail. One complicating factor is that the method used for calculating stresses is different in different FE packages and different element types.

3.3.2 Introduced crack and imaginary crack models

In this approach a notch is analysed by first introducing a sharp crack at the root of the notch (Fig. 3.3(a)). The length of the crack is assumed to be a material constant. The subsequent analysis uses fracture mechanics: we calculate the stress intensity of this notch-root crack and use this to predict failure. The method has been applied quite widely to problems in brittle fracture and in fatigue, using K_c or ΔK_{th} as the critical stress intensity parameter, respectively. Specific applications of this method will be discussed in subsequent chapters. Possibly the first use of this method was by Waddoups et al., who applied it to brittle fracture in composite materials, in a paper which is still widely quoted to this day (1971). In fatigue, the model was suggested by El Haddad et al., for the analysis of short cracks (1979), and by Klesnil and Lucas for notches (1980). Like the PM and LM, this is a method which has been discovered and re-discovered by many workers over the years.

Examining the published work in more detail, we can see that there are in fact two slightly different approaches. In the first, which I will call the 'introduced crack method', it is assumed that there is an actual crack present at the notch root. This is the case, for example, for Usami et al., studying ceramics, who suggested that fracture emanated from

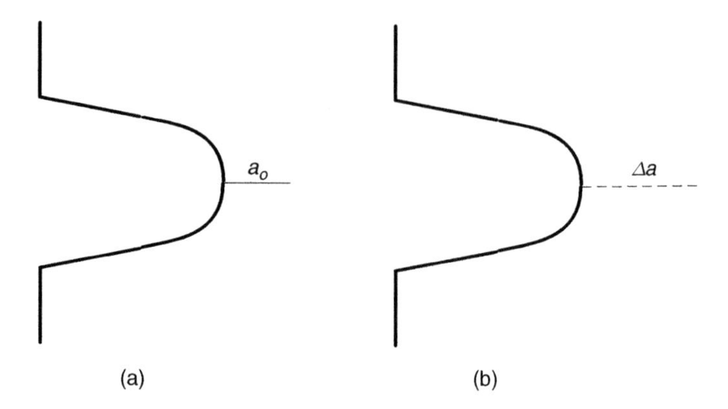

Fig. 3.3. (a) For the introduced and imaginary crack models a crack, length a_0, is placed at the root of the notch. (b) In FFM, strain-energy release is calculated during the growth of a crack increment Δa.

a small, semi-circular flaw (1986), and also for Ostash and Panasyuk who proposed that the fatigue behaviour of metals is affected by the presence of a layer of weak material on the surface (2001).

These analyses contain a fundamental theoretical problem: the introduced crack, being small in size, will not conform to a normal LEFM analysis. Size effects were mentioned briefly in earlier chapters: cracks show considerable size effects as will be revealed later (see, for example, Sections 5.2.1 and 9.2.3). Effectively, LEFM cannot be used if the length of the crack is of the same order of magnitude as L, or smaller. This leads one into a kind of circular argument: in order to use the introduced crack method to predict the behaviour of a notch, one must use a fracture criterion for the small, introduced crack. The LEFM cannot be used, and in order to be consistent one should use the *same* theory again – that is one must introduce *another* crack, at the root of the first crack and so on *ad infinitum*.

This problem can be avoided by the use of the second approach, which I call the 'imaginary crack method' (ICM). In this case the crack is not assumed to have any real, physical existence: it is imaginary. We can describe this model as follows: 'it is assumed that there is a crack at the notch root and that the propagation of this crack obeys the laws of LEFM'. This was the approach taken by El Haddad et al., for example. Waddups et al. noted that, prior to failure in composite materials, a damage zone develops at the notch root; they suggested that their notch-root crack could be assumed to be a simple representation of that damage zone, thus again avoiding the difficulties of the actual physical problem. This kind of issue arises very often in the prediction of material behaviour, and represents a fundamental division between models which attempt to predict the actual physical mechanisms of the process (which we may call 'mechanistic models') and those which, instead, proceed by using some simplifying analogy (see Chapter 4). Mechanistic models can give vital insights into the behaviour of materials, but, considering the complexity of material behaviour, non-mechanistic models will invariably give more accurate, quantitative predictions. All the methods which we consider to be part of the TCD are non-mechanistic.

3.3.3 Linking the imaginary crack method to the PM and LM

We can show that the predictions of the ICM are similar to those of the PM and LM, and in some cases identical. First consider a crack, for which (from Section 1.4) the stress intensity K can be written:

$$K = F\sigma\sqrt{\pi a} \tag{3.3}$$

Using the ICM we add a fixed amount, a_o, to the crack length, giving:

$$K = F\sigma\sqrt{\pi(a+a_o)} \tag{3.4}$$

If the crack is very long ($a >> a_o$), then this will have a negligible effect – this is obviously a necessary condition since Eq. (3.3) is the correct one for predicting the behaviour of long cracks, using LEFM. Now consider the case of a plain, uncracked specimen, for which $a = 0$. If we note that failure occurs at $K = K_c$ (for any cracked specimen) and at $\sigma = \sigma_o$ (for the plain specimen) then, making these substitutions in Eq. (3.4) we arrive at the result:

$$a_o = \frac{1}{\pi}\left(\frac{K_c}{F\sigma_o}\right)^2 \tag{3.5}$$

Thus we can see that the length of the imaginary crack, a_o, is the same as our critical distance L in Eq. (2.5), with one small difference: the parameter, F^2. For the particular case of a central, through-crack in a large plate, the value of F is unity and so $a_o = L$, the size of the imaginary crack is exactly the same as our critical distance. For many other practical cases the value of F is quite close to unity – for example, it has a value of 1.12 for an edge crack and values of the order of 0.7–0.8 for typical embedded elliptical flaws.

For the case of $F = 1$, we can also show that predictions of the effect of crack length will be identical for the ICM and the LM. The proof is as follows: the stress–distance curve for a crack can be described by the Westergaard equation (see Section 1.4):

$$\sigma(r) = \frac{\sigma}{\left[1-\left(\frac{a}{a+r}\right)^2\right]^{1/2}} \tag{3.6}$$

According to the LM, failure will occur when the average stress over a distance from $r = 0$ to $r = 2L$ is equal to σ_o, thus:

$$\frac{1}{2L}\int_0^{2L} \sigma(r)\,dr = \sigma_o \tag{3.7}$$

We can therefore predict the fracture stress, σ_f, by combining Eqs 3.6 and 3.7 and setting the nominal applied stress σ equal to σ_f. The result is

$$\sigma_f = \sigma_o\sqrt{\frac{L}{a+L}} \tag{3.8}$$

Now, to find a prediction for σ_f using the ICM, we first write Eq. (3.4) in its critical form

$$K_c = F\sigma_f\sqrt{\pi(a+a_o)} \tag{3.9}$$

and then rewrite it, noting that when $a = 0$, $\sigma_f = \sigma_o$:

$$K_c = F\sigma_o\sqrt{\pi a_o} \tag{3.10}$$

Combining these last two equations gives the result:

$$\sigma_f = \sigma_o\sqrt{\frac{a_o}{a+a_o}} \tag{3.11}$$

It can be seen that, for the case of $F = 1$ when $a_o = L$, Eq. (3.11) is identical to Eq. (3.8), thus the predictions of the LM and ICM are the same. For other values of F the two methods will coincide at the two extreme cases of $a = 0$ (plain specimen) and $a >> a_o$ (long crack). For intermediate values of crack length the two methods will give different predictions: one can see that these differences will be greatest for values of crack length close to L, at which point the difference will be of the order of F (i.e. about 10% if $F = 1.1$). We can conclude that whilst there is an important philosophical difference between the LM and the ICM, which is that the value of the critical length parameter a_o is not strictly speaking a material constant, but varies also with crack shape, nevertheless the two methods will always give similar predictions of the effect of crack length, and will be mathematically identical in the case of $F = 1$.

This result is interesting and by no means obvious, because the LM and the ICM use two fundamentally different approaches to the prediction of failure. The LM is based on an equivalence of stresses – average stress close to the crack is equated to the material's plain strength – whilst the ICM is a fracture mechanics method which relies on an equivalence of energies, the energy release rates for crack propagation, as explained in Section 1.5.

Incidentally, we can carry out exactly the same derivation in the case of fatigue, simply substituting the fatigue limit $\Delta\sigma_o$ for σ_o and the crack propagation threshold ΔK_{th} for K_c: the mathematical argument is identical. This will also apply to subsequent derivations below. This similarity between HCF and brittle fracture is obvious, but often forgotten by researchers who tend to specialise in one field to the exclusion of the other.

It is not so easy to compare the ICM to the LM and PM for the case of a notch, because the relevant equations – for the stress field near a notch and the K value of a notch-plus-crack – are different for different types of notch, and in some cases the necessary closed-form solutions do not exist. There is one trivial case in which all three solutions will be identical: that of a large, blunt notch. In this case the stress gradient near the notch root will be sufficiently low that the notch-root stress σ_{max} exists virtually unchanged over distances r of the order of L. This case is identical to that of a plain specimen loaded to $\sigma = \sigma_{max}$. There are two rather more interesting cases worth considering, as follows:

(a) Circular holes of varying size. As we saw in Section 2.7, the strength of specimens containing circular holes varies with hole radius ρ even though K_t is constant at a value of 3. In later chapters, we will see examples of this behaviour for both brittle fracture and fatigue.

(b) Long thin notches, in which $D >> \rho$ and $D >> L$. Data on the fracture behaviour of these notches is conveniently expressed in terms of the 'measured toughness', K_{cm}, which is defined as the value of K_c obtained by assuming that the notch is a crack. In the limit when $\rho = 0$, we have a crack, therefore $K_{cm} = K_c$. Such notches are discussed in more detail elsewhere (e.g. Section 5.2.2).

Figures 3.4 and 3.5 compare predictions for the two situations, using normalised values of strength and root radius. It can be seen that the solutions are not identical, but in all cases the differences are small: less than 15%. Many stress concentration features found in engineering components will approximate to one or the other of these cases. This analysis of notches has been by no means exhaustive, but can certainly give us the confidence to suggest that these three methods will give effectively similar predictions in many practical cases.

3.3.4 The finite crack extension method: 'Finite fracture mechanics'

The criterion used in this method can be stated as follows:

> "Failure will occur if there is sufficient energy available to allow a finite amount of crack growth, equal to Δa; the value of Δa is assumed to be a material constant."

This is illustrated in Fig 3.3(b); at first sight it seems similar to the ICM, but there is an important difference. In the ICM, we first inserted a crack of fixed size and then asked

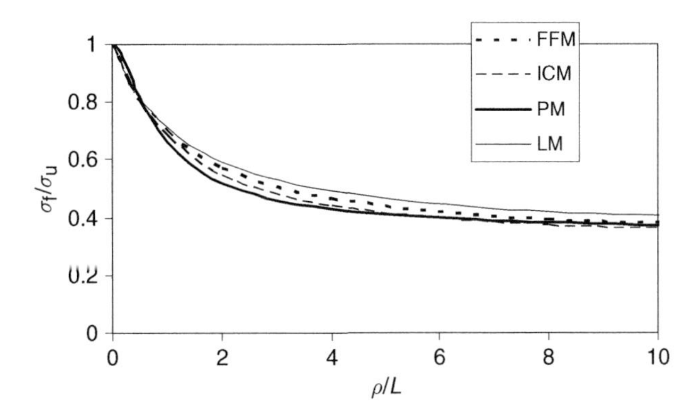

Fig. 3.4. Predictions of fracture stress σ_f (normalised by the plain specimen tensile strength σ_u) for circular holes of radius ρ (normalised by the critical distance L). Four different methods – PM, LM, FFM and ICM – give very similar results.

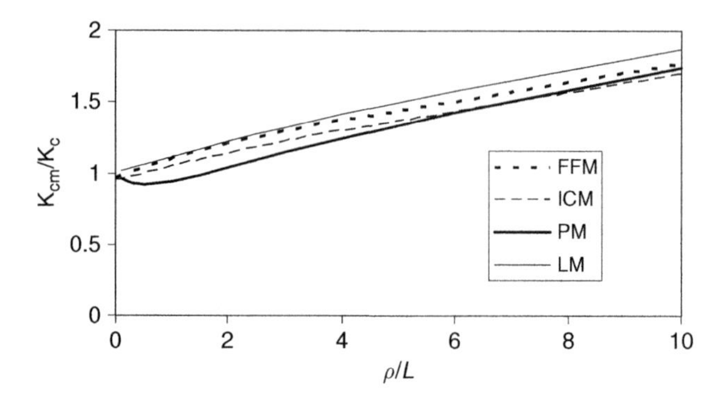

Fig. 3.5. Predictions of the measured toughness K_{cm} (normalised by K_c) for long, thin notches, as a function of notch root radius ρ (normalised by the critical distance L), using the same methods as in Fig. 3.4.

whether it was capable of growing; in the present theory, we start with no crack at the notch root. We then consider what would happen if a crack did form, specifying that the crack must be of a certain size. This is a virtual work argument, of the type used initially by Griffith in deriving the basic equations of LEFM, as outlined in Section 1.5. The difference is that in the normal Griffith approach, the amount of crack growth is assumed to be infinitesimal, allowing one to use the differential dW/da – the rate of change of strain energy with crack length. This quantity is called G, the strain-energy release rate.

By contrast, in the present model, which we have termed 'finite fracture mechanics' (FFM), the amount of strain energy released is calculated by integration. Thus, for a pre-existing crack of length a, the change in strain energy for a finite extension Δa is given by:

$$\int_{a}^{a+\Delta a} dW \tag{3.12}$$

We equate this to the amount of energy needed for crack growth, which is $G_c \Delta a$. This equation can conveniently be expressed in two other ways, either in terms of the strain-energy release rate

$$\int_{a}^{a+\Delta a} G\, da = G_c \Delta a \tag{3.13}$$

or in terms of the stress intensity (using Eq. 1.12):

$$\int_{a}^{a+\Delta a} K^2 da = K_c^{\,2} \Delta a \tag{3.14}$$

For the case of a notch the approach is just the same except that the initial crack length will be zero, so the limits of the integral are 0 and Δa. The great advantage of Eq. (3.14) is that it can be used for any notch for which a solution exists for the stress intensity as a function of crack length. Many such solutions already exist, not only for notches but for other stress concentration features; see, for example, the *Stress Intensity Factors Handbook* (Murakami, 1987).

3.3.5 Linking FFM to the other methods

We apply FFM to the case of a central through crack in an infinite body, using the expression for strain-energy release rate from Eq. (1.9):

$$G = \frac{dW}{da} = \frac{\sigma^2}{E}\pi a \tag{3.15}$$

Using Eq. (3.14) and letting $\sigma = \sigma_f$ we get the following result:

$$\sigma_f = \sqrt{\frac{G_c E}{\pi\left(a + \frac{\Delta a}{2}\right)}} \tag{3.16}$$

This is the same as the normal Griffith fracture stress (Eq. 1.10) except for the extra term $\Delta a/2$. Since $G_c E = K_c^2$, this equation is identical to Eq. (3.9) (F being equal to unity in this case) proving that the predictions of FFM are identical to those of the ICM, with the finite crack extension Δa being equal to $2L$. We already showed earlier that the ICM and LM predictions coincided, so we can conclude that, at least for this particular case of a sharp crack, the three methods give identical predictions, using critical distances which are equal to L for the ICM and $2L$ for both the LM and FFM. Again this result is not obvious, because the criteria for failure used in the three methods are quite different.

This proof, applicable only to sharp cracks, cannot easily be extended to the more general case of notches or other stress concentration features. However, we can show that FFM gives similar predictions for the two cases mentioned earlier: circular holes and long slots. The FFM predictions are also shown in Figs 3.4 and 3.5. Further work on FFM, including its use to predict fracture and fatigue in various materials, has been discussed in a recent paper (Taylor et al., 2005). Though the concept of finite crack extension has been suggested recently by other workers (Seweryn and Lukaszewicz, 2002), the solutions as expressed above in Eqs (3.13) and (3.14) – which allow the method to be used in many practical situations – were first put forward by myself and co-workers.

I regard the FFM as being a very important development, not only because it gives predictions similar to those of the other methods but, more importantly, because I believe it suggests a mechanistic explanation for the success of the PM and LM. This matter will be discussed in more detail later, in Chapter 13, after we have had the opportunity to see how the TCD works in practice.

3.3.6 Combined stress and energy methods

Another approach, investigated recently by ourselves and by some other workers, is the use of a criterion which is a combination of two of the above methods: one stress criterion and one energy criterion. For example, Hitchen et al. combined the LM with the ICM (Hitchen et al., 1994). More interestingly, Leguillon combined the PM and FFM, applying the solution to the prediction of brittle fracture in sharp V-shaped notches of the brittle polymer PMMA (Leguillon, 2002). He assumed that both the PM and FFM were necessary conditions for fracture. This means that (i) there should be sufficient energy to allow a crack to extend by Δa and (ii) there should be sufficient stress in this region of the specimen to allow the material to fracture. The virtual work argument of the FFM only tells us that the fracture process is thermodynamically possible: Leguillon argued that it still might not occur if there was not enough stress to actually break atomic bonds in the relevant region.

We considered a similar approach, but using the LM and FFM instead (Taylor and Cornetti, 2005; Cornetti et al., 2006). In both cases the mathematics becomes quite complex, so it will not be reproduced here – further details can be found in the papers. The important point to realise is that, because we have introduced a new criterion, we must relax a degree of freedom somewhere else in order to be able to solve the equations. What happens in fact is that the critical distance, Δa, is no longer a constant, but takes a value which can be calculated from the equations. In many cases this value turns out to be similar to $2L$, as before. The most important case in which it deviates significantly from $2L$ is when the size of the specimen becomes small. For example, consider a beam of height h loaded in three-point bending (Fig. 3.6). In this case there is no notch in the beam, but a stress gradient will occur anyway due to the bending moment. The figure shows predictions using the LM and FFM: the height of the beam is normalised by $2L$ and the fracture stress by σ_o. For large beams ($h >> 2L$) the predictions are similar, but as the normalised height approaches unity, the solutions diverge, becoming asymptotic in two different directions.

One can see what is happening here: as h reduces to $2L$, the LM begins to average stresses over the entire beam; since this average stress will be zero, it will become infinitely difficult to break the beam. On the other hand, the FFM is now modelling a

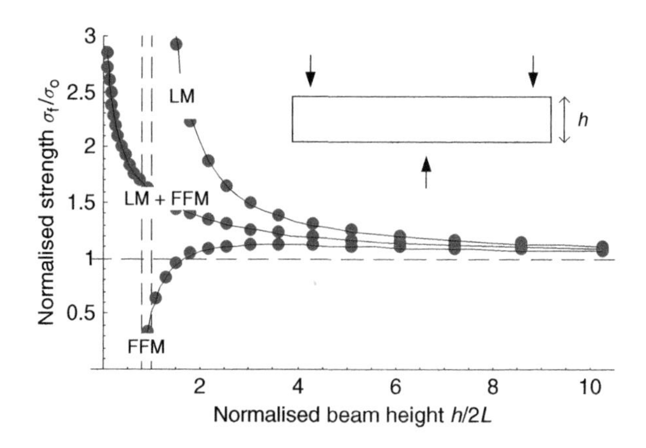

Fig. 3.6. Predictions of strength as a function of height for a beam in three point bending, using the LM, FFM and combined (LM + FFM) approaches.

crack which passes through the entire beam, so the fracture process becomes infinitely easy. Clearly neither of these solutions makes any sense: we can expect that this will always be the case – any of the above four methods will break down when the size of the specimen becomes similar to, or less than, the critical distance. However, this problem does not occur for the combined solutions, as we see from the predictions of the combined FFM/LM in the figure. The function is more well behaved, becoming asymptotic only at $h = 0$. As $h/2L$ decreases, we find that the value of Δa also decreases, remaining always less than h but approaching h as h approaches zero. This approach was successfully used to predict the failure of concrete beams, both with and without notches.

As well as providing improved solutions in some cases, these combined methods also shed light on the general question of 'Why does the TCD work?'. Detailed discussion on this matter will be reserved for Chapter 13, suffice it to say that the cases where a constant value of L can be used seem to be those in which L can be equated with a microstructural parameter such as the grain size, whereas, when L appears to be a variable quantity, it seems to be associated with a zone of damage in the material. In this respect, these combined methods have something in common with the damage-zone models (also called process-zone models) which will be discussed in the next chapter.

3.4 What is the TCD? Towards a General Definition

We have seen that these four methods – the PM, LM, ICM and FFM – are all remarkably similar. Not only do they give similar predictions, but they all use a critical distance which is simply related to L, a parameter which is itself a function of the two material constants K_c and σ_o. Furthermore, all four methods have something else in common: they are all linear, elastic continuum mechanics methods. Following on from Neuber's initial idea, we now have four methods in which continuum mechanics approaches can be used, with the fact of material inhomogeneity being introduced through a single parameter with the dimensions of length.

The similarity of these four methods leads one to suspect that they are really all expressions of the same underlying theory, despite apparent differences in their details. For this reason, I have chosen to define the TCD in such a way as to include all four of these methods.

Therefore, as a formal definition of the TCD, we can say that it is the name given to a group of methods, all of which use linear elastic analysis and a constant critical distance. Two of these methods – the PM and LM – calculate a stress value and equate it to a characteristic strength for the material; the other two methods – the ICM and FFM – use energy concepts to consider the propagation of a crack of finite size, and thus use the material parameters of G_c or K_c. Predictions obtained from the four methods are sufficiently similar that any one of them can be used in practice, the choice depending largely on convenience. For example, if the results of FEA are available, as is generally the case for industrial components, then the PM or LM will be found to be most convenient, whereas the ICM and FFM have the advantage that they can be expressed in the form of equations, at least for certain cases, allowing parametric studies to be conducted more easily.

Finally, one also has the possibility to combine one of the stress-based methods with one of the energy-based methods. These combined methods are computationally more difficult but may be appropriate in cases where the above methods break down, especially in the case of components whose size is small compared to L.

References

Cornetti, P., Pugno, N., Carpinteri, A., and Taylor, D. (2006). A coupled stress and energy failure criterion. *Engineering Fracture Mechanics* **73**, 2021–2033.

El Haddad, M.H., Smith, K.N., and Topper, T.H. (1979) Fatigue crack propagation of short cracks. *Journal of Engineering Materials and Technology (Trans.ASME)* **101**, 42–46.

Hitchen, S.A., Ogin, S.L., Smith, P.A., and Soutis, C. (1994) The effect of fibre length on fracture toughness and notched strength of short carbon fibre/epoxy composites. *Composites* **25**, 407–413.

Kinloch, A.J., Shaw, S.J., and Hunston, D.L. (1982) Crack propagation in rubber-toughened epoxy. In *International Conference on Yield, Deformation and Fracture, Cambridge* pp. 29.1–29.6. Plastics and Rubber Institute, London.

Kinloch, A.J. and Williams, J.G. (1980) Crack blunting mechanisms in polymers. *Journal of Materials Science* **15**, 987–996.

Klesnil, M. and Lukas, P. (1980) *Fatigue of metallic materials.* Elsevier, Amsterdam.

Lazzarin, P., Tovo, R., and Meneghetti, G. (1997) Fatigue crack initiation and propagation phases near notches in metals with low notch sensitivity. *International Journal of Fatigue* **19**, 647–657.

Leguillon, D. (2002) Strength or toughness? A criterion for crack onset at a notch. *European Journal of Mechanics A/Solids* **21**, 61–72.

McClintock, F.A. (1958) Ductile fracture instability in shear. *Journal of Applied Mechanics* **25**, 582–588.

McClintock, F.A. and Irwin, G.R. (1965) Plasticity aspects of fracture mechanics. In *ASTM STP 381 Fracture Toughness Testing and its Applications* pp. 84–113. ASTM, Philadelphia, USA.

Murakami, Y. (1987) *Stress intensity factors handbook.* Pergamon Press, Oxford, UK.

Neuber, H. (1936) *Forschg.Ing.-Wes.* **7**, 271–281.

Neuber, H. (1958) *Theory of notch stresses: Principles for exact calculation of strength with reference to structural form and material.* Springer Verlag, Berlin.

Novozhilov, V.V. (1969) On a necessary and sufficient criterion for brittle strength. *Prik.Mat.Mek.* **33**, 201–210.

Ostash, O.P. and Panasyuk, V.V. (2001) Fatigue process zone at notches. *International Journal of Fatigue* **23**, 627–636.

Peterson, R.E. (1938) Methods of correlating data from fatigue tests of stress concentration specimens. In *Stephen Timoshenko Anniversary Volume* pp. 179. Macmillan, New York.

Peterson, R.E. (1959) Notch-sensitivity. In *Metal Fatigue* (Edited by Sines, G. and Waisman, J.L.) pp. 293–306. McGraw Hill, New York.

Seweryn, A. and Lukaszewicz, A. (2002) Verification of brittle fracture criteria for elements with V-shaped notches. *Engineering Fracture Mechanics* **69**, 1487–1510.

Sonsino, C.M., Radaj, D., Brandt, U., and Lehrke, H.P. (1999) Fatigue assessment of welded joints in AlMg 4.5Mn aluminium alloy (AA 5083) by local approaches. *International Journal of Fatigue* **21**, 985–999.

Tanaka, K. (1983) Engineering formulae for fatigue strength reduction due to crack-like notches. *International Journal of Fracture* **22**, R39–R45.

Taylor, D. (1999) Geometrical effects in fatigue: A unifying theoretical model. *International Journal of Fatigue* **21**, 413–420.

Taylor, D. and Cornetti, P. (2005) Finite fracture mechanics and the theory of critical distances. In *Advances in Fracture and Damage Mechanics IV* (Edited by Aliabadi, M.H.) pp. 565–570. EC, Eastleigh UK.

Taylor, D., Cornetti, P., and Pugno, N. (2005) The fracture mechanics of finite crack extension. *Engineering Fracture Mechanics* **72**, 1021–1038.

Taylor, D. and Wang, G. (2000) The validation of some methods of notch fatigue analysis. *Fatigue and Fracture of Engineering Materials and Structures* **23**, 387–394.

Usami, S., Kimoto, H., Takahashi, I., and Shida, S. (1986) Strength of ceramic materials containing small flaws. *Engineering Fracture Mechanics* **23**, 745–761.

Waddoups, M.E., Eisenmann, J.R., and Kaminski, B.E. (1971) Macroscopic fracture mechanics of advanced composite materials. *Journal of Composite Materials* **5**, 446–454.

Whitney, J.M. and Nuismer, R.J. (1974) Stress fracture criteria for laminated composites containing stress concentrations. *Journal of Composite Materials* **8**, 253–265.

CHAPTER 4

Other Theories of Fracture

A Review of Approaches to Fracture Prediction

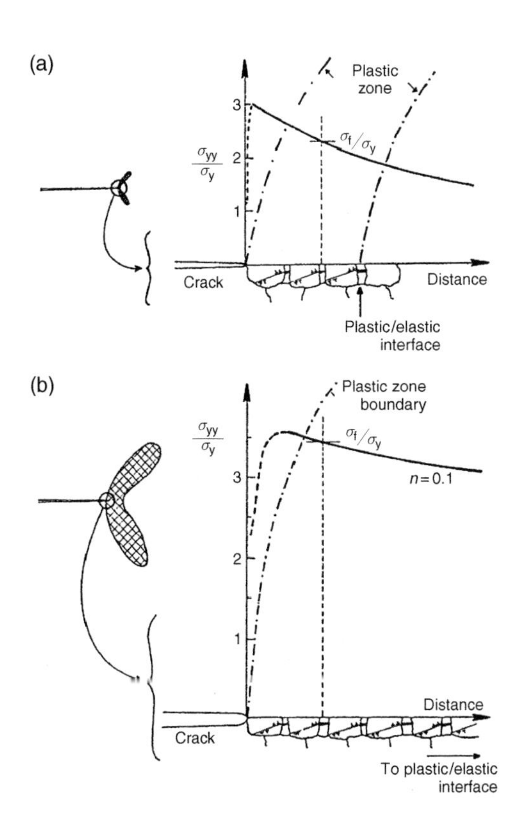

Fig. 4.1. The RKR model: an example of a mechanistic model. The figure, taken from the original paper (Ritchie et al., 1973), shows stress distributions in grains ahead of cracks at (a) low temperature and (b) higher temperature.

4.1 Introduction

Having discussed the group of theories which we refer to as the TCD, we now turn to other theoretical models which are being used to predict failure in materials. There are many such theories, a reflection of the fact that fracture processes are complex and still only partially understood. The aim of this chapter will be to introduce the main types of theory, with some specific examples, pointing out their inherent strengths and weaknesses. More examples of particular models will be presented in subsequent chapters, when we discuss the methods used to predict specific types of failure, in specific materials. We will return to a general discussion of all these theories in Chapter 13, when they will be compared with the TCD.

It is fair to say that the prediction of phenomena such as brittle fracture and fatigue has turned out to be much more difficult than anyone would have expected. There was a time, in the 1960s when fracture mechanics was becoming established, when experts in the field were saying that the major theoretical problems had been solved, and all that remained was to implement existing theories for particular materials. But as we have delved deeper into the problem, it has become more complex.

The current trend is to take advantage of modern computer power, which enables us to make predictions using complex simulations rather than by deriving analytical solutions. This has, for example, made it possible to investigate multi-factor mechanistic models, to analyse the complex shapes, of real components, and to model processes in which failure occurs gradually, by accumulation of damage. Some models are easier to apply in computer simulations than others, and this has tended to influence the direction of recent developments.

4.2 Some Classifications

A precise classification of all relevant theories is difficult, but it is useful to begin by dividing them into *mechanistic models* and *continuum mechanics models*. The starting point for a mechanistic model is the actual, physical mechanism of fracture. A theory is then constructed to try to represent this mechanism, either analytically or as a numerical simulation. Non-mechanistic models are, by definition, continuum mechanics models; fracture is assumed to occur when certain conditions are fulfilled which can be expressed in terms of continuum mechanics parameters such as stress, strain or energy. We can simplistically think of these two types of model as the 'science' and 'engineering' of fracture prediction: mechanistic models have as their aim the improved understanding of the real, physical processes involved. But, due to their complexity and uncertainty, they are not usually able to make accurate predictions of failure in real engineering structures. Continuum mechanics models contain the necessary simplifications to allow them to be used as part of the design process. Of course, both of these types of model are vital to progress in this field, and they tend to grow by feeding from each other. In fact, there is considerable overlap between the two: a mechanistic model will often use continuum mechanics assumptions at some point, whilst a continuum mechanics model may include simplified representations of real fracture processes.

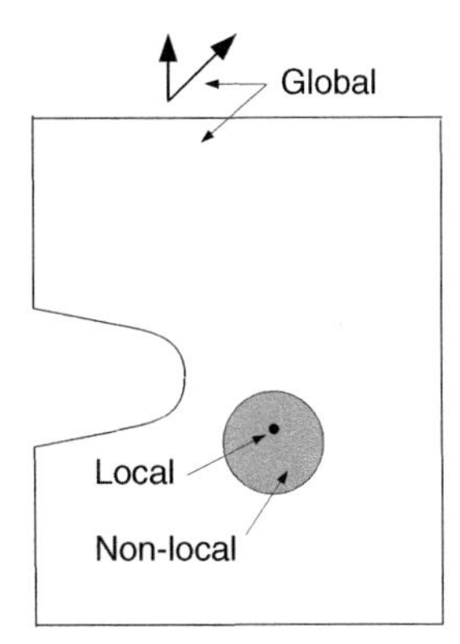

Fig. 4.2. Global, local and non-local theories.

A further classification of continuum mechanics models divides them into three types: *global, local* and *non-local* (Fig. 4.2). A global model is one which predicts fracture by considering the entire body: one example is a classic LEFM model of a body containing a crack – failure occurs at a critical value of the stress intensity, K, which is a function of the geometry and loading of the whole body. In a local model, by contrast, the failure criterion is applied at each individual point in the body. Traditional approaches to fatigue life prediction often work in this way, and have been preserved in post-processing software. Consider an FEA of a component giving the cyclic stress range, $\Delta\sigma$, at every nodal point. The number of cycles to failure for each point can be calculated using $\Delta\sigma$ and the appropriate stress–life curve for the material. The important feature is that the calculations at a particular point use only data obtained at that point, and not elsewhere in the body.

Local models are useful because they can be implemented easily in complex bodies, and can be interfaced with statistical models. They sometimes work well, usually in cases where the stress gradient (i.e. the rate of change of stress with distance) is small, but often they are too simplistic to be reliable. Many models, such as, for example, the Beremin model for cleavage fracture in steels (see Section 7.3.2) began life as local models but, as they developed, took on non-local characteristics to enable them to make realistic predictions. Non-local models, like local models, make calculations at each individual point in the body; the difference is that the calculations use data not only from that point but also from other points, often in a prescribed surrounding volume of material. Such models can take account of stress-gradient effects, but at the price of increased complexity.

Whilst this classification into three groups is a useful way to think about these models, it can, in practice, be difficult to decide which category a particular model belongs to. For example, our PM in the TCD appears at first sight to be a local model because it uses only stress information at a point, but in fact it is non-local, because the location of the point (at a distance $L/2$ from the point of maximum stress) brings in some information from elsewhere in the body. Furthermore the basic philosophy of the local and non-local models is that they work by analysing all points in the body, rather than concentrating attention on a particular stress-concentration feature as we do in the TCD.

4.3 Mechanistic Models

A key feature of any mechanistic model is simplification. At the outset, one should realise that a perfect description of all aspects of a fracture process is impossible, especially considering the behaviour of the material at the microstructural level. An effective model will identify the key features of the process to be considered and will describe those features as simply as possible in analytical or numerical form. This is not easy, because in order to be more realistic one is tempted to include more and more features of the real process, adding more and more degrees of freedom to the model. These models always tend to work – the problem is that one can have so many degrees of freedom that it becomes possible to predict anything, simply by changing the constants in the equations.

An example of a simple and effective model is the Ritchie Knott and Rice (RKR) model of brittle fracture in steels (Ritchie et al., 1973), which is discussed in Section 7.3.2. A simple assumption – that failure is initiated at cracked carbides in grain boundaries – leads to a model which is relatively easy to implement and capable of predicting several different phenomena. This model is rather similar to our PM (see Fig. 4.1), the only real difference being that it uses an elastic–plastic stress distribution rather than an elastic one. Indeed this may explain the success of the PM in this particular case. However, on a more fundamental level the two theories are very different: the RKR starts from a specific mechanism and tries to describe it in mathematical terms (using continuum mechanics along the way), whilst the PM is essentially a continuum mechanics prediction, introducing a length constant, which subsequently turns out to be similar in magnitude to the grain size.

There are many other mechanistic models in use, some of which will be described in later chapters. At this stage, one further example will suffice, which is the prediction of fatigue in metallic materials. It is well known that fatigue is a two-stage process: in Stage 1, a crack initiates, for example through intense plasticity on a shear band; in Stage 2, the crack propagates, increasing its length on every cycle. These two stages have distinctly different characters: for example, in Stage 1, cracks are often found on planes of maximum shear (and therefore at 45° if the loading is axial tension), whilst in Stage 2, the crack will generally propagate on a plane normal to the maximum tensile stress. The boundary between the two stages is blurred by the presence of short-crack behaviour, in which crack propagation occurs more quickly than would be predicted by LEFM. In some cases, Stage 1 may be effectively bypassed due to pre-existing cracks or flaws in the material. These features have been incorporated into many different models, with varying degrees of sophistication, some of which are discussed in Chapter 9. Whilst the behaviour of long cracks can be quite accurately modelled using LEFM, the Stage 1 and short-crack behaviour has not proved so amenable to description. Some very interesting numerical simulations are now being constructed in which all the individual grains in a sample can be specifically modelled.

4.4 Statistical Models

Statistical models of fracture can be traced to the work of the Swedish engineer Weibull. The basic concept is that the strength of a material sample is not a single-valued, deterministic quantity but rather a variable, stochastic quantity. This idea is naturally linked to the real properties of materials which we can expect will vary from place to place on a microstructural level: another important reason for this variation is the existence of manufacturing flaws and other imperfections, scattered at random throughout the material.

In general, then, one can introduce statistical aspects into any model, but the particular approach developed by Weibull has one further assumption, that of a 'weakest link' process. Given that material varies in quality from place to place, one can assume that, for a body experiencing a uniform stress, failure will occur from the worst place. This assumption is valid for the fracture of a brittle ceramic material, for example, which will fail from the largest flaw present in the sample. The fatigue limit of a metallic material is also amenable to this kind of prediction, since fatigue will occur from the weakest material point (or worst crack-initiating defect); other properties, such as the yield strength, for example, cannot be predicted in this way because, whilst some parts of the specimen will certainly yield before others, σ_y is characterised by a general spread of plasticity throughout the specimen.

Some functional form must be assumed for the stochastic variables; in some cases this information can be found from measurement data, for example the size distribution of inclusions in the material, if these are known to initiate failure. More often the form of the distribution is assumed at the start, the most popular version being that originally proposed by Weibull, whereby the cumulative probability of failure P_f for a particular volume of material under stress σ is expressed in terms of two constants: σ^* (essentially a measure of material strength) and b (which describes the degree of scatter in the distribution), (Weibull, 1939) thus:

$$P_f = 1 - \exp\left[-\left(\frac{\sigma}{\sigma^*}\right)^b\right] \tag{4.1}$$

This approach is particularly useful for predicting statistical size effects, as will be discussed in Chapter 13. In principle, the approach can predict the probability of failure of a component, which is very useful industrially. The most common difficulty is the problem of predicting very low levels of failure probability, often required for safety-critical components, which implies accurate modelling of the extremes of the distribution. An important use of statistical methods has been their incorporation into models of brittle cleavage fracture by Beremin and others (e.g. Beremin, 1983), which is discussed in Chapter 7.

4.5 Modified Fracture Mechanics

The LEFM (Section 1.5) has been extremely successful, within a certain range of problems which can be briefly described as 'long, sharp cracks with small plastic zones'. This has tempted many workers to try to extend the validity of fracture mechanics through modifications to the theory. Three particular modifications can be mentioned here, which are known by the acronyms EPFM, NSIF and CMM.

Elastic plastic fracture mechanics (EPFM) arose from the desire to use fracture mechanics in situations of greater plasticity, where the crack-tip plastic zone is comparable in size to the crack length or specimen dimensions, including the extreme case where the crack is growing in a completely plastic strain field, for example at the root of a notch. Two new parameters – the J integral and the crack-opening displacement – were devised. Chapter 7 contains more discussion of these matters: the crucial point is that increased plasticity often coincides with a change in fracture mechanism, from one of instantaneous, unstable crack growth to one in which complete failure is preceded by the gradual development of cracking or other forms of damage. This means that, in principle at least, failure becomes a property not only of the local conditions near the crack (stress, strain, strain energy etc.) but also of the conditions remote from the crack such as the type of loading and constraint on the body. EPFM is currently being employed to predict component failure, generally through computer simulations, but its more obvious use is in the measurement of toughness in materials for comparative purposes.

The notch stress intensity factor (NSIF) method and the crack modelling method (CMM) approaches arose as attempts to apply LEFM to problems in which the relevant feature was not a crack. Many features occur in components which, whilst they are not actually cracks, nevertheless create local stress fields that are more or less similar to those created by cracks. The NSIF method focuses on a particular type of feature: the sharp V-shaped notch. This notch is defined as having zero root radius and a notch opening angle (α in Fig. 1.7), which is greater than zero. As mentioned in Chapter 1, solutions for the stress field ahead of this type of notch were first obtained by Williams, who showed that, for distances r very much less than the notch length D, they can be written in the following general form (Williams, 1952):

$$\sigma(r) = \Psi r^{-\lambda} \tag{4.2}$$

In the case of a crack, the exponent λ has the value $1/2$ and the constant Ψ becomes $K/(2\pi)^{1/2}$. The value of λ is constant for any given angle α – as the angle increases, λ decreases, eventually becoming zero as α approaches 180° giving us a plain specimen. Several workers have noted that the value of Ψ has the same function as K, so this has been termed NSIF (Boukharouba et al., 1995). Examples of the use of this parameter are given in later chapters (e.g. Sections 6.2.3, 7.3.2 and 9.6); though the approach is theoretically restricted to notches of zero root radius, it does open up to analysis some important practical applications, such as welded joints (Lazzarin et al., 2003), in which the root radius is close to zero.

I developed the CMM in an attempt to apply LEFM to fatigue problems involving sharp notches and other sharp stress concentration features. This followed on from the work of Smith and Miller, who showed that a notch behaves like a crack, from the point of view of HCF, provided its root radius is small enough (Smith and Miller, 1978). A similar effect will be demonstrated for brittle fracture in later chapters. Thus, for Smith and Miller, a sharp notch of length D could be modelled as a crack of the same length.

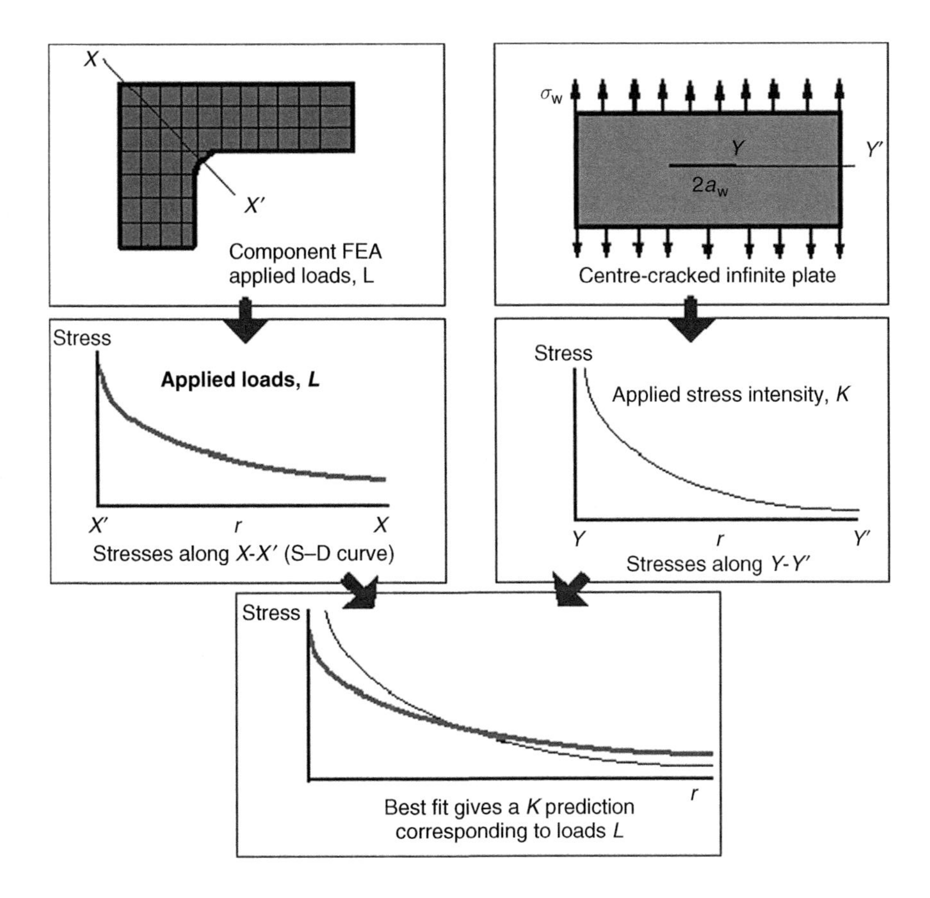

Fig. 4.3. The crack modelling method (CMM).

I extended this approach to consider stress concentration features of any geometry by using the local stress fields. Employing a curve-fitting approach, the stress–distance curve for any feature can be compared to the stress–distance curve for a crack. By optimising the comparison, we identify the crack which is most similar to the feature of interest. The methodology is represented in Fig. 4.3. Some other tests are needed to confirm that the feature in question is sharp enough to qualify as crack-like, but once this is done the method is capable of being interfaced to FEA and can give accurate predictions (Lawless and Taylor, 1996; Taylor, 1996; Taylor et al., 1997).

4.6 Plastic-Zone and Process-Zone Theories

From an early stage in the development of fracture mechanics it was realised that the high stresses near the crack tip would give rise to two phenomena: plastic deformation and damage. Plastic deformation predominates in metals, where it is an important toughening mechanism. In brittle materials, where plasticity is limited or non-existent, damage arises in the form of microscopic cracks, delaminations and so on. In some brittle materials these damage zones can be very extensive and may significantly increase toughness: these materials – examples of which are fibre composites and concrete – are often referred to as *quasi-brittle*. For convenience we will refer to this plastic zone or damage zone as the *process zone*. Estimation of the size of the process zone, and of stress–strain conditions within it, is clearly of importance in understanding these toughening mechanisms. Indeed, some theories of fracture and fatigue have proposed simply that failure will occur when the process zone reaches a certain fixed size. Such models are still used today, though there is no clear theoretical argument to explain why the process-zone size should be constant at fracture.

However, even the estimation of process-zone size and shape turns out to be a surprisingly difficult problem to solve. Initial work by Dugdale for metals and Barenblatt for brittle materials (Barenblatt, 1959; Dugdale, 1960) was developed by Hillerborg and others (Hillerborg et al., 1976) to create a variety of models which are now referred to as 'process zone models' or 'cohesive zone models'. These models are very widely used today: they will be referred to again in subsequent chapters so it is appropriate to say a few words here to outline the basic ideas. Lawn provides a more thorough description of the underlying theory (Lawn, 1993).

The approach is generally restricted to *2D* problems; the process zone is assumed to exist as a line of length d extending from the crack tip (Fig. 4.4). In reality, of course, the process zone will occupy an area (in *3D* a volume) rather than a single line, but this assumption makes the analysis more tractable. For this reason, the method is sometimes referred to as the *fictitious crack* or *cohesive crack* method, since the process zone has been reduced to a crack-like line. The behaviour of material within the zone is represented by a stress–displacement curve as shown in the figure: here the stresses and displacements are in the direction perpendicular to the line. The initial part of the curve represents elastic separation of material across the process zone. The maximum stress σ_p gives the effective strength of the material; the subsequent decrease of stress with displacement is known as the *softening curve*. The length of the process zone is not a fixed value, rather it is assumed that all material along a line of infinite length has the properties given by the stress–displacement curve so that, as the applied load is increased, the amount of material which experiences damage (i.e. which is stretched beyond the maximum point) will likewise increase. For a high enough applied load the process zone will propagate right through the specimen, indicating failure.

These models have achieved great popularity, partly because they can be implemented within FEAs by introducing special elements with the properties given by the

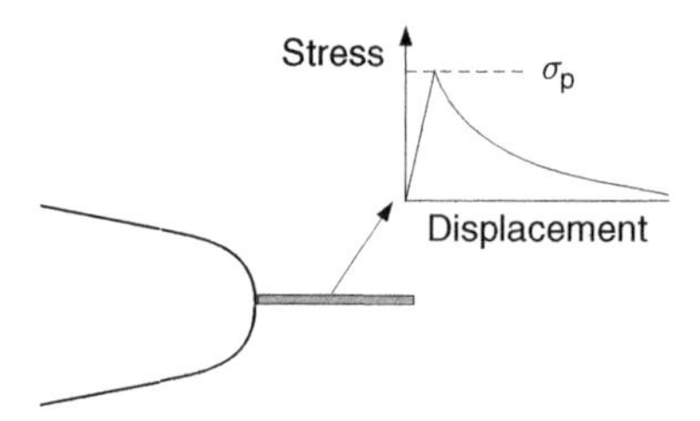

Fig. 4.4. The process zone model.

stress–displacement curve. They have been applied to a range of problems in brittle fracture – especially for quasi-brittle materials (for recent reviews, see Bazant, 2004; Carpinteri et al., 2003) and also in fatigue. Though initially used for bodies containing pre-existing cracks, the theory is now applied also to notches and even to plain specimens in which stress gradients occur, for example, due to bending. The approach could be described as partially mechanistic, since it starts from the idea of a process zone and the failure of material within this zone. However, in most cases no attempt is made to derive the shape of the stress–displacement curve from the actual mechanism of failure. The normal procedure is an empirical one in which the shape of this curve is varied until a good fit is obtained to the available fracture data. In practice, it has proved very difficult to link the parameters of the curve back to any physical mechanisms of deformation and failure in the material.

Whilst the size of the process zone just prior to failure is not a constant, nevertheless it is often of the same order of magnitude as the function $(K_c/\sigma_u)^2$, and therefore this is often used as a general scaling parameter. It is interesting to note that this function is very closely related to the critical distance L in the TCD. The relationship between the TCD and the process zone models will be discussed in Chapter 13.

4.7 Damage Mechanics

In the damage mechanics approach, the level of damage in a material element is represented by a scalar function $\boldsymbol{D}$, which varies from 0 (undamaged material) to 1 (failed material). This approach avoids the necessity to model the physical damage in a realistic way. The main advantage is that it becomes possible to include gradual changes in damage within a complex simulation of the failure of a component or structure. Damage is assumed to be a function of local stress or strain, and of time or number of cycles, so it is possible to incorporate the effects of both fatigue and creep, as well as gradual damage accumulation under increasing monotonic loads in quasi-brittle materials such as composites (see Section 8.6). A particularly useful feature is that the value of $\boldsymbol{D}$ in an element can be linked to other material properties such as elastic modulus or strength, which will tend to decrease as $\boldsymbol{D}$ increases: when $\boldsymbol{D} = 1$, the element can be deleted from the simulation, or given a stiffness of zero. This allows one to model changes in the stress field – for example, a damaged element may, in some circumstances, unload itself, passing stress on to adjacent, undamaged elements.

Damage mechanics models, when combined with FEA, can produce some very realistic simulations. Their main limitations are the simplistic treatment of damage as a single quantity: in reality different types of damage (e.g. microcracking, delamination) may have different effects and may interact in complex ways. Another problem is their sensitivity to factors in the numerical model such as mesh density. They are normally implemented as part of a local approach, the damage in each element depending only on the stress–strain history of that element. Damage mechanics models are not generally used to predict cracking processes because they are not well adapted to deal with stress singularities.

4.8 Concluding Remarks

What emerges most obviously from this chapter is the fact that there are a considerable number of different theoretical approaches, which can be applied to the solution of the same set of problems in the fracture and fatigue of materials. The very fact that there are so many different methods implies that we are still unsure as to what is the best approach in many cases. However, in some ways these approaches can be complementary; for example, mechanistic models help to shed light on the real physical mechanisms of failure, and this should in turn help us to design better continuum mechanics models. That having been said, it is worth remembering that some continuum mechanics models – notably LEFM – are successful despite the fact that they do not incorporate physical mechanisms: LEFM works just as well to describe crack growth by cleavage, for example, as by void coalescence. This is because it describes something which is essential in both mechanisms: the necessary energy for crack growth.

This chapter completes a set of four introductory chapters in which we have discussed the TCD and other theories of fracture prediction, especially in relation to brittle failure under monotonic loading, and fatigue failure under cyclic loading. In the following chapters, we will consider how these approaches are applied in particular cases, starting with monotonic fracture in various different materials, moving on to fatigue failure and subsequently to failures associated with surfaces in contact.

References

Barenblatt, G.I. (1959) The formation of equilibrium cracks during brittle fracture. General ideas and hypothesis, axially symmetric cracks. *Prikl Mat Mekh* **23**, 434–444.

Bazant, Z.P. (2004) Quasibrittle fracture scaling and size effect. *Materials and Structures* **37**, 1–25.

Beremin, F.M. (1983) A local criterion for cleavage fracture of a nuclear pressure vessel steel. *Metallurgical Transactions A* **14A**, 2277–2287.

Boukharouba, T., Tamine, T., Nui, L., Chehimi, C., and Pluvinage, G. (1995) The use of notch stress intensity factor as a fatigue crack initiation parameter. *Engineering Fracture Mechanics* **52**, 503–512.

Carpinteri, A., Cornetti, P., Barpi, S., and Valente, S. (2003) Cohesive crack model description of ductile to brittle size-scale transition: Dimensional analysis vs renormalization group theory. *Engineering Fracture Mechanics* **70**, 1809–1839.

Dugdale, D.S. (1960) Yielding of steel sheets containing slits. *Journal of the Mechanics and Physics of Solids* **8**, 100–108.

Hillerborg, A., Modeer, M., and Petersson, P.E. (1976) Analysis of crack formation and crack growth in concreteby means of fracture mechanics and finite elements. *Cement and Concrete Research* **6**, 777–782.

Lawless, S. and Taylor, D. (1996) Prediction of fatigue failure in stress concentrators of arbitrary geometry. *Engineering Fracture Mechanics* **53**, 929–939.

Lawn, B. (1993) *Fracture of brittle solids*. Cambridge University Press, Cambridge.

Lazzarin, P., Lassen, T., and Livieri, P. (2003) A notch stress intensity approach applied to fatigue life predictions of welded joints with different local toe geometry. *Fatigue and Fracture of Engineering Materials and Structures* **26**, 49–58.

Ritchie, R.O., Knott, J.F., and Rice, J.R. (1973) On the relationship between critical tensile stress and fracture toughness in mild steel. *Journal of the Mechanics and Physics of Solids* **21**, 395–410.

Smith, R.A. and Miller, K.J. (1978) Prediction of fatigue regimes in notched components. *International Journal of Mechanical Science* **20**, 201–206.

Taylor, D. (1996) Crack modelling: a technique for the fatigue design of components. *Engineering Failure Analysis* **3**, 129–136.

Taylor, D., Ciepalowicz, A.J., Rogers, P., and Devlukia, J. (1997) Prediction of fatigue failure in a crankshaft using the technique of crack modelling. *Fatigue and Fracture of Engineering Materials and Structures* **20**, 13–21.

Weibull, W. (1939) The phenomenon of rupture in solids. *Proceedings Royal Swedish Institute of Engineering Research* **153**, 1–55.

Williams, M.L. (1952) Stress singularities resulting from various boundary conditions in angular corners of plates in extension. *Journal of Applied Mechanics* **19**, 526–528.

CHAPTER 5

Ceramics

Brittle Fracture in Engineering Ceramics, Building Materials, Geological Materials and Nanomaterials

5.1 Introduction

In previous chapters I have described the TCD in general terms and compared it with other theories. This chapter will be the first of several which consider the application of the TCD in a specific field, in this case the prediction of brittle fracture in ceramic materials. We begin with this topic because it is theoretically the easiest to understand, offering the simplest and most direct demonstration of the ability of the TCD to predict experimental data. The approach, in this and the following chapters, will be first to outline the general problem posed by a particular class of materials, secondly to demonstrate the accuracy, and any shortcomings, of the TCD predictions when compared against the available data, and finally to discuss alternative methods of prediction, placing the TCD in the context of other work in the literature.

The term 'ceramics' covers a broad range of materials, from traditional building materials, pottery (Colour Plate 1) and geological materials, through high-specification engineering ceramics to the new materials currently being developed for micro- and nano-scale devices. Properties which make ceramics suitable for a wide range of applications include high hardness, good wear resistance and thermal and chemical stability. As regards their mechanical properties, ceramic materials have two important features: high inherent strength and low toughness. These properties arise directly from the nature of the atomic bonding in these materials: 3D lattices of ionic and covalent bonds confer not only the potential for high strength, but also an inability to undergo plastic deformation, which severely limits toughness.

The result is that these materials react strongly to the presence of any form of stress concentration such as a notch, crack or defect. To take a specific example, an engineering ceramic such as silicon nitride may have an inherent strength as high as 1000 MPa; the inherent strength is a quantity which will be defined precisely below, suffice it to say that it is associated with the strength of material containing minimal defects. The same material has a fracture toughness (K_c) of the order of 5–8 MPa(m)$^{1/2}$; this means that a crack or defect as small as 1 mm has the effect of reducing the tensile strength to about 100 MPa. By contrast, most metallic materials, having much higher toughnesses, would be completely unaffected by a crack of this size. Polymers also have low toughness values but since they also have lower strengths (usually less than 100 MPa) the effect of small defects in reducing strength is less of a practical problem.

Because these materials are so sensitive to defects, and because such defects will invariably occur – arising either during processing or in subsequent use of the component – some form of defect tolerance analysis, such as LEFM, is required. In practice however, the difficulty of identifying and measuring the defect responsible for failure has led to an alternative approach, of a statistical nature. In this approach, material strength is considered to be a statistical quantity, usually described by the Weibull equation, which gives the cumulative failure probability, P_f, as a function of the applied stress, σ, with (in its simplest form) two material constants σ^* and b, thus:

$$P_f = 1 - \exp\left[-\left(\frac{\sigma}{\sigma^*}\right)^b\right] \tag{5.1}$$

This approach has some theoretical difficulties which will be discussed below: its practical difficulties include the need for a large amount of test data with which to define the constants, and the fact that the results also depend on the size of the test specimen. These size effects raise enormous problems in the case of building materials such as concrete, which are used in very large section sizes which can have considerably lower strength than any specimen that can conveniently be tested. The same is true for natural materials, such as rocks and ice, which exist in very large volumes. At the opposite end of the scale, the requirement for very small components for nanotechnology devices has lead to the use of brittle ceramic materials such as silicon and carbon in quantities so small that vacancies at the atomic level constitute significant defects.

5.2 Engineering Ceramics

In recent decades, techniques have been developed for producing ceramic materials with very high levels of purity and greatly reduced porosity, allowing them to be used for important load-bearing applications such as engine components. Examples are silicon nitride (Si_3N_4), silicon carbide (SiC) and alumina (Al_2O_3). At the same time, the toughness of these materials has been addressed through intensive research, worthwhile because even a modest increase in toughness has the effect of greatly expanding their range of application. Consequently we know a great deal about the mechanisms of cracking in these materials, and the various ways in which crack growth can be hindered, leading to increased toughness. An important finding is that, unlike metals which achieve toughness largely due to the plastic deformations that occur ahead of the crack tip, in ceramics the important toughening mechanisms are mostly those which act *behind* the crack tip, such as bridging of the crack faces by uncracked ligaments and fibres of material. An exception to this, and an example of a material specifically designed with toughness in mind, is partially stabilised zirconia (PSZ), which achieves improved

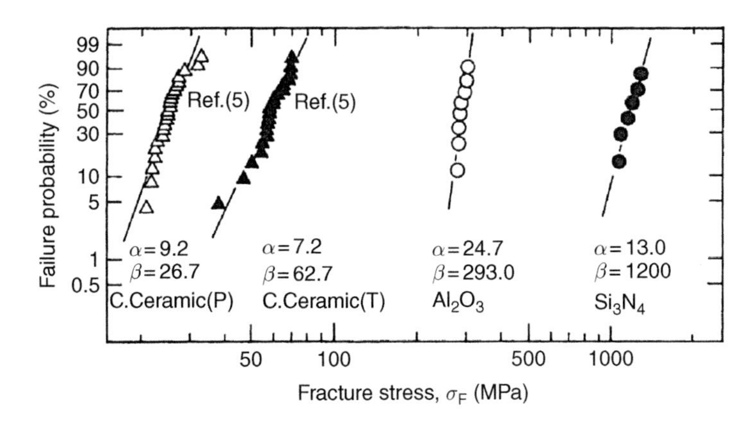

Fig. 5.1. Data from Ando et al. (1992); Weibull distribution of fracture stress in various ceramic materials.

toughness through a transformation reaction driven by stress ahead of the crack tip. For a thorough treatment of these issues the reader is directed towards Brian Lawn's excellent book *Fracture of Brittle Solids* (Lawn, 1993).

Though the Weibull approach is still widely used in assessing engineering components made from ceramic materials, in practice the amount of scatter in mechanical properties is often quite small, thanks to improved methods of processing and quality assurance. For example, Figs 5.1 and 5.2 show some typical data on the measured strength and toughness of several materials (Ando et al., 1992). The Weibull exponent (b in Eq. 5.1, denoted α on these figures) is generally greater than 10, giving a relatively steep curve with little difference between the 10 and 90% probability levels. However, examination of a larger dataset in Fig. 5.3 (Usami et al., 1986) shows an important deviation: at high values of the fracture stress the data fall on a straight line, indicating that they conform to the Weibull equation, but there is a long 'tail', at which the fracture stresses are lower than would be predicted from the Weibull approach.

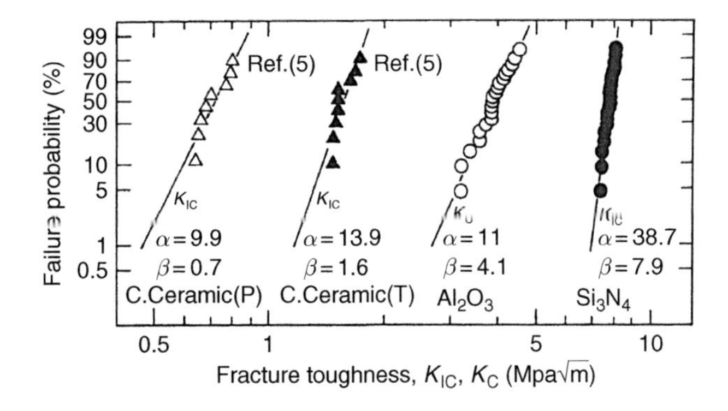

Fig. 5.2. Weibull distribution of fracture toughness for the same materials as Fig. 5.1.

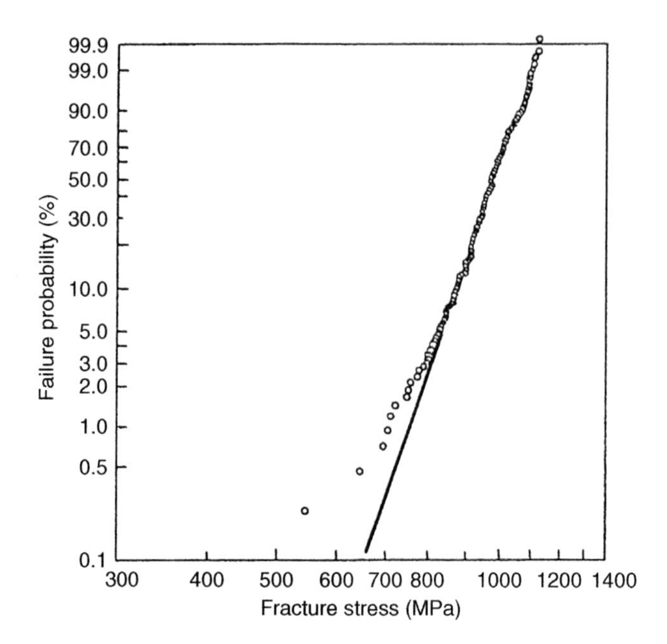

Fig. 5.3. Weibull plot of the strength of Si_3N_4 specimens (Usami et al., 1986) This large dataset shows a 'tail' at low strengths.

These weak specimens will be the ones containing relatively large defects. The same effect may happen to the materials in Figs 5.1 and 5.2 but cannot be seen due to the small dataset used. These results emphasise the potential weaknesses of the statistical approach and the importance of developing a robust analysis of defect tolerance for these materials.

5.2.1 The effect of small defects

Figure 5.4 shows typical experimental data (Kimoto et al., 1985) measuring the strength of specimens of silicon carbide containing small flaws; here the fracture stress, σ_f, is plotted as a function of defect size. Various techniques can be used to introduce these defects, including machining of narrow slots and grooves, cracking induced by contact from a hard indenter, and the identification of naturally occurring defects such as pores and machining marks. In practice, the method used to introduce the defects has a negligible effect, unless residual stresses are introduced, which occurs, for example, if a sharp indenter is used.

When examining this data, the first thing to note is that there is relatively little scatter: the points all fall onto a single curve, with variation of the order of 10%, at least some of which can be attributed to errors in the measurement of failure load and, especially, of defect size. This emphasises the fact that, when defect size is accounted for, these materials display properties which have as little scatter as other classes of materials such as metals.

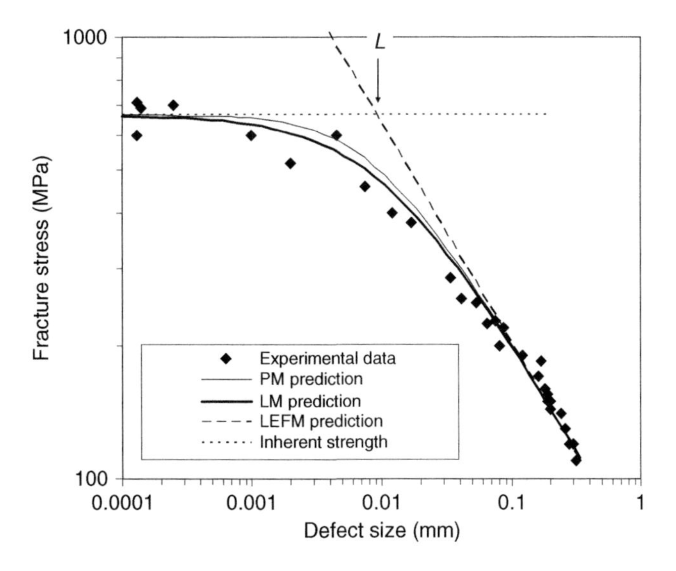

Fig. 5.4. Fracture stress σ_f as a function of defect size in SiC: data from Kimoto et al., 1985; predictions using LEFM and using the TCD (PM and LM). The value of L is given by the intersection of the LEFM line and the line corresponding to the inherent strength.

For large defect sizes the data fall onto a straight line on this logarithmic plot, giving the expected LEFM behaviour for cracks, in which σ_f is linked to crack length, a, through the fracture toughness, K_c, using the standard equation of fracture mechanics:

$$K_C = F\sigma_f\sqrt{\pi a} \tag{5.2}$$

As noted previously in Chapter 1, we will *not* use the convention whereby the plane strain fracture toughness (in Mode I) is denoted by K_{IC} and the plane stress value is denoted by K_c; instead we will use K_c throughout, specifying the degree of constraint where necessary. In practice plane strain conditions prevail in almost all cases for ceramics. In Fig. 5.4, and also in all the similar figures in this chapter, the actual defect size has been modified to give an equivalent size, which is defined as the length of a central, through-thickness crack which would give the same value of K for the same applied stress. Since this through-crack would have an F value of 1 the equivalent size, a_{eq} can be found knowing the actual length a and shape factor F, for the physical defect:

$$a_{eq} = F^2 a \tag{5.3}$$

For defects with 3D shapes such as pores, the shape is imagined to be projected onto a plane normal to the tensile axis, to give an equivalent planar defect. Thus, for example, a spherical pore of radius a will be modelled as a circular crack, which has an F value of 0.64, giving an equivalent length of $a_{eq} = 0.41a$. This modification is convenient because it allows us to compare different shapes of defect on the same graph. In fact we would expect slightly different predictions from the TCD depending on the F value but, as noted previously, any differences in the results will be small, so small that they will tend to be lost in the scatter in the experimental data.

As the defect size is reduced, the measured strength increases, but deviates from the straight-line behaviour of LEFM. For very small defects, σ_f approaches a constant value. It is this value which we will call 'the inherent strength of the material', σ_u. This can be thought of as the strength of material which does not contain any significant defects. It should not be imagined that the material is entirely defect-free: it will certainly still contain defects, such as microscopic pores and inclusions, but it is clear from the graph that these defects must be of such a size that they do not individually act to reduce the material's strength. Drawing a horizontal line on the graph to represent σ_u, we can see that there are three regimes of behaviour: (i) small defects, for which the strength is $\sigma_f = \sigma_u$; (ii) large defects, for which the strength conforms to LEFM, therefore:

$$\sigma_f = \frac{K_C}{\sqrt{\pi a_{eq}}} \tag{5.4}$$

and; (iii) defects of intermediate size, whose strengths are lower than would be predicted using either the constant-stress or constant-K arguments used for the other two categories. These defects constitute a major problem area for damage-tolerance theories: we shall see later that similar problems arise in other types of failure prediction, especially in fatigue where it has received much attention.

Also shown on Fig. 5.4 are predictions made using the TCD: both the PM and the LM. In this case these predictions can be made very simply in analytical form, using the relevant equation for stress $\sigma(r)$ at a distance r from the tip of a central through-crack in an infinite plate (Westergaard, 1939):

$$\sigma(r) = \frac{\sigma}{\left[1 - \left(\frac{a}{a+r}\right)^2\right]^{1/2}} \tag{5.5}$$

This equation accurately describes $\sigma(r)$ at all distances from the crack tip in an infinite body loaded by a uniform nominal tensile stress σ. We need to use this equation, rather than the simplified form more commonly used in fracture mechanics (Eq. 1.5) because the simplified equation applies only when $r << a$; when using the PM and the LM for small defects we need to consider distances similar to, and larger than, a. For both the PM and the LM we define the critical distance, L, as explained in previous chapters, by:

$$L = \frac{1}{\pi}\left(\frac{K_C}{\sigma_u}\right)^2 \tag{5.6}$$

The material considered in Fig. 5.4 has a fracture toughness of $3.7\,\text{MPa(m)}^{1/2}$ and a strength of 667 MPa, giving $L = 0.01$ mm. Using the PM, we set $r = L/2$ and $\sigma(r) = \sigma_u$ in Eq. (5.5), to give $\sigma = \sigma_f$, thus:

$$\sigma_f = \sigma_u\left[1 - \left(\frac{a}{a+L/2}\right)^2\right]^{1/2} \tag{5.7}$$

To use the LM, we require the average stress over a given distance 0 to r, $\sigma_{av}(r)$ which is found by integrating Eq. (5.5), to give

$$\sigma_{av}(r) = \sigma\sqrt{\frac{2a+r}{r}} \tag{5.8}$$

Setting $\sigma_{av}(r) = \sigma_u$, $r = 2L$ and $\sigma = \sigma_f$ in Eq. (5.8) gives the prediction:

$$\sigma_f = \sigma_u\sqrt{\frac{L}{a+L}} \tag{5.9}$$

As Fig. 5.4 shows, the prediction lines for these two methods are quite similar; they necessarily tend to the same values at each end of the curve, when they merge with the straight-line predictions. In between they do separate slightly but in this case both give reasonable predictions, within the scatter of the experimental data.

It is clear that the TCD has been very successful here, giving an accurate estimate of the effect of defect size on strength throughout the entire range of defect sizes. It is worth pointing out, in relation to both this and much of the subsequent data in this chapter, that the prediction is an absolute one in the sense that it contains no adjustable parameters whose values might be changed to obtain a better fit to the data. The values of K_c and σ_u are the only material parameters we require. On this type of plot the value of σ_u essentially fixes the position of the left hand end of the curve, whilst the straight line based on K_c establishes the position of the right hand end. It is useful to note that L can be found by the intersection of the two straight lines, as indicated on the figure. Results in the middle of the plot show the greatest deviation from the straight-line predictions at values of defect size around L. Thus, even without making any quantitative predictions, we can use L to make a useful qualitative judgement about the behaviour of a defect: if the defect size is significantly less than L, we can expect that the defect will have a negligible effect, the strength being approximately that of the defect-free material. Alternatively, if defect size is significantly larger than L, then the defect will behave like a sharp crack, and the normal equations of fracture mechanics will apply.

Figures 5.5–5.9 show further examples of this kind of data, for a variety of different engineering ceramics and different shapes and sizes of defect. Figure 5.5 shows results on Si_3N_4 and Al_2O_3 (Ando et al., 1992); Fig. 5.6 shows data on Sialon (Kimoto et al., 1985); Figure 5.7 shows two further datasets on Si_3N_4 and one on Al_2O_3 (Kimoto et al., 1985 and Taniguchi et al., 1988). Figure 5.8 shows a composite series of results for Si_3N_4, collected by Usami and co-workers (Usami et al., 1986) from various sources; all the materials in this last figure had approximately the same K_c value, whilst σ_u varied from one set of samples to another. These variations resulted from differences in processing conditions, with a clear trend towards a decreasing strength with increasing grain size. In fact, Usami et al showed that most of the variation in this data could be removed if the defect size was normalised by the grain size, d. Interestingly, this also implies that L should be a function of d: in this case the best prediction could be obtained by setting $L = 4d$, though values for individual materials varied in the range d–$10d$. The relationship between L and microstructural dimensions is a matter which we will return to in later discussions.

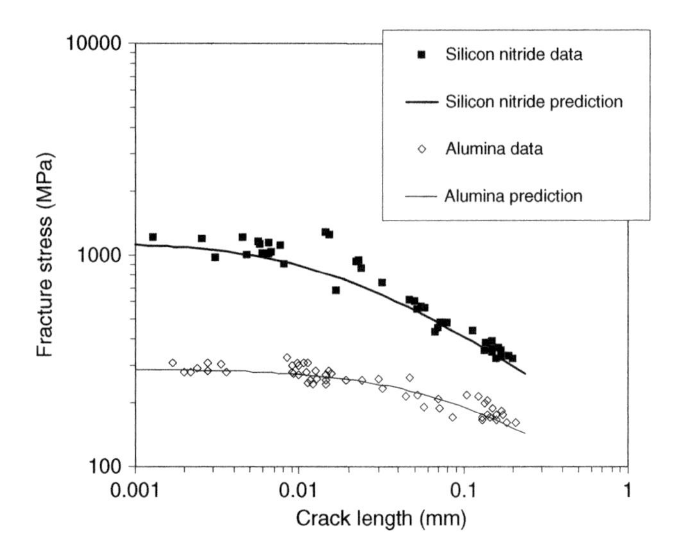

Fig. 5.5. Data from Ando et al. (1992), with predictions using the LM.

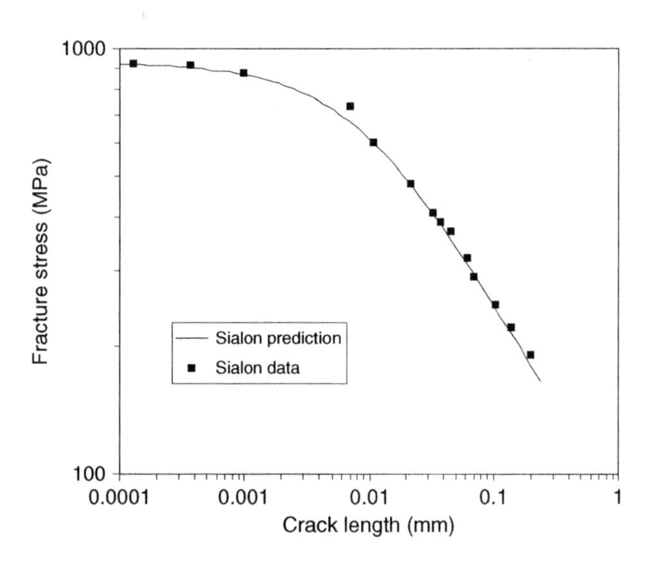

Fig. 5.6. Data on sialon, from Kimoto et al. (1985), with predictions using the LM.

In all of the data presented in Figs 5.4–5.8, it is clear that the TCD provides an accurate prediction of the results: predictions from the LM are shown on the figures, though the PM was also quite accurate. Finally, Fig. 5.9 shows results for soda-lime glass (Kimoto et al., 1985), which fall on the LEFM line indicating a constant value of $K_c = 0.63\,\text{MPa(m)}^{1/2}$ down to lengths as small as $6\,\mu$m. No L value can be determined from this data; this is probably because the material is amorphous, having no grain structure or other microstructural features.

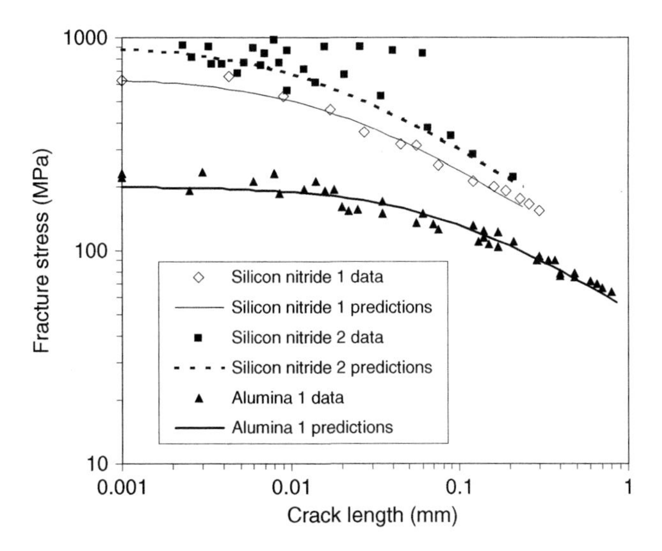

Fig. 5.7. Data from Kimoto et al., 1985 on alumina and two different silicon nitrides; predictions using the LM.

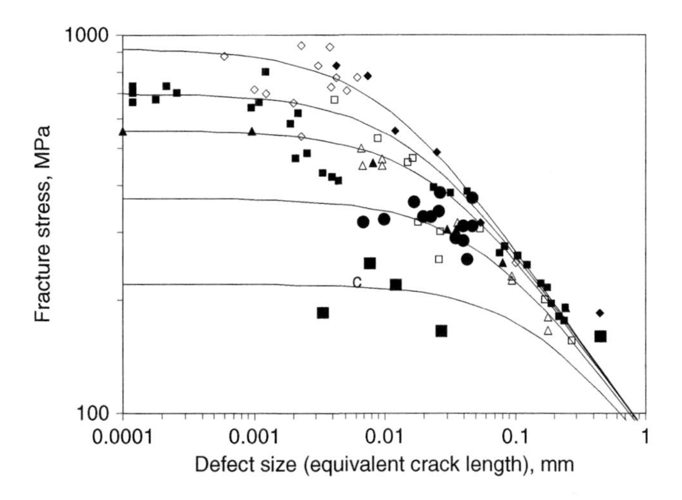

Fig. 5.8. Data collected by Usami et al., (1986) on silicon nitride. Different symbols indicate different material sources: prediction lines using the LM.

The data from the above graphs can be plotted in a different way, by calculating the K value at failure for each defect. This has been done in Fig. 5.10 for the data of Fig. 5.4. The results can be thought of as a series of experimentally obtained values of K_c, but of course K_c should be a material constant. Clearly a valid result for K_c (in this case equal to $3.7\,\mathrm{MPa(m)}^{1/2}$) can only be obtained from relatively long cracks. So the information presented in Fig. 5.4 in terms of strength can here be reinterpreted in terms of toughness: we can say that the measured toughness of a cracked body varies with the crack length, approaching K_c for relatively large cracks and approaching zero for very small cracks.

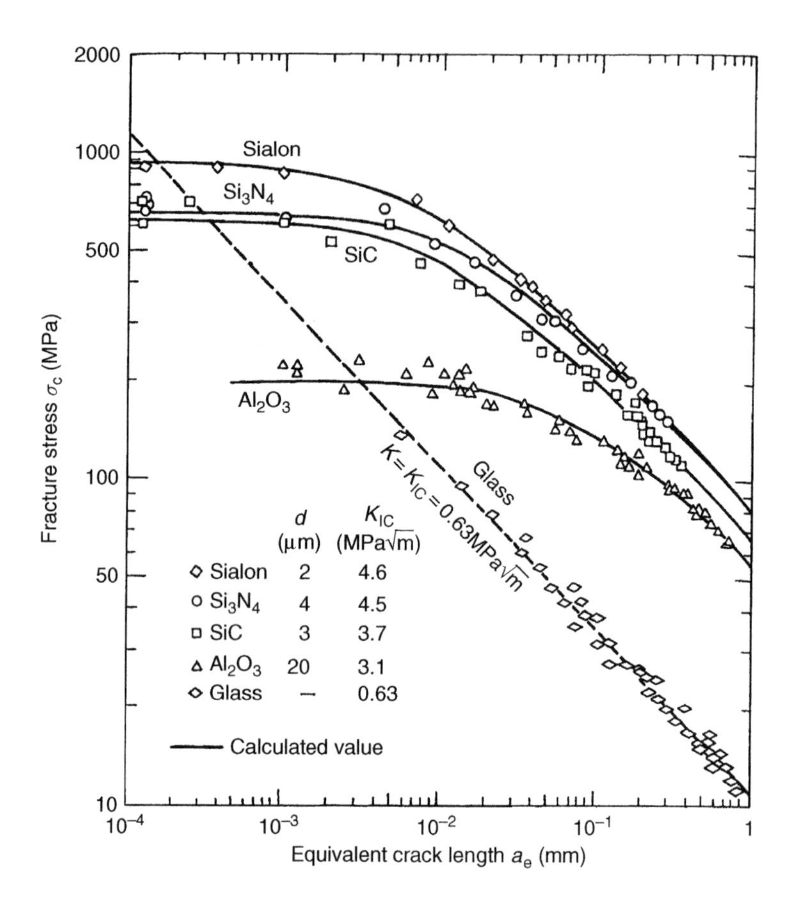

Fig. 5.9. Data on glass compared to other materials (Usami et al., 1986). The dashed line indicates the LEFM prediction.

Again the defining length parameter is L: in order to obtain a valid result for K_c the cracks used in the test specimens should be significantly longer than L.

It was only possible to obtain these types of results, including data for very small defects, because of the high quality of the manufacturing process. One can imagine that if these specimens had been produced using less stringent manufacturing standards, then they would have contained many larger defects, such as internal porosity or surface machining marks. These defects, which we might call 'natural' defects, would clearly have limited the strength of the material in cases where any introduced defects (such as machined notches) were smaller than the natural defects. In this situation the strength/size graphs above would have looked rather different, as Fig. 5.11 shows schematically.

Failure at low values of defect size would be characterised by a large scatter band, representing failure from the natural defects. This scatter band would intersect the line corresponding to the introduced defects at some point. An important issue here, which will become much more important in later discussions on polymeric and metallic

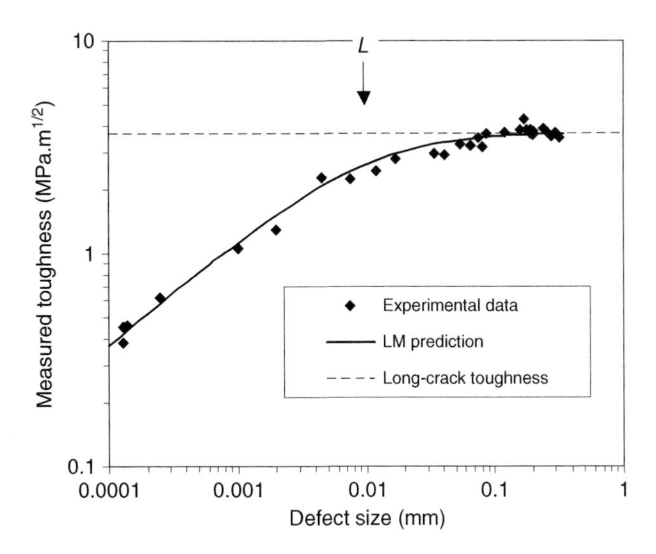

Fig. 5.10. The data of Fig. 5.4 replotted in terms of the measured value of K_c.

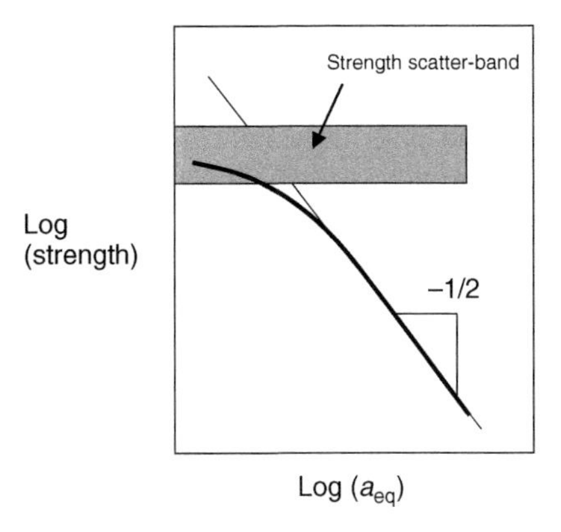

Fig. 5.11. The effect of scatter in plain-specimen strength due to naturally occurring defects. The curved line shows the expected relationship between strength and defect size.

materials, is that the measured strength of the 'plain' specimens (i.e. those with no introduced defects), as well as showing a lot of scatter, is now a function of the size of the natural defects. We would no longer be able to use this σ_u value when making our predictions since it would be lower than the true, defect-free value. However, we could still deduce the value of the inherent strength by ignoring the data in the scatter band and fitting a curve to the data from larger defect sizes. With sufficient knowledge about the nature and distribution of the natural defects we could then go on to predict the scatter band also. Therefore, in the general case we need to make a distinction between the strength of the material as measured using plain specimens, σ_u, and the true, inherent strength, which we will call σ_o. In this chapter on ceramics, the terms can be used interchangeably (if we assume that the specimens used to measure σ_u were free from large defects) but we shall see that in subsequent chapters, on polymers and metals, this distinction between σ_u and σ_o becomes much more significant.

The data and predictions shown above were concerned only with relatively small defects, less than 1 mm in size. This adequately covers the microscopic defects which are likely to be found in these engineering ceramics, which are made to high processing standards. Larger defects cannot necessarily be analysed in the same way because in the above predictions we have assumed that the defects were crack-like, that is they had sharp root radii. This allowed us to use Eq. (5.5), which strictly speaking is only valid for a crack, that is for a linear defect with zero root radius. If, however, we consider defects or, more importantly, design features such as holes and corners, which are relatively rounded or blunt in shape, having larger root radii, then this equation will no longer adequately describe the stress field near the defect. These types of features will be considered in the next section.

5.2.2 Notches

Figure 5.12 shows typical experimental results obtained from tests on specimens of a ceramic material containing notches (Tsuji et al., 1999). In this case the notches used were macroscopic in size, that is the notch length was significantly larger than L. The material tested was alumina. The aim of these tests was to investigate the effect of the notch root radius, ρ.

There are two reasons for studies of this kind. First, engineering components contain design features similar to these notches, which cause stress concentration and thus act as points of potential failure. Secondly, notched specimens similar to these might be used in tests to measure K_c. In practice, it is difficult to introduce macroscopic cracks of controlled length into ceramic materials; they are so brittle that any crack tends to

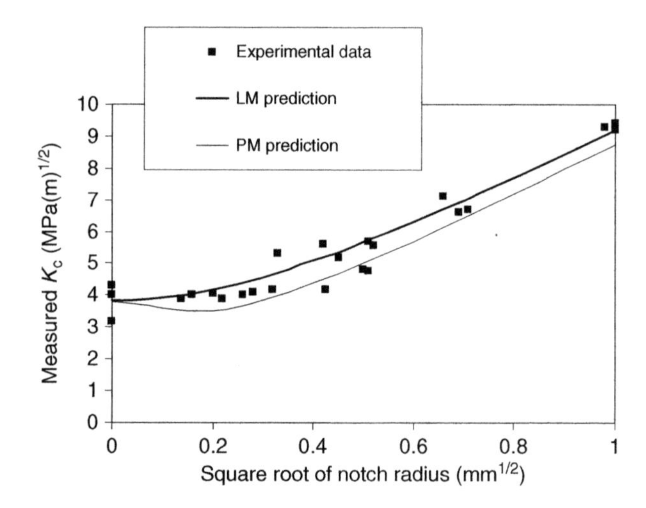

Fig. 5.12. Data on measured toughness of alumina as a function of notch radius (Tsuji et al., 1999). Predictions using the PM and LM.

propagate unstably, right through the specimen. Metallic materials can be pre-cracked by fatigue loading but for most ceramics this is not usually an option because the range of stress levels over which fatigue occurs is too small. Cracks of controlled length can be introduced through contact with an indenter, such as the diamond pyramid used for measuring Vickers hardness, but this also tends to introduce residual stresses into the material around the crack which have to be taken into account in estimating stress intensity (Lawn, 1993). On the other hand, a long, sharp slot can be machined into the specimen relatively easily, causing relatively little residual stress, but this slot will inevitably have a finite root radius.

The data of Fig. 5.12, in which pre-notched specimens were loaded to failure, is presented in terms of the 'measured K_c' value, which is the value of K_c calculated by assuming that the notch is a sharp crack, that is using the standard fracture mechanics equation (Eq. 5.2) with a equal to the notch length D, and F given its appropriate value for the particular shape of notch and specimen. Of course, this 'measured K_c' value is, strictly speaking, only equal to the true fracture toughness of the material when the notch is a crack, that is when $\rho = 0$. It is clear from Fig. 5.12 that the true K_c value (in this case $3.8\,\mathrm{MPa(m)^{1/2}}$) is found not only for sharp cracks but also for all notches up to some critical value of ρ, which in this case is about 0.1 mm. The data have been plotted in terms of the square root of ρ, this being a convention amongst researchers in this field because it has been noticed that, for root radii above the critical value, the measured K_c value is approximately proportional to the square root of the radius. Also shown in the figure are predictions made using the TCD, employing both the PM and the LM once again. For these predictions, a simple analytical solution can be obtained, starting from the equation for the stress as a function of distance for a long, narrow slot (Creager and Paris, 1967), which can be written as follows:

$$\sigma(r) = \frac{K}{\sqrt{2\pi x}}\left(1+\frac{\rho}{2x}\right) \tag{5.10}$$

Here K is the stress intensity value for a crack of the same length (Eq. 5.2 above) and x is distance measured from a point halfway between the notch tip and its centre of radius. This can be rewritten in terms of the distance, r, measured from the notch tip (and therefore from the point of maximum stress) by noting that $x = r + \rho/2$, giving:

$$\sigma(r) = \frac{K}{\sqrt{\pi}}\frac{2(r+\rho)}{(2r+\rho)^3} \tag{5.11}$$

We can apply the PM, as before, by setting $\sigma(r) = \sigma_u$ and $r = L/2$. This will give us a value for K at failure which is the measured K_c; to avoid confusion we will call this K_{cm}. The result is

$$K_{cm} = \sigma_u\sqrt{\pi}\left[\frac{\sqrt{(L+\rho)^3}}{L+2\rho}\right] \tag{5.12}$$

If we remember that $K_{cm} = K_c$ when $\rho = 0$, then we can obtain the following result:

$$\frac{K_{cm}}{K_c} = \frac{(1+\frac{\rho}{L})^{3/2}}{(1+2\frac{\rho}{L})} \tag{5.13}$$

This emphasises that the difference between the measured toughness and the true toughness is a function of the root radius, normalised by the critical distance L. To make predictions using the LM, we calculate the average stress over the distance $r = 0$ to $2L$ by integrating Eq. (5.11). The result is

$$\sigma_{av} = \frac{K}{2L\sqrt{2\pi}}\left(2\sqrt{\frac{\rho}{2}+2L} - \frac{\rho}{\sqrt{\frac{\rho}{2}+2L}}\right) \tag{5.14}$$

Proceeding as before, we obtain the following surprisingly simple result for the ratio between measured and true toughness values:

$$\frac{K_{cm}}{K_c} = \sqrt{\frac{\rho}{4L}+1} \tag{5.15}$$

These analytical solutions are very useful but their limitations should be borne in mind. The Creager and Paris equation is only valid for long, thin slots ($D >> \rho$); when used for smaller, rounder notches it tends to underestimate the local stress (for example, it would predict a stress concentration factor of 2 for a circular hole, instead of the correct result of 3). It is also only valid for considering distances $r << D$, so it should not be used for physically small notches where D is similar to, or less than, L. Finally it assumes infinite body dimensions (width, length, thickness), so corrections would be needed if any of these dimensions was of the same order of magnitude as D. In cases where this equation is not valid the option always exists to obtain the stress data using FEA, which in any case will be almost essential when considering real components.

However, the equations derived above are accurate for most types of test specimen used to obtain notch strength data for ceramics. It can be seen that the PM and LM methods give good predictions for the data on Fig. 5.12. The LM gives a slightly better fit to the whole set of data, whilst the PM emphasises more clearly the almost horizontal portion at low root radii. In fact the curve for the PM prediction actually dips slightly, giving a minimum value at $\rho = L/2$, at which the measured toughness is predicted to be 8% lower than K_c. Whether or not this decrease actually occurs is difficult to tell, given that there will always be some scatter in the experimental data. At $\rho = L$, the PM curve lies slightly above K_c (by just 4%) and, at the same point, the LM curve predicts an increase of 12%, so this is a convenient choice for the critical root radius above which the notch is effectively no longer behaving like a sharp crack.

Figures 5.13–5.17 show further examples of this kind of data, for various materials, using the LM to make predictions throughout. Figures 5.13 and 5.14 show results on Si_3N_4 and SiC respectively (Takahashi et al., 1985); though the total amount of data is rather small in these cases, the values of the input properties K_c and σ_u were given in the publications (in fact these are the same materials for which small-defect results were already presented above in Section 5.2.1); accurate predictions could be made in both cases.

Figure 5.15 shows data on Al_2O_3 with a grain size of 10 μm (Bertolotti, 1973) and electrical porcelain, which is also an alumina-based material but made to a lower

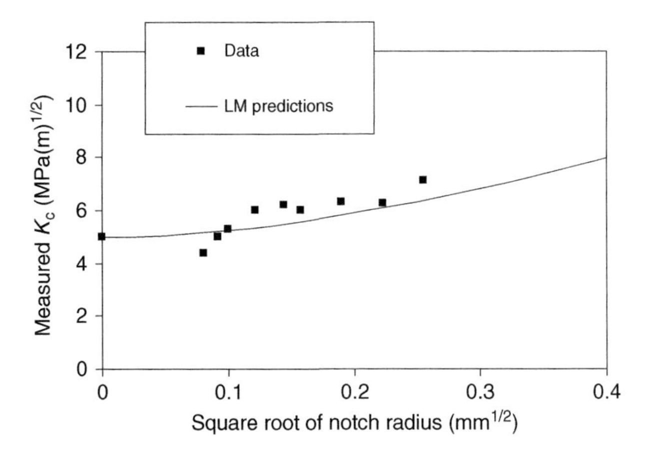

Fig. 5.13. Data on Silicon Nitride from Takahashi 1985; predictions using the LM.

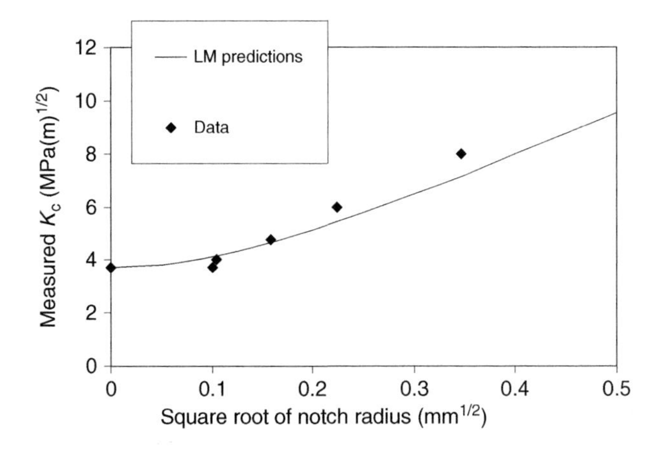

Fig. 5.14. Data on Silicon Carbide from Takahashi 1985; predictions using the LM.

processing standard and having a slightly larger grain size of 15 μm (Clark and Logsdon, 1974). In these cases some material property data was missing, so the predictions were necessarily more speculative in nature. For the Al_2O_3 material, the value of K_c was known, but not the value of σ_u, so L could not be calculated from first principles. Choosing a value 30 μm (which is three times the grain size) gave good predictions using the LM. For the porcelain neither K_c nor σ_u were known: values chosen to give the best fit were 1.1 MPa(m)$^{1/2}$ for K_c and 200 μm for L, which interestingly is 13 times the grain size. This suggests that whilst L may be related to grain size (as noted above in relation to the data of Usami on small defects) it is also affected by other factors.

Figure 5.16 shows data on a relatively tough ceramic: magnesia partially stabilised zirconia (MgPSZ), which had a grain size of 40 μm (Damani et al., 1996). In this case the material constants were not known, and there was no data below the critical

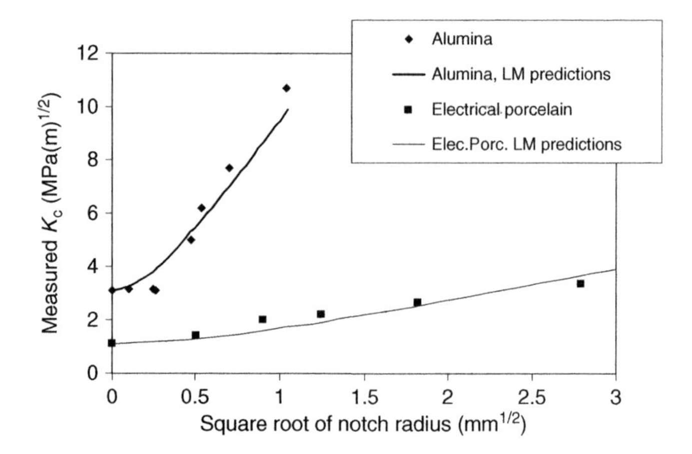

Fig. 5.15. Data on alumina (Bertolotti, 1973) and electrical porcelain (Clark and Logsdon, 1974): predictions using the LM.

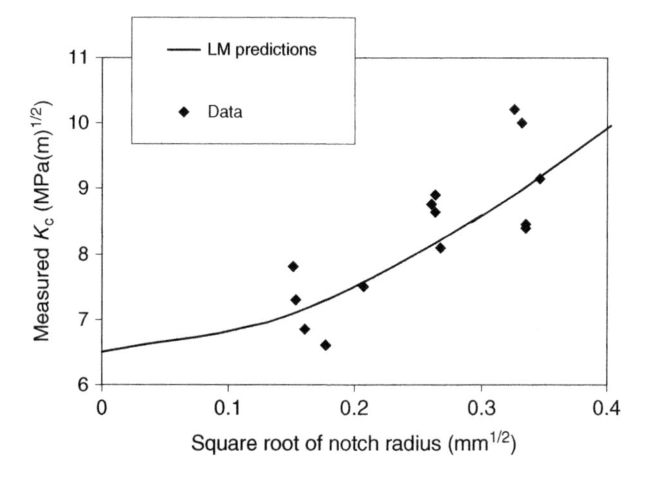

Fig. 5.16. Data on MgPSZ (Damani et al., 1996); predictions using the LM.

root radius, so the analysis is necessarily speculative. Reasonable predictions could be obtained using a K_c value of 6.5 MPa(m)$^{1/12}$ and an L value of 30 μm, which is slightly less than the grain size. Finally Fig. 5.17 displays results for the same alumina material shown in Fig. 5.12, but tested at an elevated temperature of 1000 °C; clearly the TCD also works under these circumstances.

As a final, and slightly unusual, example of an engineering ceramic material, we consider nuclear graphite. Polycrystalline graphite is used in the cores of nuclear reactors; these cores consist of a series of components joined together using keyways. Stresses may be set up at the sharp corners of these keyways due to internal shrinkage and temperature changes; these stresses sometimes cause cracking. To study this problem, Zou and

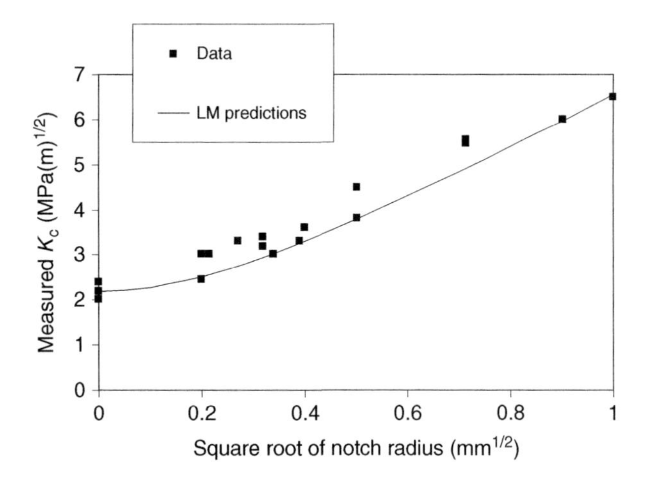

Fig. 5.17. Data on the same material as in Fig. 5.12, but tested at 1000 °C (Tsuji et al., 1999); LM predictions.

co-workers made specimens containing features similar to those found in service: a channel section and an L-shaped specimen (Zou et al., 2004). Figure 5.18 shows the L-shaped specimen, which was loaded in tension until failure occurred by cracking at the corner. Four different values were used for the corner radius: 0, 1, 2 and 4 mm.

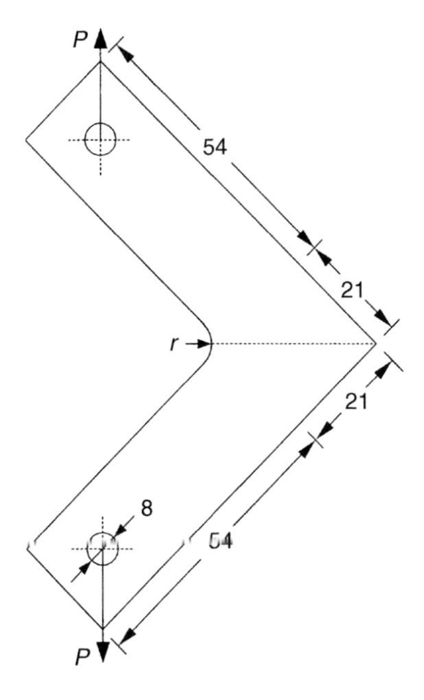

Fig. 5.18. The geometry of the L-shaped specimen used by Zou et al. in testing nuclear graphite. Dimensions are in millimetres. The root radius (labelled r in the diagram) was varied between 0 and 4 mm.

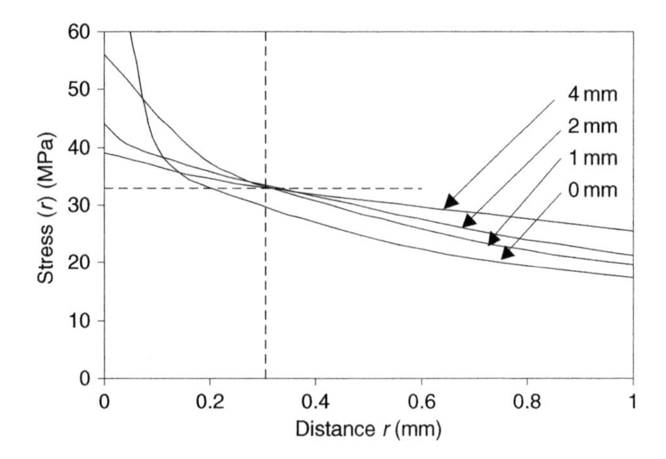

Fig. 5.19. Stress–distance curves at the relevant failure loads for L-shaped specimens of nuclear graphite with four different root radii (0, 1, 2 and 4 mm), from Zou et al. (2004). The two dashed lines indicate the material strength and the distance $L/2$ respectively, intersecting at the critical point.

Separate testing established the mechanical properties, including a plane strain fracture toughness of $1.44\,\text{MPa(m)}^{1/2}$ and a plain specimen strength of 33 MPa, from which we can calculate an L value of 0.61 mm. This is certainly a very different material from the high-strength ceramics considered above, with a much lower σ_u and consequently a much larger L.

These workers carried out FEA to establish the stress fields in their specimens at the applied loads corresponding to failure. Figure 5.19 reproduces these curves; we can see immediately that the TCD will give accurate predictions of these failure loads, because all four curves intersect at the point which corresponds to the distance $r = L/2$, and the stress value equal to σ_u. In fact the accuracy with which the curves for the three non-zero radii intersect at exactly the correct point is quite uncanny. The curve corresponding to zero radius falls slightly below the others, but even in this case the difference in terms of the stress level at the critical distance is only about 10%. In practice, the corner radius for this specimen type must have been slightly greater than zero, so its true stress/distance curves would have been slightly higher anyway.

This example illustrates a number of points. First, the method can be used to make predictions for geometrical shapes which are not simple notches – in this case a right-angle corner – and for which no simple analytical function exists for the stress analysis. Secondly, the procedures developed for high-strength ceramics still apply to this much weaker material, and thirdly, if FEA results are available then predictions can be made very quickly and easily.

5.2.3 *Large blunt notches*

So far we have considered two different types of stress concentration: small defects and long, thin notches. The first type was characterised by small values of both length

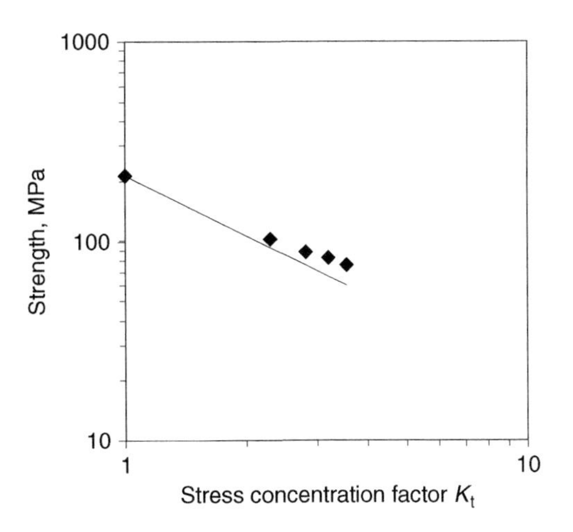

Fig. 5.20. Data points showing the strength of notched specimens of alumina (Wang et al., 1995); the line is a prediction using σ_u/K_t which works well for these large, blunt notches.

(D) and root radius (ρ): in fact, we assumed that ρ was equal to zero in making the predictions. The second type was characterised by large, macroscopic values of length ($D >> L$) but relatively small values of ρ, often of the same order of magnitude as L. These notches will have high values of the stress concentration factor, K_t, since this is related to the ratio $(D/\rho)^{1/2}$. Another category of notches which is of practical interest are those which are relatively large and blunt, that is both D and ρ are much larger than L. In this case, predictions made using the TCD will be similar to predictions made by simply dividing the plain specimen strength by K_t. This is because, if $L << \rho$, then the point at which we are examining the stress field will be, relatively speaking, very close to the notch root. Another way of saying this is that the stress gradient near the notch will be sufficiently low that the stress at $r = 0$ will be very similar to the stress at $r = L/2$ (or to the average stress over $2L$). A case in point is illustrated in Fig. 5.20 which reproduces some data on alumina (Wang et al., 1995). These workers used plain specimens and specimens containing notches of depth 3 mm and root radius 0.5, 0.75, 1.0 and 1.5 mm, giving K_t factors ranging from 1 to 3.53. The data (on measured strength at the 50% probability level) can be predicted reasonably well by a line which corresponds to σ_u/K_t; this line slightly underestimates the strength at the higher K_t values, but the error is small (20%). We can anticipate this result if we consider the stress distribution near these notches. For the sharpest notch, Eq. (5.11) is valid; if we assume an L value of 0.05 mm, which is typical for alumina, then the stress at $r = L/2$ will be lower than that at $r = 0$ by only 9%. For the blunter notches the difference will be even smaller. Thus the TCD predicts that there will be a class of notches for which the simple K_t factor is an accurate, slightly conservative, guide to performance.

5.2.4 Discussion: other theories and observations

The above data and predictions have demonstrated that the TCD, especially in the form of the LM, is capable of predicting the effect of stress concentrations of all kinds, from microscopic defects and cracks to large notches. In this section, I will discuss some of the other approaches which have been used to predict this kind of data, and some useful observations made by other workers. In Chapters 3 and 4, we already examined a variety of prediction methods, showing that many of them have something in common with the TCD whilst others stem from quite different philosophies of material behaviour. Since the merits and limitations of these various methods were already discussed in those chapters, we will not cover all this ground again; in what follows, I will present some examples of the application of these methods to engineering ceramics.

Statistical methods, especially the Weibull approach, are still commonly used for the assessment of notches (Bruckner-Foit et al., 1996; Hertel et al., 1998; Hoshide et al., 1998; Wang et al., 1995); I would venture to suggest that this is inappropriate in most cases, especially for relatively sharp or small features. A Weibull analysis will certainly make predictions of the correct type – that is it will predict, for example, that the change in strength accompanying a change in root radius is not, in general, as great as the change in K_t factor. This happens, in the Weibull analysis, because whilst the local stress is increasing, the volume of material under stress is decreasing due to the greater stress gradient. But in order to make this prediction one must necessarily extrapolate, assuming that the constants in the equation apply to higher stresses, and smaller volumes, than can be demonstrated experimentally. More importantly, this approach takes no account of the changing sensitivity of the material to defects of different sizes. We have seen clearly, in many examples above, that small defects, less than L in size, have very little effect on strength compared to larger defects whose effect can be described by standard LEFM. This may explain the anomalies observed in statistical data such as shown earlier, in Fig. 5.3, where the data deviate from the Weibull line at low strengths. These failures will have occurred due to particularly large defects, which are clearly having a much greater effect on strength than the Weibull line, based mainly on the smaller defects, would have predicted. Some recent publications (e.g. Hertel et al., 1998) attempt to introduce the complexities of defect size effects into the traditional Weibull analysis. This approach certainly has some merit, but my own opinion is that the statistical aspect can be dispensed with, at least for many of the better produced engineering ceramics whose behaviour can be expected to be largely deterministic.

Turning then to other deterministic theories, several workers have attempted to predict the effect of notches and small defects using a modified LEFM approach in which a small crack is introduced at the notch tip. In Chapter 3, we classified this approach into two types, which we called the 'introduced-crack' and 'imaginary-crack' models, according to whether the crack is assumed to actually be present, or only imaginary. Examples of introduced crack models are those of Usami et al. (1986) and Damani et al. (1996). Usami et al. assumed that the crack forms due to the failure of a single grain located at the notch tip. This was assumed to be a particularly large grain, which fails because it is relatively weak. The resulting K value of the combined notch-plus-crack was then calculated, failure occurring when $K = K_c$. The calculation of K was greatly simplified by using only the stress at a single point, located at the furthest extent of the crack from the notch. This means, of course, that the resulting theory is essentially the same as our PM, except for the shape factor (F) of the crack and the way in which the critical distance is chosen. In the end, Usami et al. decided on a value of twice the average grain size for their crack length, presumably because this gave the best fit to the experimental data. The resulting predictions were, unsurprisingly, very similar to those of the PM.

These workers made the mistake which is commonly made when developing the introduced-crack theory, namely they did not allow for the fact that their introduced crack is itself a short crack, and therefore will have a different K_c value, lower than the true, long-crack value. Damani et al. did make some allowance for the short-crack effect in their work but, as I explained in Chapter 3, there is an underlying inconsistency in these models which cannot be avoided. Their introduced-crack model was used to predict various sets of data on the effect of ρ on measured K_c but unfortunately they had to estimate the values of important constants such as K_c and the size of the introduced crack which, they argued, could either be a fractured grain (as in Usami's approach) or else a machining mark. Hoshide and Inoue used what is essentially an imaginary-crack model, though they suggested a possible source for the crack in the form of a weak surface layer, presenting some possible evidence for this in the form of acoustic emission results (Hoshide and Inoue, 1991). Other workers have also used empirical models which are essentially the same as the imaginary-crack model (Keith and Kedward, 1997; Suo et al., 1993).

Other workers have developed various types of process zone theory to explain the behaviour of cracks and notches. These methods were also discussed in general terms in Chapter 4. The simplest approach, in which failure is assumed to occur when the process zone reaches a critical size, was used by Ando and co-workers (Ando et al., 1992; Tsuji et al., 1999). As we saw in Chapter 4, this approach will inevitably give predictions similar to those of the TCD, because of the similarity between L and the process zone size. Ando et al. presented evidence to show that, in engineering ceramics, the size of the process zone is a function of the plain-specimen strength, σ_u:

$$r_p = \frac{\pi}{8}\left(\frac{K}{\sigma_u}\right)^2 \tag{5.16}$$

This equation is clearly very similar to our equation for L (Eq. 5.6), the constant $\pi/8$ differing from our $1/\pi$ by only a factor of 1.2. We will see in subsequent chapters that this is not the case for metals or for polymers, but it seems to be a reasonable assumption for very brittle materials. Tsuji et al. (1999) carried out detailed microscopic examinations of the fracture surface and found that, for their notched specimens, failure was initiated not at the surface of the notch but rather at a point some distance away from the notch root. Figure 5.21 shows an example of their findings, in this case for alumina. Initiation sites included cavities and large grains. They found that the average distance of the initiation site from the notch root was 20.3 μm, which is very similar to the calculated value of $L/2$ for this material: 26.2 μm, providing a strong justification for the use of the PM in this case.

More sophisticated models take into account the various microscopic processes that will occur around the tip of the crack or notch, which will affect the local stress field and may help or hinder the process of crack growth. Lawn (1993) considers these various

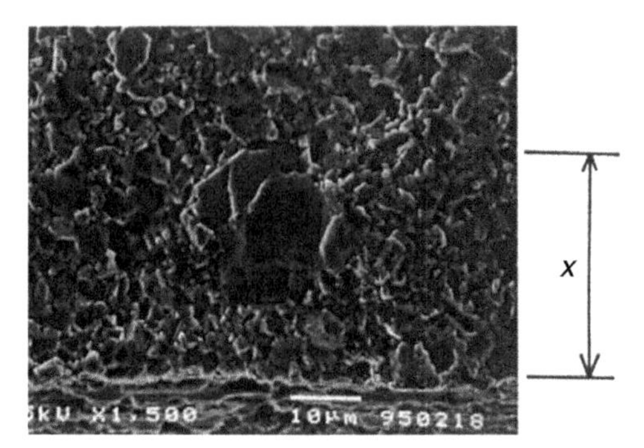

Fig. 5.21. An SEM photograph from Tsuji et al., 1999, showing an initiation site located a distance X from the notch root, in alumina.

processes in some detail, attempting in the case of a crack to predict quantitatively their effect on the resulting toughness. For example, the zone ahead of the crack is often characterised by the presence of many small cracks. These microcracks, usually no larger than individual grains, are a form of damage but in fact have a slight toughening effect because they use up energy and reduce local stresses by changing the material's stiffness. Potentially much larger effects occur behind the crack tip; for example, bridging of the crack faces by interlocking grains and unbroken ligaments of material plays a major role in increasing toughness (Yuan et al., 2003); such a mechanism will clearly be relatively ineffective at small crack sizes. Much work still has to be done to allow quantitative predictions based on the real mechanisms of crack initiation and growth in these materials.

Finally, several workers have developed process-zone models of the continuum mechanics variety, which were already discussed in general terms in Chapter 4. It is interesting to note that some researchers (Suo et al., 1993) concluded that the approach was too complex to use in practice and settled instead for an empirical equation which is in fact identical to the ICM discussed earlier, rewritten in terms of net section stress.

5.3 Building materials

The materials used in civil engineering structures such as concretes and mortars form another large class of ceramic materials. The main difference between these and the engineering ceramics that we have discussed above is that building materials have much larger microstructural features (e.g. aggregates) and contain larger defects. Despite the difference in the scale at which they are used, similar problems arise: concrete structures will contain stress concentration features, and tests must be conducted on notched specimens to determine their toughness. A particular problem – and one that has preoccupied many researchers in recent times – is the dependence of material strength on specimen size. The measured value of σ_u will decrease with increasing size of test specimen. This is not surprising, in fact all brittle materials will display such scaling effects, but it is a matter of particular importance for building materials because they are used in extremely large structures such as dams – structures which are orders of magnitude larger than any specimen that can realistically be made and tested. A number of complex and elegant mathematical models have been developed to predict these scaling effects (Bazant, 2004) (Carpinteri and Cornetti, 2002). The problem is a complex one and can be expected to have at least two elements, which it is convenient to call 'statistical' and 'geometrical'. Statistical size effects will arise due to the increasing probability of large defects in larger volumes of material; geometrical size effects will occur in any situation where a stress gradient is present, and so it is these which we are predicting when using the TCD. Stress gradients are present at notches of course, but also in plain specimens if they are tested in bending or torsion rather than in axial tension.

Figure 5.22 shows some experimental data on the measured strength of concrete beams tested in three-point bending (Karihaloo et al., 2003). Data are plotted as a function of the height of the beam, h, using plain beams and also beams containing notches of three different lengths, characterised by given values of the ratio notch length to beam height, a/h.

It is clear that there is a significant reduction in strength for increasing beam size, which is more marked in the notched specimens than in the plain ones. The figure shows predictions using the TCD (Cornetti et al., 2005); in this case we used the FFM approach. This approach was described previously, in Section 3.3.4; it is one of the theories in the TCD group, and gives predictions of notch fracture which are usually very similar to those of the LM. Here we obtained very good predictions for the notched specimens: the prediction for the plain specimens was somewhat inaccurate, especially for small specimens where the beam height became similar in magnitude to the critical distance L. We found that the plain-specimen data could be more accurately predicted using another

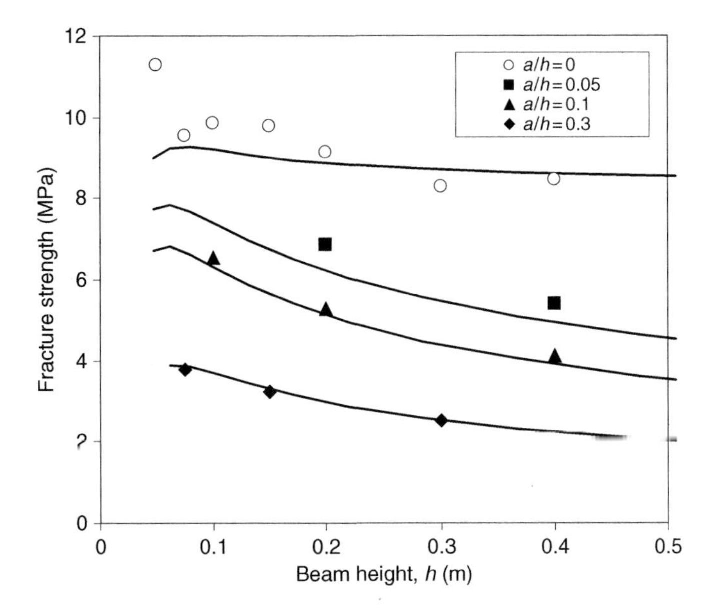

Fig. 5.22. Fracture strength of plain and notched concrete beams: data from Karihaloo (2003); predictions using the TCD (FFM method).

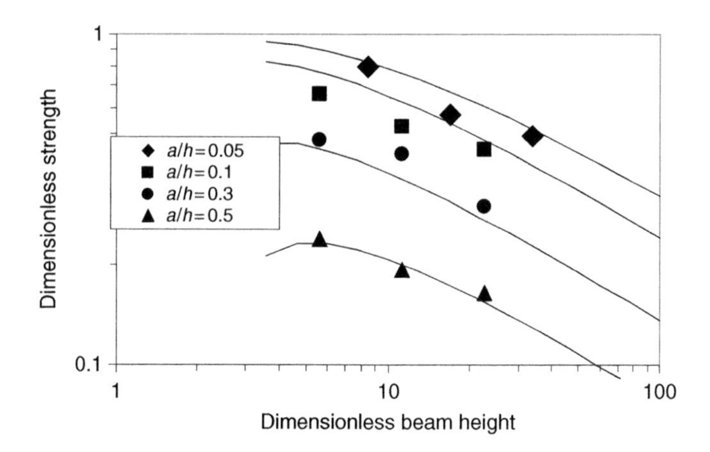

Fig. 5.23. Data and FFM predictions for the fracture strength of notched cement paste.

variant of the TCD (Taylor and Cornetti, 2005) in which two different TCD methods are combined, (see Section 3.3.6). These various methods will be discussed more fully in Chapter 13.

The value of L in this case was 10.7 mm. This distance is much larger than we encountered above for the engineering ceramics; mathematically it arises because whilst the K_c value of concrete is the same order of magnitude as that of engineering ceramics, its σ_u value is much smaller. Physically, the value of L is similar to microstructural features such as aggregate particles. Figure 5.23 shows further data from Karihaloo, in this case for hardened cement paste, which had an L value of 5.6 mm; again the FFM gave good predictions.

5.4 Geological Materials

We can expect that the TCD should be suitable for predicting the fracture of rocks and minerals, since they are also essentially brittle, ceramic materials. Given their low strengths and relatively coarse structures we might expect relatively large L values, similar to those of concretes and other building materials. In fact there is some evidence of L values as large as several metres in sea ice (Dempsey et al., 1999); this evidence comes from what must surely be the largest test specimens ever made: square sheets of floating ice with sides from 0.5 to 80 metres long!

Ito and Hayashi used the PM in their work on hydraulic fracturing of rocks, following on from previous work by Lajtai. In fact these are the only specific references to the use of the PM which I have been able to find for any ceramic material (Ito and Hayashi, 1991; Lajtai, 1972). These workers derived the idea from the work of Whitney and Nuismer on composite materials, which will be discussed in detail in Chapter 8 (Whitney and Nuismer, 1974). They used this approach to predict the fracture of a wellbore, which is a hole drilled into a rock and pressurised with fluid. This required an analysis to predict the stresses around the hole, taking account of the permeability of the surrounding rock, and a failure criterion for the material in the form of an effective stress of a type commonly used in this field. The critical distance $L/2$ for the PM was derived from measured values of material toughness and strength using just the same equation that we have developed here. Values obtained for two different types of rock (Kofu andesite and Honkomatsu andesite) were 6.8 mm and 3.2 mm respectively, of the same order of magnitude as we found for concrete and mortar above.

5.5 Nanomaterials

We bring this chapter to a close by mentioning the materials which are currently being developed for use in microscopic devices, the so-called 'micro-electromechanical systems' (MEMS) and 'nano-electromechanical systems' (NEMS). There has, for some time now, been interest in the idea of developing machines on a very small scale, and in recent years this has begun to be technologically feasible. Indeed we now find MEMS devices such as microscopic switches in many common domestic items. The materials from which these are made fall into the broad class of ceramics as regards their mechanical properties. A material commonly used in MEMS devices, for example, is silicon, in both its single-crystal and polycrystalline forms.

Figure 5.24 shows data on the fracture strength of microscopic specimens of single crystal silicon, containing very small notches, whose lengths varied from 0.02 μm (i.e. 20 nm) up to 0.5 μm (Minoshima et al., 2000). Due to the method of manufacture, the notch root radius decreased with increasing depth, so the shallowest notch had a radius of 0.26 μm (and therefore a K_t factor of about 1.6) whilst the deepest notch had a root radius of only 15 nm (giving $K_t = 12.5$). Unnotched specimens were also tested, and as can be seen from the figure they showed a very large amount of scatter, about an average value of 6 GPa. The figure shows a prediction which I made using the PM; this gives a reasonable fit considering the amount of scatter in the data. The use of either

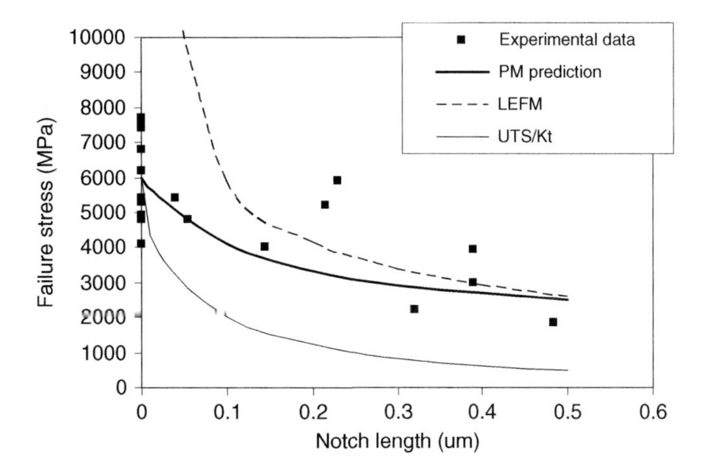

Fig. 5.24. Data on the fracture strength of single crystal silicon containing very small notches. PM predictions are compared with predictions using LEFM and using σ_u/K_t.

LEFM (assuming the notches to be long cracks) or σ_u/K_t (as would be valid for large blunt notches) gave very poor predictions, as we would expect.

The measured value of σ_u for this material was very high – much higher than that of any of the engineering ceramics considered above. Even greater strengths, of the order of tens of Giga Pascals, can be obtained for materials made in the form of very small fibres. These nano-fibres, for example carbon nanotubes, have been suggested as structural materials of the future, because their measured strengths are much larger than those of steel. They may indeed find important future uses, but it is important to remember that these high strengths occur because the specimen size is so extremely small.

A material such as carbon, whether in the form of diamond, graphite or nanotubes, is inherently brittle; unless its low toughness is specifically addressed by material modifications it will always be susceptible to defects: the microscopic specimens considered here are simply too small to contain large defects.

Pugno and Ruoff have used the TCD (in the form of the FFM, to which they gave the name Quantised Fracture Mechanics) to attempt to predict the strength of nanomaterials, including carbon nanotubes, SiC nanorods and Si_3N_4 whiskers (Pugno and Ruoff, 2004). Their argument was that, in this form, the appropriate value of L would be the atomic spacing. The measured strength of nanotubes tends to show a lot of scatter: Pugno and Ruoff argued that the reason for this scatter was that the specimens contained defects in the form of atomic vacancies Thus different levels of strength should occur if the defect consisted of one atom, two atoms, three atoms and so on. Figure 5.25 reproduces data from Yu et al. who measured the tensile strength of carbon nanotubes: the 19 samples tested had strengths varying from 63 GPa to 11 GPa (Yu et al., 2000) The figure also shows predictions using the LM, assuming that these atomic vacancy defects can be considered to be cracks: Pugno and Ruoff obtained slightly different predictions by assuming the defects to be elliptical holes. The defect size (i.e. the number of missing atoms, n) was unknown experimentally, so I have chosen values which best fit the prediction line: nevertheless the prediction is impressive in that it is able to explain

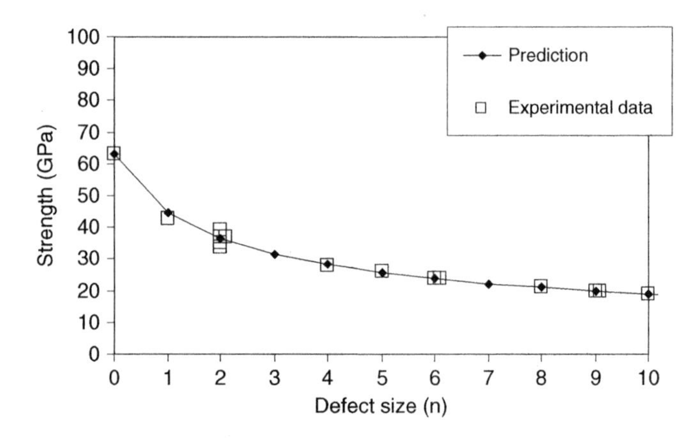

Fig. 5.25. Measured fracture strengths of carbon nanotubes (Yu et al., 2000); predictions using the LM, assuming the samples contained defects in the form of vacancies of size n atoms.

the large amount of the scatter and the tendency of the measured strengths to cluster around certain values. These predictions also demonstrate just how rapidly the strength of nanotubes will decrease if they contain defects of even a few atoms in size, underlining the point made above about the poor defect tolerance of these materials.

5.6 Concluding Remarks

We have seen in this chapter how the TCD methods (PM, LM and FFM) can be successfully used to predict the failure of brittle materials of all kinds, from building materials and rock through high-specification engineering ceramics to nanomaterials. Table 5.1 below summarises the properties of all the materials which have been considered above.

Table 5.1. Values of L, σ_u and K_c for various ceramic materials

Material	L	σ_u (MPa)	K_c (MPa.m$^{1/2}$)	Reference
Nanomaterials	(atomic spacing?)			(Pugno and Ruoff, 2004)
Silicon (single crystal)	0.12 μm	6000	3.7	(Minoshima et al., 2000)
Sialon	8 μm	920	4.6	(Kimoto et al., 1985)
Silicon carbide	9.8 μm	666	3.7	(Kimoto et al., 1985)
Silicon carbide	11 μm	620	3.7	(Usami et al., 1986)
Silicon nitride	11.8 μm	920	5.6	(Ando et al., 1992)
Silicon nitride	14.4 μm	1160	7.8	(Ando et al., 1992)
Silicon nitride	15.3 μm	650	4.5	(Ando et al., 1992)
Silicon nitride	26 μm	550	5.0	(Usami et al., 1986)
Silicon nitride (5 materials)	9.4 – 164 μm	920 –220	5.0	(Usami et al., 1986)
MgPSZ	(20 μm)	(896)	(7.1)	(Damani et al., 1996)
Alumina (at 1000 °C)	31.8 μm	215	2.2	(Tsuji et al., 1999)
Alumina	22 μm	(373)	3.1	(Bertolotti, 1973)
Alumina	52.4 μm	297	3.83	(Tsuji et al., 1999)
Alumina	76 μm	200	3.1	(Ando et al., 1992)
Alumina	76.6 μm	290	4.5	(Ando et al., 1992)
Electrical Porcelain	150 μm	(51.6)	1.12	(Clark and Logsdon, 1974)
Nuclear Graphite	610 μm	33	1.44	(Zou et al., 2004)
Honkomatsu Andesite	3.2 mm	12.1	1.32	(Ito and Hayashi, 1991)
Kofu Andesite	6.8 mm	11.1	1.62	(Ito and Hayashi, 1991)
Hardened Cement Paste	5.6 mm	4	0.53	(Karihaloo et al., 2003)
Concrete	10.7	8.28	1.52	(Karihaloo et al., 2003)
Sea Ice	(several metres?)			(Dempsey et al., 1999)

Note: Brackets indicate approximate or speculative values.

Our main concerns here have been in predicting the reduction in strength caused by crack-like defects and by introduced notches, since an understanding of these areas will allow these kinds of materials to be used with confidence in engineering situations. Also, we have emphasised here the ability of the TCD to predict the existing experimental data. We have not given much thought as to why this method works so well, in such a wide variety of circumstances. Indeed the same approach will be taken in several subsequent chapters, in which we consider different types of materials (polymers, metals and composites), different types of failure (fatigue, fretting) and the complexities of multiaxial loading and of complex component geometries. Some of these materials and applications areas will require modifications to the basic TCD methods shown in this chapter. It is only when we have demonstrated satisfactorily that we can predict the data in all these situations that we will turn to the theoretical questions surrounding the TCD.

References

Ando, K., Kim, B.A., Iwasa, M., and Ogura, N. (1992) Process zone size failure criterion and probabilistic fracture assessment curves for ceramics. *Fatigue and Fracture of Engineering Materials and Structures* **15**, 139–149.

Bazant, Z.P. (2004) Quasibrittle fracture scaling and size effect. *Materials and Structures* **37**, 1–25.

Bertolotti, R.L. (1973) Fracture toughness of polycrystalline alumina. *Journal of the American Ceramic Society* **56**, 107–117.

Bruckner-Foit, A., Heger, A., and Munz, D. (1996) On the contribution of notches to the failure probability of ceramic components. *Journal of the European Ceramic Society* **16**, 1027–1034.

Carpinteri, A. and Cornetti, P. (2002) Size effects on concrete tensile fracture properties: An interpretation of the fractal approach based on the aggregate grading. *Journal of the Mechanical Behaviour of Materials* **13**, 233–246.

Clark, W.G. and Logsdon, W.A. (1974) The applicability of fracture mechanics technology to porcelain ceramics. In *Symposium on Fracture Mechanics of Ceramics, Vol.2* pp. 843–861. Plenum, New York.

Cornetti, P., Pugno, N., and Taylor, D. (2005) Strength predictions via finite fracture mechanics. In *Proceedings of the 11th International Conference on Fracture* p. 73. ESIS, Turin, Italy.

Creager, M. and Paris, P.C. (1967) Elastic field equations for blunt cracks with reference to stress corrosion cracking. *International Journal of Fracture Mechanics* **3**, 247–252.

Damani, R., Gstrein, R., and Danzer, R. (1996) Critical notch-root radius effect in SENB-S fracture toughness testing. *Journal of the European Ceramic Society* **16**, 695–702.

Dempsey, J.P., Adamson, R.M., and Mulmule, S.V. (1999) Scale effect on the in-situ tensile strength and failure of first-year sea ice at Resolute, NWR. *International Journal of Fracture, special issue on fracture scaling* 9–19.

Hertel, D., Fett, T., and Munz, D. (1998) Strength predictions for notched alumina specimens. *Journal of the European Ceramic Society* **18**, 329–338.

Hoshide,T. and Inoue, T. (1991) Simulation of anomalous behaviour of a small flaw in strength of engineering ceramics. *Engineering Fracture Mechanics* **38**, 307–312.

Hoshide, T., Murano, J., and Kusaba, R. (1998) Effect of specimen geometry on strength in engineering ceramics. *Engineering Fracture Mechanics* **59**, 655–665.

Ito, T. and Hayashi, K. (1991) Physical background to the breakdown pressure in hydraulic fracturing tectonic stress measurements. *International Journal of Rock Mechanics and Mineral Science and Geomechanics Abstracts* **28**, 285–293.

Karihaloo, B.L., Abdalla, H.M., and Xiao, Q.Z. (2003) Size effect in concrete beams. *Engineering Fracture Mechanics* **70**, 979–993.

Keith, W.P. and Kedward, K.T. (1997) Notched strength of ceramic-matrix composites. *Composites Science and Technology* **57**, 631–635.

Kimoto, H., Usami, S., and Miyata, H. (1985) Relationship between strength and flaw size in glass and polycrystallline ceramics. *Japanese Society of Mechanical Engineers* **51–471**, 2482–2488.

Lajtai, E.Z. (1972) Effect of tensile stress gradient on brittle fracture initiation. *International Journal of Rock Mechanics and Mineral Science and Geomechanics Abstracts* **9**, 569–578.

Lawn, B. (1993) *Fracture of brittle solids*. Cambridge University Press, Cambridge.

Minoshima, K., Terada, T., and Komai, K. (2000) Influence of nanometre-sized notch and water on the fracture behaviour of single crystal silicon microelements. *Fatigue and Fracture of Engineering Materials and Structures* **23**, 1033–1040.

Pugno, N. and Ruoff, R. (2004) Quantized fracture mechanics. *Philosophical Magazine* **84**, 2829–2845.

Suo, Z., Ho, S., and Gong, X. (1993) Notch ductile-to-brittle transition due to localised inelastic band. *Journal of Engineering Materials and Technology* **115**, 319–326.

Takahashi, I., Usami, S., Nakakado, H., Miyata, H., and Shida, S. (1985) Effect of defect size and notch root radius on fracture strength of engineering ceramics. *Journal of the Ceramics Society of Japan* **93**, 186–194.

Taniguchi, Y., Kitazumi, J., and Yamada, T. (1988) Bending stress analysis of ceramics based on the statistical theory of stress and fracture location. *Journal of the Japanese Society of Materials Science* **38–430**, 777–782.

Taylor, D. and Cornetti, P. (2005) Finite fracture mechanics and the theory of critical distances. In *Advances in Fracture and Damage Mechanics IV* (Edited by Aliabadi, M.H.) pp. 565–570. EC, Eastleigh UK.

Tsuji, K., Iwase, K., and Ando, K. (1999) An investigation into the location of crack initiation sites in alumina, polycarbonate and mild steel. *Fatigue and Fracture of Engineering Materials and Structures* **22**, 509–517.

Usami, S., Kimoto, H., Takahashi, I., and Shida, S. (1986) Strength of ceramic materials containing small flaws. *Engineering Fracture Mechanics* **23**, 745–761.

Wang, F., Zheng, X.L., and Lu, M.X. (1995) Notch strength of ceramics and statistical analysis. *Engineering Fracture Mechanics* **52**, 917–921.

Westergaard, H.M. (1939) Bearing pressures and cracks. *Journal of Applied Mechanics A* 49–53.

Whitney, J.M. and Nuismer, R.J. (1974) Stress fracture criteria for laminated composites containing stress concentrations. *Journal of Composite Materials* **8**, 253–265.

Yu, M.F., Lourie, O., Dyer, M.J., Moloni, K., Kelly, T.F., and Ruoff, R. (2000) Strength and breaking mechanism of multiwalled carbon nanotubes under tensile load. *Science* **287**, 637–640.

Yuan, R., Kruzic, J.J., Zhang, X.F., DeJonghe, L.C., and Ritchie, R.O. (2003) Ambient to high-temperature fracture toughness and cyclic fatigue behaviour in Al-containing silicon carbide ceramics. *Acta Materialia* **51**, 6477–6491.

Zou, Z., Fok, S.L., Oyadiji, S.O., and Marsden, B.J. (2004) Failure predictions for nuclear graphite using a continuum damage mechanics model. *Journal of Nuclear Materials* **324**, 116–124.

CHAPTER 6

Polymers

Brittle Fracture in Polymeric Materials

6.1 Introduction

This chapter deals with the failure under monotonic loading of polymeric materials containing stress concentrations. The general approach is similar to that taken in the previous chapter, on ceramics, but we will find some important differences, necessitating a major modification to the TCD.

Polymers, though still relatively new materials by historical standards, are increasingly being used in load-bearing applications where the prevention of failure is of crucial importance. Two examples from my own work in failure analysis will serve to illustrate this. Colour Plate 2 shows part of a child's car seat, which was made from a thin shell of moulded PVC. The design involved several slots through which passed the straps of the seatbelt. During a car accident the high forces in the straps, combined with the stress concentration effect of the slot, caused a brittle fracture in the material. This loosened the straps, releasing the child from the seat, with fatal consequences. Figure 6.1 shows the brittle fracture of a polymer resin which was used in a car component. The fracture, which was preceded by a small amount of slow crack growth, caused a failure in the timing system which precipitated a complete seizure of the engine whilst the car was travelling at high speed. A serious accident was narrowly avoided. In this case the failure was due to unusually high stresses arising from poor tolerances in the surrounding components.

In addition to these types of applications, that is critical, load-line components, polymers are used in great volumes as casings on equipment such as computers and household goods, for which the major mechanical requirement is resistance to impact. This chapter will mostly be concerned with fractures initiated under monotonic loading at relatively low strain rates, but we will also mention the effect of notches under impact situations.

Fig. 6.1. Brittle fracture in a polymeric car component.

The study of fracture in polymers really came of age in the 1980s with the publication of two excellent books: Kinloch and Young's *Fracture Behaviour of Polymers* and Williams' *Fracture Mechanics of Polymers* (Kinloch and Young, 1983; Williams, 1984). These publications came at a crucial time, summarising the work done in the previous two decades on the application of the new science of linear elastic fracture mechanics (LEFM), to polymeric materials. Though much work has been done since, these books still provide a very useful perspective on the subject.

Almost all polymers will display classic brittle fracture behaviour provided the temperature is low enough but, unlike ceramics, they will usually display some form of plastic or non-linear deformation before failure, at least in the most highly stressed region. The general deformation and fracture behaviour of polymers is much more complex than that of ceramics and metals, for two main reasons. First, there are a larger number of mechanisms available by which polymers can achieve permanent or temporary deformation. Many polymers will undergo plastic deformation through a yielding process which is normally called 'shear yielding' to indicate that it is controlled by shear stress. That is also true for metals of course, though in the case of polymers hydrostatic stress does have a minor role in encouraging plastic deformation. But polymers also display a mechanism known as crazing, which is not found in other classes of materials. Crazes form by the accumulation of microscopic voids (driven by hydrostatic stress) and develop into supported cracks, that is cracks which have small fibrils of material spanning their faces. Thanks to this support, crazes require more stress to grow. Craze growth is controlled by the tensile stress normal to the faces of the craze. Eventually, if the stress is high enough, the craze will break down into a normal crack, but one that always has a craze at its tip. Crazes are a form of damage, but they also have a toughening effect, because multiple crazing near the crack tip or notch root can consume energy and reduce local stresses in the same way that plastic deformation can. Some polymers (e.g. PMMA at room temperature) are sufficiently brittle that the first craze which forms immediately propagates, causing a brittle fracture even in an unnotched specimen. Others (e.g. polystyrene) exhibit multiple crazing behaviour which can effectively take the place of plasticity as a general deformation mechanism.

The second major complexity in the deformation and fracture behaviour of polymers is the effect of temperature and, linked to it, the effect of time or strain rate. These materials are sensitive to changes in temperature in the vicinity of room temperature. Raising the temperature tends to suppress brittle fracture, initially encouraging the crazing and yielding mechanisms and, at higher temperatures, allowing some polymers to undergo extremely large amounts of deformation by a drawing mechanism. Other polymers retain brittle behaviour even at temperatures approaching disintegration, but for any polymer the operating temperature is always a crucial feature. Likewise, increasing strain rate encourages brittle behaviour, suppresses yielding and increases elastic stiffness. In fact the shear yielding mechanism should really be thought of as a mixture of plastic (i.e. permanent) and elastic deformation, since it often occurs by the movement of chain segments which can, given time, return to their original positions. For these reasons, the behaviour under high-speed impact loading can be very different from that at slower strain rates, so materials which are designed to resist impact, such as high-impact polystyrene (HIPS), may perform poorly under static loading.

It is universally true that polymers have low toughness: typical values for K_c are in the range 1–3 $MPa(m)^{1/2}$. However, small defects, which so dominate the behaviour of ceramic materials (see Chapter 5), are less important here because polymers also have relatively low values of strength and stiffness, so they must necessarily be used at much lower applied stresses, effectively increasing the critical defect size. Also, as we shall see later, their plastic and non-linear deformation behaviour leads to a situation in which some types of stress concentration, especially small defects and blunt notches, have no effect whatsoever on strength.

In this chapter we will examine the accuracy of the TCD in predicting the effect of notches, cracks and other stress concentrations in several different polymers, including PMMA, polycarbonate (PC), polyvinylchloride (PVC), polystyrene (PS), and high-impact polystyrene (HIPS).

6.2 Notches

6.2.1 *Sharp notches*

We will begin by looking at some data on PC (Tsuji et al., 1999) which is of the same type as that examined in the previous chapter (Section 5.2.2), namely tests on long, relatively sharp notches having lengths D very much greater than their root radii, ρ. The parameter recorded is the 'measured K_c', that being the value of fracture toughness calculated assuming that the notch is a crack of the same length.

The method of analysis is exactly the same as described in Section 5.2.2 (Eqs 5.10–5.15). In summary, we use the following equation to calculate the critical distance, L as a function of K_c and the material's tensile strength σ_u:

$$L = \frac{1}{\pi}\left(\frac{K_c}{\sigma_u}\right)^2 \tag{6.1}$$

For this type of notch we can use the Creager and Paris formula (Eq. 5.10) for the stress as a function of distance, giving, for the point method (PM), the value of the measured toughness, K_{cm}, as:

$$\frac{K_{cm}}{K_c} = \frac{\left(1+\frac{\rho}{L}\right)^{3/2}}{\left(1+\frac{2\rho}{L}\right)} \tag{6.2}$$

In this case the material constants were $K_c = 3.47\,\mathrm{MPa(m)}^{1/2}$, $\sigma_u = 70.2\,\mathrm{MPa}$ giving $L = 0.78\,\mathrm{mm}$. Figure 6.2 shows the experimental data and also the prediction using this value of L. It is clear that the prediction is very poor; it is necessarily correct at $\rho = 0$ because there it must correspond to K_c, but for higher values of the root radius the prediction line is much lower than the data and remains almost constant for the whole range of ρ values studied. Clearly something is wrong. However, as the figure also shows, we can achieve a very good prediction if we use a different value of L. Trying various values of L, we obtain a best fit using $L = 0.061\,\mathrm{mm}$, much smaller than that calculated from Eq. (6.1). Clearly the value of K_c must remain unchanged, otherwise the prediction will be incorrect at $\rho = 0$. This implies, from Eq. (6.1), that the stress value used must differ from σ_u. We call this new stress value σ_o; it can be found by

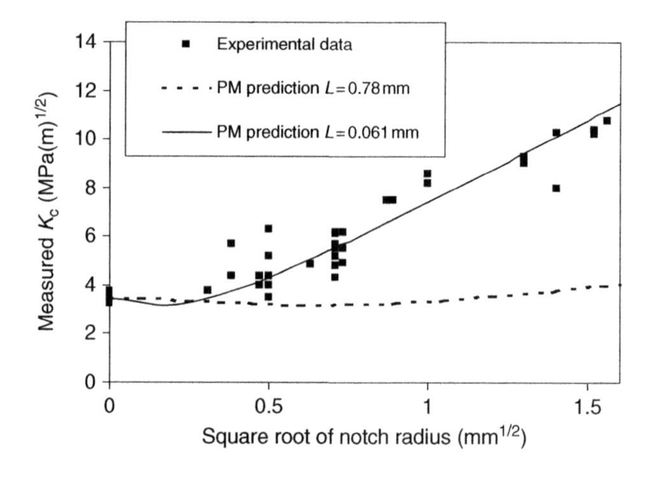

Fig. 6.2. Data on the measured K_c of notched specimens of PC (Tsuji et al., 1999). Predictions using the PM, with two different values of L: 0.78 mm (calculated from Eq. 6.1) and 0.061 mm (which gives the best fit to the data).

rearranging Eq. (6.1), using the normal value of K_c along with the new value of L, found by fitting to the data:

$$\sigma_o = \frac{K_c}{\sqrt{\pi L}} \tag{6.3}$$

The value of σ_o in this case turns out to be 250 MPa, larger than σ_u by a factor of 3.56. We could, alternatively, have obtained this result by choosing a value of σ_o which gave the best fit to the data, calculating L accordingly: the result would have been the same. Clearly something has changed: we can no longer make predictions using σ_u in the way that we did for ceramic materials. But nevertheless we can still use the TCD to make accurate predictions, once we know the appropriate values of the constants.

Exactly the same analysis was described by Kinloch, Williams and co-workers in a series of articles in the 1980s (Kinloch and Williams, 1980; Kinloch et al., 1982; Kinloch et al., 1983). These workers studied various epoxy resins: they found that accurate predictions could be obtained using the PM in this modified form, finding suitable values of L and σ_o by comparison with experimental data. Figures 6.3 and 6.4 show examples of their results.

Interestingly their purpose in carrying out this work was not to assess the performance of notches and defects in components: their aim was to predict the variation in toughness which occurs in these materials with temperature and strain rate. The idea was that, due to local yielding, a certain amount of crack tip blunting will occur prior to failure, turning even a sharp crack into a notch with a finite value of ρ, which can be estimated from standard fracture mechanics theory. In Figs 6.3 and 6.4, the open points are calculated

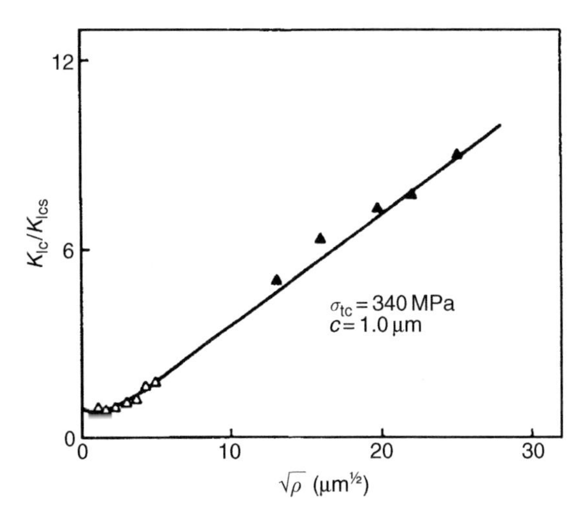

Fig. 6.3. Data from Kinloch et al., 1983. Measured K_c (normalised by the value for a sharp crack) as a function of $\rho^{1/2}$, for unmodified epoxy. Solid symbols are for cracks with drilled holes, open symbols are for sharp cracks, ρ being the calculated amount of crack blunting prior to fracture. The line shows a prediction using the PM.

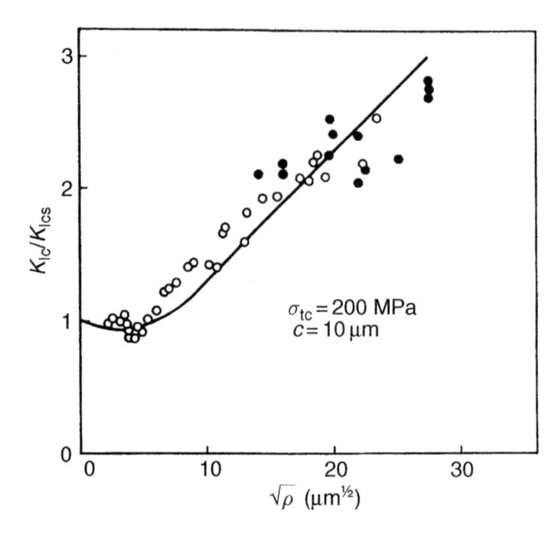

Fig. 6.4. Further data and predictions from Kinloch et al.: as Fig. 6.3 but for a rubber-modified epoxy.

values of ρ for the blunted cracks. But in order to confirm the validity of their approach, these workers also made some specimens in which the crack tips were deliberately blunted, by drilling a small hole at the end. Various other materials were also tested and characterised in this way (Kinloch and Young, 1983; Williams, 1984).

The only other use of TCD-type approaches such as the PM and LM which I have been able to find in the literature is a series of recent papers concerned with sharp V-shaped notches of zero root radius (Carpinteri and Pugno, 2005; Grenestedt et al., 1996; Leguillon, 2002; Seweryn and Lukaszewicz, 2002). These types of notches will be considered in Section 6.2.3. It is interesting to note that, in searching the literature, I was unable to find any recent reference to the application of this method to notches of finite root radius. This is very surprising, given that the groundwork was laid in the early 1980s, and considering the large number of polymeric components for which such an analysis would be relevant.

Our own results (of which more will be said below) and those of Kinloch, Williams and co-workers, all indicate that the appropriate value of σ_o is invariably higher than the material's yield and fracture strengths. For example, Kinloch and Young 1983 showed results for various different polymers in which σ_o takes values of 3–4 times σ_y. For this reason they suggested that σ_o represented a kind of constrained yield strength. Under conditions of plane strain (which will occur in many polymers if the section thickness is large enough – see Section 6.4 below) the stress near the crack tip rises to values of the order of $3\sigma_y$ due to constraint effects. However, as we shall see in further data below, this relationship does not always hold; we have found smaller values of the ratio σ_o/σ_y, sometimes less than 2, and values as high as 5.5 have been found in polymers (Kinloch et al., 1982) and 10 in some metals (see Chapter 7). It seems that there is no simple, analytical method for finding the value of σ_o from first principles. This issue will be discussed at greater length, in this chapter and later on in this book, as it is a matter of some importance. At this stage, however, we can take a purely practical view, noting that the appropriate values of the two constants which we need, σ_o and L, can be found provided we have test data from notched specimens with two different radii. But will these constants provide accurate predictions, not only for the results on relatively long, sharp notches as in Figs 6.2–6.4, but for all types of notches, of any possible size and shape? This issue will be addressed in the next section.

6.2.2 A wider range of notches

In my own laboratories, we carried out a series of tests, the aim of which was to examine the validity of the TCD when applied to stress concentrations of a wide variety of sizes and shapes. We used tensile specimens made from sheet material of constant thickness. Various features were introduced into the specimens, as follows:

(a) *Sharp notches*: These were single edge notches of dimensions similar to those used in the tests described above (e.g. Fig. 6.2), having high ratios of D/ρ. The ρ value chosen was the minimum that could conveniently be made using standard machining techniques. Typical dimensions for these notches were $D = 3$ mm, $\rho = 0.1$ mm.

(b) *Medium notches*: These were similar to the sharp notches but with a larger root radius. Various values were used, in the range 0.2–4 mm, giving stress concentration factors in the range 4–10.

(c) *Blunt notches*: Some very blunt notches were used to test the theory at low K_t. With depths of 0.5–1.5 mm and root radii of 12–50 mm, these notches had K_t factors of 1.5–2.25.

(d) *Holes*: Central circular holes with diameters in the range 1–3 mm. In some cases these holes were drilled at an angle of 45° or 70° to the specimen surface.

(e) *Hemispheres*: Small surface depressions of approximately hemispherical shape were made to simulate porosity and other small manufacturing defects. These had diameters in the range 0.45–3 mm.

(f) *Fillets*: specimens containing a reduction in width with a 90° fillet.

Four materials were tested: PS, HIPS, PMMA and bone cement, which is a low-strength form of PMMA used in surgical operations such as hip joint replacement. Tensile tests were carried out in deformation control at a rate of 5 mm/minute. More experimental details can be found elsewhere (Taylor et al., 2004). Table 6.1 shows the measured values of σ_u (from the plain specimens) and K_c (from the sharp notches, which had root radii less than the critical value). The values obtained were typical for these materials as reported elsewhere. Table 6.1 also shows, for completeness, the value of L which would be calculated using σ_u in Eq. (6.1). This is denoted L_u and is included here only for comparative purposes, since it was not used in the analysis. Both the PM and the LM were used, but since we obtained predictions of similar accuracy from both methods only the PM predictions are reported here.

Table 6.1. Material property values for the polymers tested

Material	K_c (MPa(m)$^{1/2}$)	σ_u (MPa)	L_u (mm)	σ_o (MPa)	L (mm)
PS	1.8	41.9	0.59	57.6	0.42
HIPS	0.9	19.2	0.70		
PMMA	2.23	71.5	0.31	146	0.107
Bone cement	1.6	52	0.30	104	0.154

Note: Valid results for L and σ_o could not be obtained for HIPS from the specimens tested.

As noted above, in order to obtain values for the parameters L and σ_o we need results from notched specimens of two different types. Previously, when working with ceramics (in Chapter 5) we calculated L using data from a plain specimen (to obtain σ_u) and a cracked specimen (to obtain K_c). For polymers we can continue to use the cracked specimen, but now the plain specimen will give no useful information; we need to substitute a notched specimen.

We chose, rather arbitrarily, to use the circular hole of diameter 3 mm as our second specimen type. In principle, one could use any notch geometry, with the limitation that the stress concentration factor must be greater than the ratio σ_o/σ_u (we will return to this point in later discussions): we chose the circular hole because it represents a relatively large root radius (therefore being very different from the sharp notch) and because specimens can easily be made in a reproducible manner.

FEA was used to model all the specimens, because there are no simple analytical solutions for these geometries and in any case FEA would be the normal method for analysing industrial components. The method of finding L and σ_o was as follows: stress/distance curves were drawn for the sharp notch and 3 mm hole specimens using loads applied to the FE models which corresponded to the experimentally determined failure loads for these specimens. In practice, since a linear-elastic analysis was carried out, it was easy to find this data from the FEA simply by scaling. The maximum principal stress was used, measured along a line drawn from the point of maximum stress in a direction normal to the applied load. Plotting the two curves on the same axes, the point at which they intersect gives the values of L and σ_o. This is illustrated in Fig. 6.5 for the PMMA material. The curve corresponding to the plain specimen (which of course is simply a horizontal line) is also drawn, for comparison. If the PM can be used as a means of prediction, then the corresponding curves for all notches should pass through the same point. In fact, as we shall see later, this will not be the case for very blunt notches or, obviously, for the plain specimen. An alternative approach to find L and σ_o would have been to plot the curves for all the different notches and to find the point on the graph corresponding to the geometric centre of all the points of intersection. This would be a 'best-fit' approach to all the available data. However, we chose to use only the results from two specimens because, whilst this approach will be less accurate from a scientific point of view, it emphasises the fact that relatively little testing is needed in order to use the TCD and it simulates the kind of industrial situation in which, following a small amount of laboratory testing, the method could be applied when designing components.

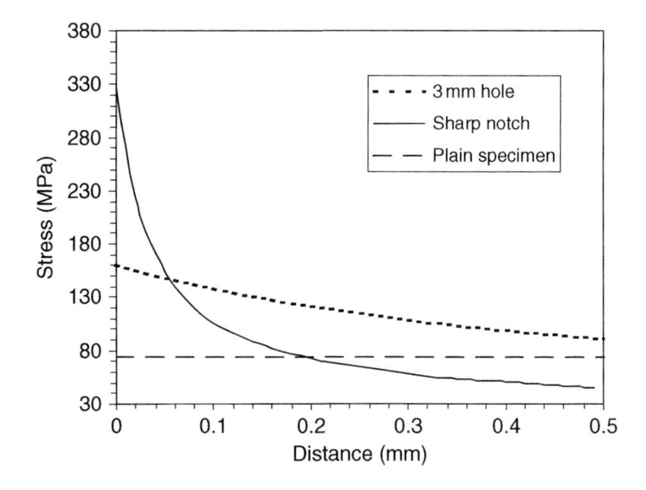

Fig. 6.5. Stress–distance curves corresponding to the failure loads for three specimens. The values of L and σ_o are found at the point of intersection of the lines for the 3 mm hole and the sharp notch.

Table 6.1 shows the resulting values L and σ_o for the PS, PMMA and bone cement. For the HIPS material, it was not possible to calculate L, or indeed to use the TCD, for reasons which will be explained later. Leaving this material aside, the other three failed in a brittle fashion, by the sudden growth of cracks. It was generally possible to distinguish the point of crack (actually craze) initiation on the specimen: for the plain specimens and the notches and holes with large root radii there was usually only one point of initiation (or, for the holes, one on each side) whilst for the sharp notches there was clear evidence of multiple initiation all along the notch root. Load/deformation curves showed generally linear behaviour, but there was usually some amount of curvature indicating plasticity or other non-linear deformation before failure.

Figure 6.6 presents the results for PMMA in terms of the nominal, net-section stress at failure: also shown are the predictions using the PM. Figure 6.7 presents similar results for PS. It is clear that in the great majority of cases the predictions have been successful. We regard the prediction as being a success if it has an error of less than 20%, reasoning that errors of the order of 10% will arise in both the experimental testing and the stress analysis, so it would be practically impossible to achieve greater accuracy. In general, the accuracy was very good; with the exception of the blunt notches, which we will return to shortly, the worst error was 25% and most predictions fell within 10% of the experimental value. Figure 6.8 shows prediction errors for PMMA from our own tests plus a further large set of data obtained by other workers (Gomez et al., 2000) who used single edge notch specimens loaded in both tension and three-point bend. They used a very wide range of notch depths, root radii and notch-depth/specimen-width ratios. We found that the constants for this material were only slightly different from those deduced for our own PMMA ($\sigma_o = 136\,\text{MPa}$, $L = 0.06\,\text{mm}$). As the figure shows, the prediction errors for all these notches were very acceptable, and were not significantly affected by the root radius or, for that matter, any other notch dimension.

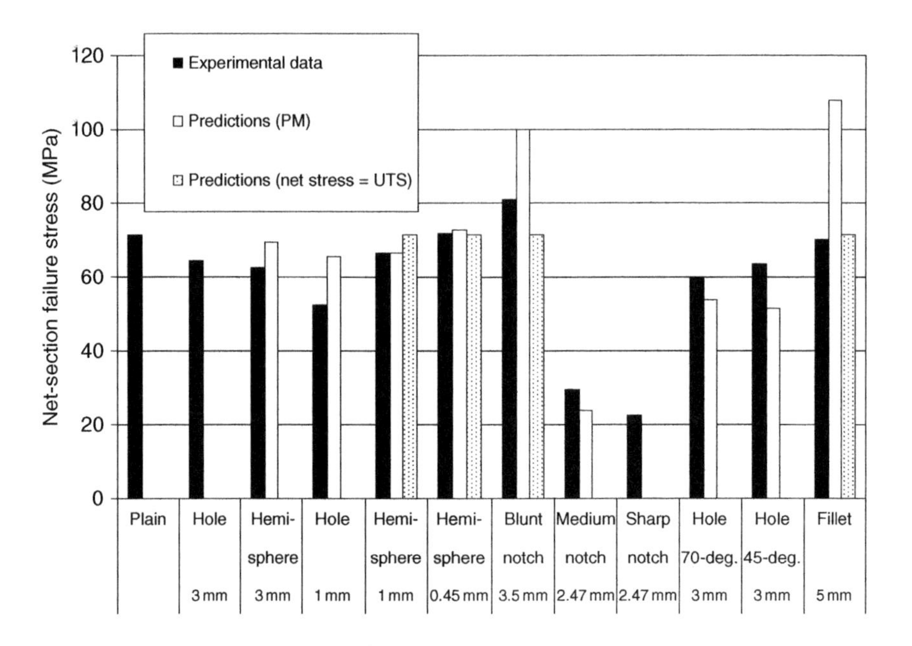

Fig. 6.6. Strength of PMMA containing various stress concentration features; predictions either using the PM or simply by net-stress = UTS (σ_u).

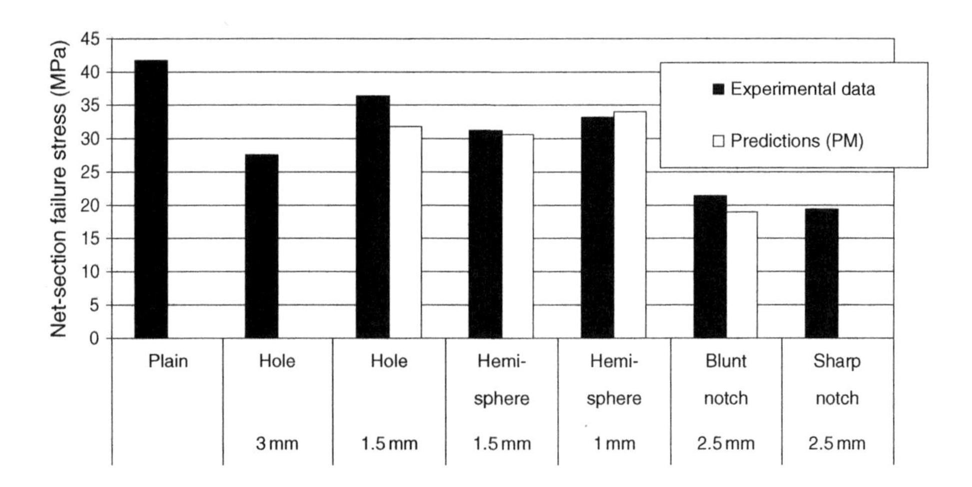

Fig. 6.7. Strength of PS containing various stress concentration features; predictions using the PM.

The bone cement material was also analysed successfully, though the material constants differed from those of commercial PMMA: the details are not included here but have been published elsewhere (Taylor et al., 2004).

It is worth noting that good predictions were possible even for the case of the holes drilled at different angles to the specimen face. These features create quite complex

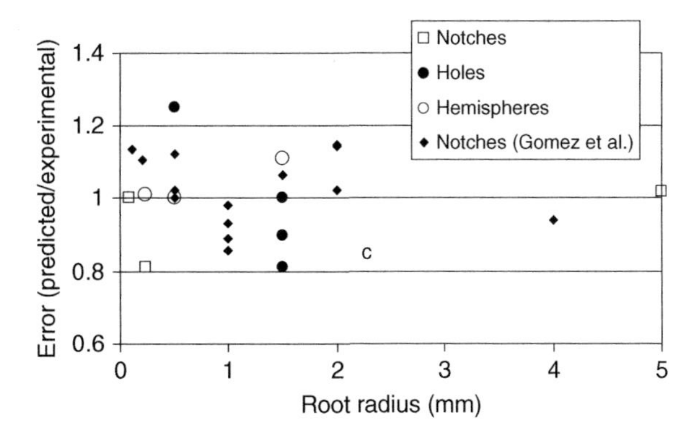

Fig. 6.8. Prediction errors (defined as the ratio between predicted and experimental failure stress) for PMMA, using data from our own work and another study (Gomez et al., 2000).

3D stress fields in contrast to the essentially 2D nature of the more typical specimens. The application of the TCD to features of 3D geometry and complex loadings presents some challenges in terms of its implementation and interpretation; these issues will be dealt with in more detail in later chapters, concerned with engineering components (Chapter 12) and multiaxial loading (Chapter 11).

This approach is clearly working well, but it is obvious that the use of a value of σ_o greater than σ_u is going to cause problems for certain types of notches, specifically those for which K_t is less than σ_o/σ_u. In those cases, we will predict a nominal fracture stress, in tension, which is greater than σ_u; this is clearly impossible since at σ_u failure can occur elsewhere in the specimen. The situation is illustrated schematically in Fig. 6.9, which is a graph similar to that of Fig. 6.2, but instead plotting the fracture stress σ_f rather than the measured K_c and extending the ρ axis to larger values. As ρ reaches infinity we have a plain specimen. The original form of the TCD, using L_u and taking

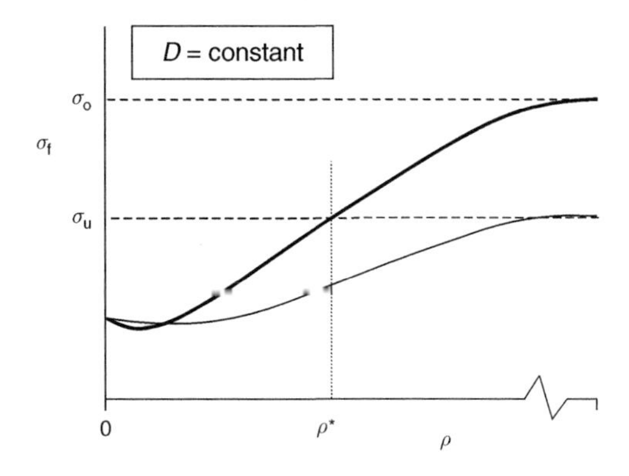

Fig. 6.9. Schematic showing TCD predictions of fracture stress as a function of root radius, at constant notch depth. The lower line is the prediction using σ_u as the critical stress; the upper line uses σ_o.

σ_u as the critical stress, will give a prediction line which tends to σ_u at high ρ, and this is satisfactory for predicting the behaviour of very brittle materials like ceramics.

The new form of TCD will give a higher prediction line, one which necessarily crosses the horizontal line corresponding to σ_u at some finite radius, which we can call ρ^*. We can view the situation in a different way (Fig. 6.10) by plotting K_t on the horizontal axis instead of ρ and using a logarithmic scale. Now the plain specimen corresponds to $K_t = 1$ (at the left hand end) and sharp, crack-like notches occur at high K_t, approaching infinity. Note that in both of these plots the fracture stress of the sharp notches will depend not only on ρ (or K_t) but also on the notch length, so in general there will be a series of lines which will tend to converge, at high ρ and low K_t, to a straight line corresponding to $\sigma_f = \sigma_o/K_t$.

We can see that now there are two separate predictions occurring: the TCD prediction and a prediction which is simply $\sigma_f = \sigma_u$. It is not obvious from a theoretical point of view how these two predictions will interact: we must have experimental data to discover what will happen in the region where the two predictions intersect. Figure 6.11 shows a graph similar to Fig. 6.10, with data from the above tests on PMMA and bone cement, plotting only those points with relatively low K_t factors. The value of σ_f has been normalised by dividing by σ_u. We can use the same prediction lines for both materials because, though they had different values for their material constants, the ratio σ_o/σ_u was the same in both cases (actually 2.04 for PMMA and 2.0 for bone cement). It can be seen that the experimental data lie close to the prediction lines, even at points near the intersection point. This implies that the two predictions can be made independently, the correct prediction being the one which gives the lowest value for σ_f in each case. Another way to express this idea is to say that there exist 'non-damaging notches': features which concentrate stress but which have no effect on strength; these notches are defined by $K_t < \sigma_o/\sigma_u$. It is obviously very useful to be able to identify features of this type when they occur, since the designer can use them with no fear that their

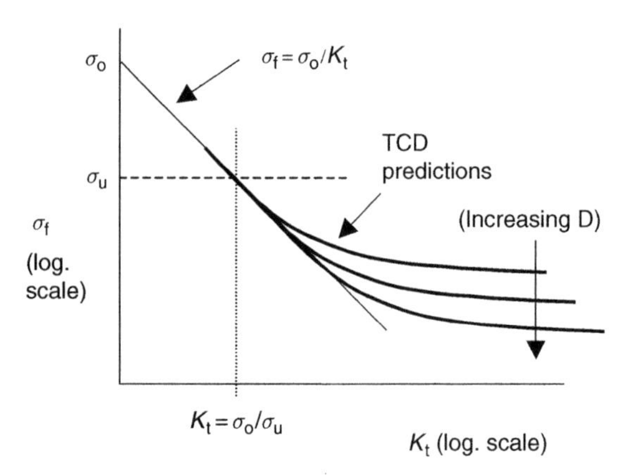

Fig. 6.10. As Fig. 6.9 but plotting K_t instead of ρ and using logarithmic scales. The curved lines show TCD predictions for various notch depths D, which tend to the line $\sigma_f = \sigma_o/K_t$ at low K_t. Strength cannot rise above σ_u, therefore non-damaging notches are predicted when $K_t < \sigma_o/\sigma_u$.

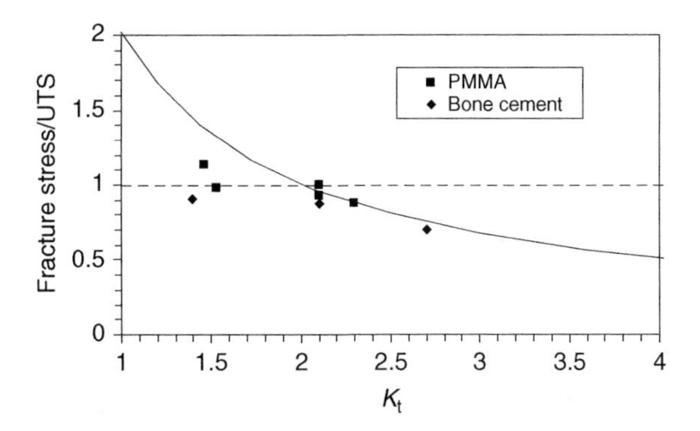

Fig. 6.11. Experimental data (σ_f/σ_u) for stress-concentration features of low K_t in commercial PMMA and Bone Cement; prediction lines as in Fig. 6.10. Note that the data follow the lower of the two lines, even at points close to the intersection point.

presence will compromise the strength of the component. However, a word of caution is needed here. The above tests were all carried out using specimens loaded in pure tension; this type of loading in fact rarely occurs in real components, which generally experience more complex, multiaxial loading modes. In particular there is always some degree of bending or torsion present, which will tend to set up a stress gradient even in a body containing no geometrical stress-raisers. Since polymers are frequently used in the form of thin sheets, out-of-plane bending will often occur.

To investigate this we carried out some tests on PMMA using specimens of the same geometry as described above but loaded in out-of-plane bending, creating a through-thickness variation in stress from tension on one face to compression on the other. We tested plain specimens and specimens containing a 3 mm diameter hole, which creates a rather complex 3D stress field. Figure 6.12 shows the geometry and loading: Table 6.2 summarises the results. Again the TCD was able to predict these results with good accuracy. It is interesting that we were able to predict even the plain bending test: clearly this is a preliminary result which would need further investigation, but it is very encouraging as it implies that the problem described above in relation to notches of low K_t in tensile stress fields will not generally arise in the great majority of engineering components.

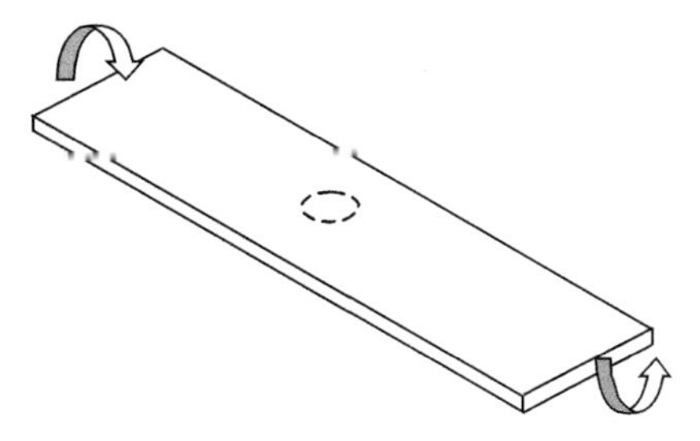

Fig. 6.12. Specimens of PMMA sheet, with or without a central hole, loaded in out-of-plane bending. Table 6.2 shows the results.

Table 6.2. Out-of-plane bending experiments: results and predictions

Feature	σ_f (MPa) Experimental	σ_f (MPa) Predicted	Error %
Plain specimen	124	146	16%
3 mm diameter hole	84.4	66.4	21%

Note: σ_f here is the maximum nominal stress at the surface.

Several other workers have generated experimental data which confirms the above approach (Inberg and Gaymans, 2002a; Nisitani and Hyakutake, 1985; Zheng et al., 2003): the first two references will be considered below in Section 6.4, because their results were obtained under conditions of varying constraint. Zheng et al. (2003) tested two types of PMMA, which they referred to as 'commercial PMMA glass' and 'oriented PMMA', using notched specimens with a wide range of K_t values. We found that the TCD can successfully be used to predict all this data. Interestingly the 'commercial PMMA glass' behaves almost like a classic brittle ceramic material, σ_o being only slightly larger than σ_u whereas their 'oriented PMMA' has quite similar material constants to the PMMA materials tested by ourselves and others such as Gomez et al. This illustrates that the ratio σ_o/σ_u can take various values, from unity upwards.

6.2.3 V-Shaped notches

V-shaped notches are defined as those having zero root radius but a finite opening angle, θ. A crack is of course just the limiting case of $\theta = 0$. The basic theory surrounding these notches was discussed in Chapter 3, and again in Chapter 4 in relation to the notch stress-intensity factor (NSIF) method. We recall that the local stress field for this kind of notch ($\sigma(r)$ for distances $r << D$) can be described by:

$$\sigma(r) = \Psi r^{-\lambda} \tag{6.4}$$

This is similar to the stress field for a crack, and reverts to this form at $\theta = 0$, when $\lambda = 0.5$ and Ψ has the same meaning as K. For $\theta > 0$, λ is a function of θ. In practice, it remains almost constant at 0.5 up to $\theta = 90°$ but then decreases with increasing θ, becoming zero when $\theta = 180°$ at which point we have a plain specimen. The existence of a relatively simple form for the stress field has led several workers to investigate this type of notch in detail, developing theories which can be explored mathematically, such as the NSIF theory. Consequently quite a lot of experimental data has been generated for these kinds of notches. The PM and LM have been successfully used by other workers (Carpinteri and Pugno, 2005; Grenestedt et al., 1996) to predict the fracture strengths of such notches, in PVC foam and PMMA: Fig. 6.13 shows one example.

Other workers considering V-shaped notches have used theories related to the TCD; for example, Leguillon used a combined stress–energy approach of the type mentioned in Chapter 3 (Leguillon, 2002) whilst Seweryn and Lukaszewicz used various TCD-type models (Seweryn and Lukaszewicz, 2002).

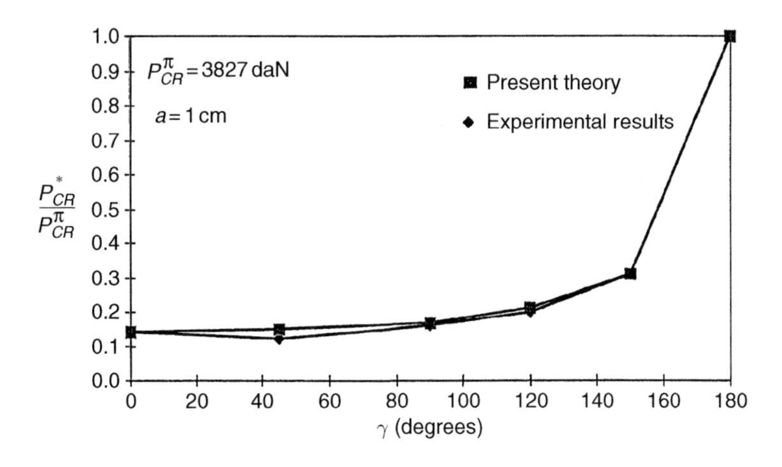

Fig. 6.13. Data and predictions for V-shaped notches (Carpinteri and Pugno, 2005).

It is very good to know that the TCD also works for these kinds of notches, though in practice it would rarely be possible to apply the analytical expression of Eq. (6.4) to the case of an engineering component. Normally FEA would be used, but a problem would be encountered, which also occurs when cracks are modelled in FEA, namely the creation of a stress singularity: a point at which the elastic stress theoretically reaches infinity. Thus it is impossible to converge the solution at the notch tip: increasing mesh refinement only increases the predicted stress at that point. However, the TCD can still be used because convergence does occur at every other point, including the critical point at $r = L/2$. This is in fact a major practical advantage of the TCD. This issue, and especially its consequences for the creation of simplified FE models of components, is discussed further in Chapter 12.

6.3 Size Effects

Returning to the discussion above (Section 6.2.2) on the behaviour of notches of various shapes and sizes, we now consider very small notches and small cracks. In the previous chapter (Section 5.2) it was shown that small cracks and defects in ceramic materials had fracture stresses which were lower than would be predicted by normal LEFM methods, the fracture stress tending to the value of σ_u as the crack length approached zero, rather than to infinity as would be predicted by LEFM. This behaviour could be predicted by the TCD; now, considering the way in which the TCD has been applied in this chapter, we can expect somewhat different behaviour from polymers. Just as we found that there were certain large, blunt notches which would be non-damaging, so we can also expect that there will be a class of small cracks and defects which likewise will have no effect on the tensile strength of the specimen. This arises again due to the difference between σ_o and σ_u, which will lead to situations in which the predicted fracture stress (using the TCD) is greater than σ_u.

In fact, we already saw this behaviour in some of the data from our own testing work, described above. The PMMA specimens containing small hemispheres (0.45 mm diameter) had fracture strengths which were very similar to those of the plain specimens (Fig. 6.6); in fact some of the fractures did not initiate at the hemispheres, but elsewhere in the specimens. This result is anticipated by the TCD: in small defects the critical point at $r = L/2$ is so far away from the notch that the stress at that point is similar to the nominal applied stress. Given a value of σ_o/σ_u greater than unity, the notch will essentially not exist as far as the TCD is concerned.

We can also expect the same behaviour from sharp cracks: there should be a crack length below which the crack has no effect on strength. In practice, these crack lengths will be small: of the same order of magnitude as L. There has been very little experimental data generated concerning such cracks in polymers, in contrast to the large amount of such data for ceramics. This is presumably because these small defects have a much more detrimental role in ceramics. However, the effect of small processing defects, inclusions, fatigue cracks and so on in polymers is also a matter worthy of study. A good example of this is orthopaedic bone cement, which, when used in surgical operations such as the artificial hip and knee joints, invariably contains defects such as bubbles of air and evaporating monomer, and casting porosity. There has been a lot of research into the behaviour of these defects since it has been noticed that they influence the long-term failure of surgical implants and prostheses (Culleton et al., 1993), and much effort has been put into devising techniques for reducing or eliminating them.

Berry carried out a series of tests to measure the fracture strength of PMMA and PS as a function of crack length, preparing small through-thickness cracks in tensile specimens by carefully machining material away from specimens containing larger cracks (Berry, 1961a; Berry, 1961b). For PMMA, he found that LEFM predictions were accurate down to the smallest crack length which he could generate, which was 0.05 mm. This is not surprising considering that we found L values of 0.06–0.11 mm for this material. For PS, Berry found that LEFM could be used down to a crack length of about 1 mm, but for shorter cracks a different behaviour applied. The fracture stress became constant, equal to the plain-specimen tensile strength, and in most cases the fracture did not initiate at the crack, but elsewhere. Fig. 6.14 reproduces some of this data for PS. Unfortunately it is not possible to make a prediction using the TCD because we have no data from notched specimens with which to calculate L and σ_o. The PS which we tested had a similar strength to that of Berry but a much lower toughness, so the values deduced for that material do not necessarily apply. However, it is interesting to see the form of the PM prediction using different values of σ_o/σ_u, as shown on Fig. 6.14. The prediction using $\sigma_o/\sigma_u = 1$ clearly underestimates the short crack data. Good predictions can be obtained for values of $\sigma_o/\sigma_u > 2$, for which the prediction lines stay close to the LEFM line, at least up to the point where it is intersected by the line $\sigma_f = \sigma_u$. Cracks behave either as normal, long cracks or else are completely non-damaging. This is interesting and merits further investigation, since this behaviour of short cracks in a polymer is quite different from their behaviour in ceramics (Chapter 5) and also different from the behaviour of short fatigue cracks, as we shall see in Chapter 9.

Incidentally, some of the first plain specimens of PMMA which we made contained small surface marks caused by a clamp which was used to hold the specimens during machining. These marks took the form of circular depressions about 1 mm across: though very shallow they had quite sharp edges. It is interesting to note that whilst these marks

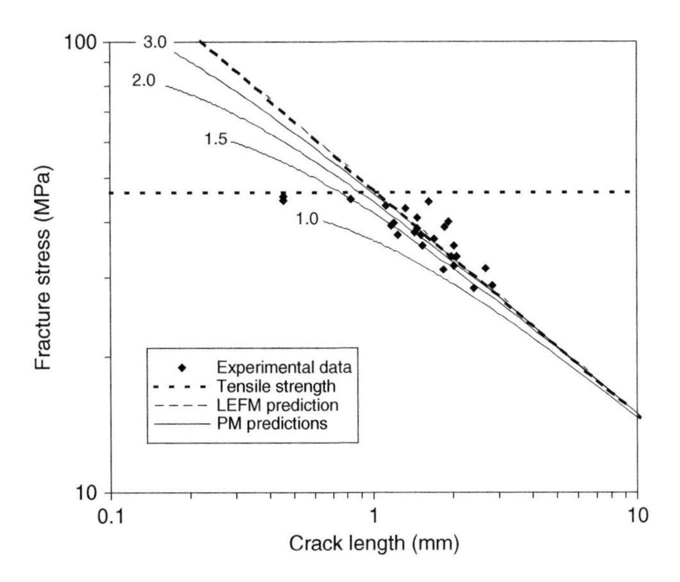

Fig. 6.14. Short crack data on PS (Berry, 1961b); predictions using the PM with various values of σ_o/σ_u.

did not cause any reduction in the strength of the specimens, nevertheless they provided the initiation sites for the failures. This illustrates a fact which always should be borne in mind when analysing failed components: just because a defect is found at the point of initiation of the failure does not necessarily mean that the defect has reduced the strength of the component.

6.4 Constraint and the Ductile–Brittle Transition

The concept of constraint was alluded to in Chapter 1, where it was mentioned that the toughness of materials is strongly affected by the amount of constraint in the region the crack tip. The general concept of constraint, though simple enough to understand in principle, turns out to be very difficult to handle in detail, as regards predicting the effect of a generalised 3D state of stress and strain on material strength and fracture. In this section, we will consider only the relatively tractable problem of out-of-plane constraint, by which we mean the effect of changing specimen thickness on the state of stress and strain in the thickness direction. Most of the specimens from which data have been obtained in this, and the preceding, chapter were essentially flat samples – plates or bars – containing through-thickness notches of constant depth. Conditions of plane stress always occur at the front and back surfaces, but if the specimen thickness, B, is sufficiently large then plane strain conditions exist throughout most of the thickness, that is the majority of the crack (or notch) tip experiences plane strain. Since the fracture toughness K_c is considerably lower in plane strain than in plane stress, fracture initiates from inside the specimen and the plane-strain K_c value is the relevant one.

The brittle fracture of ceramic materials almost always occurs under plane strain conditions, even for very thin specimens. For polymers, however, plane strain conditions can be lost in sheets of thickness less than a few millimetres. The process by which plane strain conditions are lost is essentially one of plasticity: the plane-stress regions at the surfaces give rise to relatively large plastic zones which, if they grow sufficiently to spread through the thickness and touch each other, effectively relieve the constraint and establish plane-stress conditions throughout. Since polymers are often used in the form of thin sheets, the degree of constraint is an important factor in predicting their mechanical behaviour.

This matter will be dealt with in more detail in the next chapter, on metals; suffice it to say that all the data presented above in this chapter was obtained using specimens of sufficient thickness to ensure plane strain conditions. In fact the plane-strain/plane-stress transition in polymers is almost always associated with a change from brittle to ductile behaviour, so brittle fracture generally occurs under plane strain conditions. A good example, in which there is a comprehensive set of data showing the effect of constraint on notch behaviour, is found in the work of Nisitani and Hyakutake, who tested double-edge-notched samples of PC in three different thicknesses: 1, 2 and 5 mm (Nisitani and Hyakutake, 1985). They showed that, as the notch root radius was increased, a transition occurred from brittle fracture to ductile fracture at some critical radius ρ_t, the value of which varied with specimen thickness B. They were able to observe the plastic zones in this transparent material. Brittle fracture was characterised by the sudden initiation and unstable growth of a crack; at the time of fracture the plastic zone was small, typically no larger than ρ. In ductile fracture the plastic zone grew much larger, spreading right across the width of the specimen to meet the plastic zone from the notch opposite. Final separation only occurred later, at a stress value similar to that of the plain specimen, that is in ductile fracture the notches were non-damaging.

Figures 6.15 and 6.16 show their data. In Fig. 6.15, the brittle fracture results are plotted, using a rather different method of presentation from that used previously. The vertical axis gives the value of stress *at the notch root* at the time of failure. This stress was calculated using linear–elastic analysis, so plasticity is not allowed for in this calculation. The stress used here is, by definition, equal the nominal fracture stress σ_f, multiplied by K_t. On the horizontal axis, Nisitani and Hyakutake plotted the inverse of the root radius, $1/\rho$. All the data points shown are for cases of brittle fracture: it is clear that there is no effect of specimen thickness, and our PM prediction describes the data very well. Also shown on the graph is a series of three lines corresponding to the predicted onset of plane-stress conditions. The theory used to draw these lines is explained in more detail in Chapter 7 (Section 7.2.2): it corresponds to a prediction of the point at which the plane-stress plastic zones spread through the thickness. The three lines correspond to the three different values of sheet thickness that were used: the points where they intersect the PM prediction are expected to be the values of ρ_t – the brittle–ductile transition radius – for each sheet thickness. As Fig. 6.16 shows, these predictions agree reasonably well with the experimentally measured values.

A second example of this kind of behaviour is provided by the work of Inberg and co-workers (Inberg and Gaymans, 2002a; Inberg and Gaymans, 2002b; Inberg et al., 2002), who also tested notched specimens of PC in tension. They, however, used a higher loading rate, of 1 m/sec, in order to simulate impact conditions.

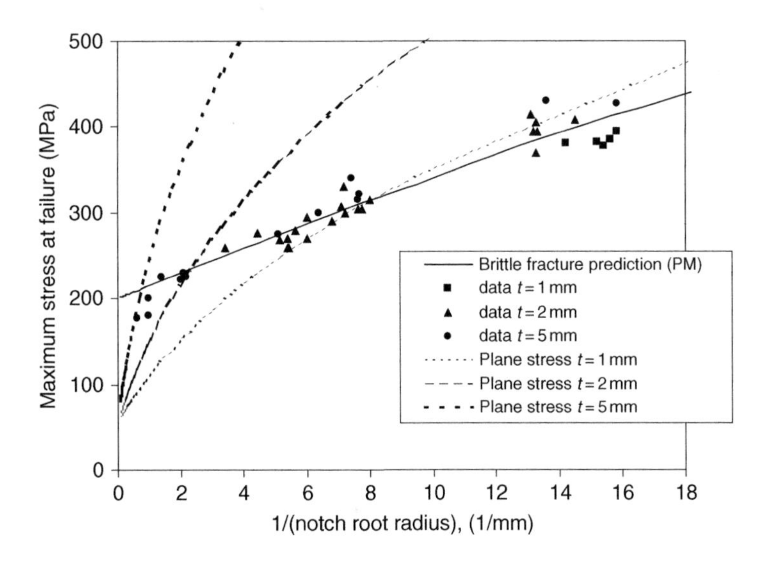

Fig. 6.15. Brittle fracture data for PC specimens with three different thicknesses (Nisitani and Hyakutake, 1985), with PM prediction line. Also lines predicting the onset of plane stress conditions for each thickness.

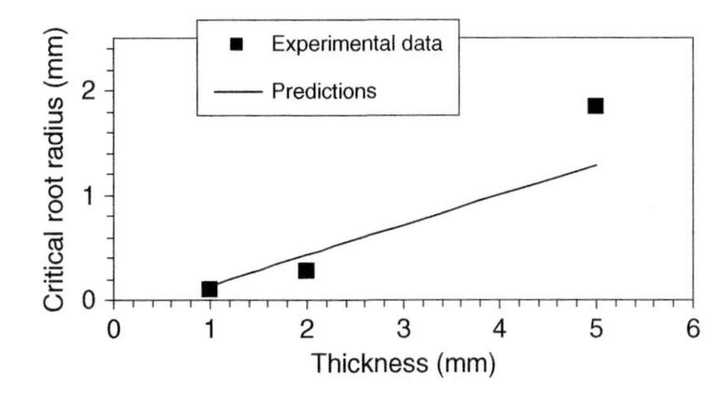

Fig. 6.16. Experimental and predicted values of the critical root radius for the ductile–brittle transition in PC (data from Nisitani and Hyakutake, 1985). Predictions using the method described in Section 7.2.2.

Despite the high loading rate, their results are numerically very similar to data from PC obtained from conventional tensile tests at lower deformation rates of the order of millimetres per second, as reported above; so the increased testing rate seems to have had no effect on either K_c or σ_u. Figure 6.17 shows fracture stress as a function of root radius, along with a prediction using the PM. It is clear that the prediction does not describe very well the behaviour of the blunter notches: in particular, those with root radii of 0.25 mm and 0.5 mm fail at significantly higher stresses than predicted. The reason for this becomes clear when we insert the line that corresponds to the onset of plane-stress conditions. This shows that the three blunter notches are failing in plane stress and are therefore predicted to be ductile failures, which in fact they were. The

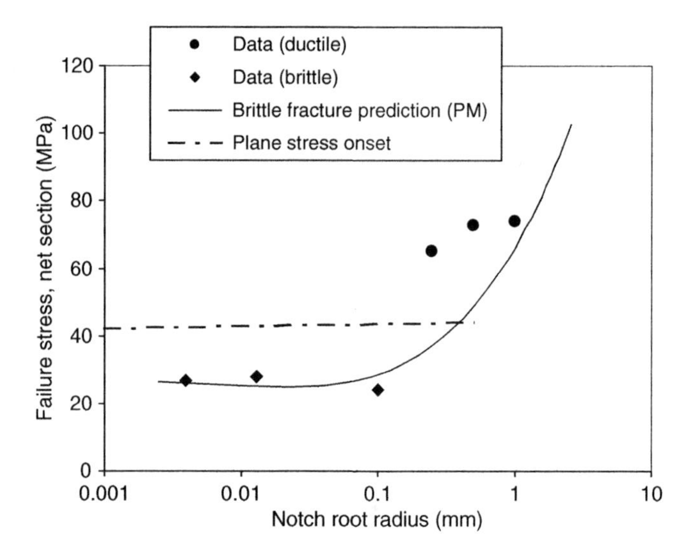

Fig. 6.17. Experimental data on PM at high loading rate (Inberg and Gaymans, 2002a). Predictions using the PM plus a line indicating the stress above which plane stress conditions will occur.

failure stress of these three notches is almost constant, close to the plain specimen strength of the material.

Finally, we consider some results from two rubber-toughened polymers: HIPS, which was tested as part of our own investigations reported above (Table 6.1), and PC toughened with ABS (Inberg and Gaymans, 2002a). The addition of small particles of a rubbery material is often used to improve the toughness of polymers, and especially to increase their resistance to impacts, a property which will be considered in the next section. Our tensile tests on HIPS used the same loading rate (5 mm/min) and the same types of notches as described above for PMMA and PS; the specimen thickness was 3 mm. The plain specimen strength was 19.2 MPa: when we calculated the nominal strengths of the other specimens using the net-section stress in each case, we found that there was almost no change: strengths varied from 14.57 to 19.44 MPa. Even a very sharp notch of length 5 mm only reduced strength by 24%. This 3 mm thick sheet was certainly too thin to establish plane strain conditions in the material, hence all the failures were ductile. We were just able to achieve plane strain when testing a 13 mm-diameter bar with a circumferential notch, from which we measured a fracture toughness of $K_c = 0.92\,\mathrm{MPa(m)}^{1/2}$. Even this test was not in strict conformance with the standards for plane-strain toughness testing, because the notch length was too short, but the result is similar to other values quoted in the literature (Fleck et al., 1994). The PC/ABS material tested by Inberg and Gaymans (2002) at higher strain rates also showed a very narrow range of strengths (53–69 MPa) for a wide range of notch root radii (0.004–1 mm), as shown in Fig. 6.18. The estimate of plane-stress onset conditions shows that all these specimens were failing in plane stress. This emphasises once more the considerable insensitivity to notches displayed by polymers when they are failing in a ductile manner, as will often occur in practice when they are used in the form of thin sheets.

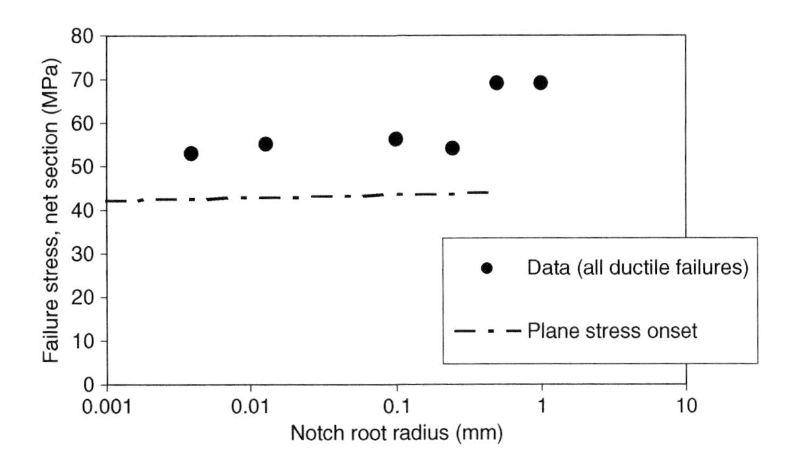

Fig. 6.18. Data from the same source as Fig. 6.17, in this case for rubber-toughened PC/ABS material (Inberg and Gaymans, 2002a). All failures were ductile, as predicted by the position of the plane-stress onset line.

6.5 Strain Rate and Temperature Effects

As noted above, the mechanical properties of polymers are greatly affected by changes in temperature and strain rate within the ranges frequently encountered in service. The study of these effects constitutes a very large subject in itself and one which has been examined in an extensive body of research. The two excellent books which I mentioned earlier (Kinloch and Young, 1983; Williams, 1984) provide a very good introduction to this subject.

In practice, the toughness of polymers is usually measured in impact tests. Such tests are certainly important but they provide quite different data from conventional K_c tests: for example, the HIPS mentioned above has a much higher impact toughness than unmodified PS, but its K_c value is actually lower than that of PS. This implies that, under loads which are static or applied only slowly, PS will always be stronger than HIPS, using either notched or unnotched samples. This difference arises because the impact test measures an essentially different property: the total amount of energy absorbed during specimen failure. A material will achieve a high value in this test, even if its K_c is low, by having a large area under the stress–strain curve (i.e. a large amount of elastic and/or plastic deformation before failure). In rubber-toughened materials an important mechanism is the multiple crazing occurring in the matrix material between the rubber particles, which acts to absorb energy in a manner similar to plastic deformation.

A further difficulty with the impact test is that the energy parameter which is obtained cannot be normalised in such a way as to make it independent of specimen size. Therefore, unlike K_c and the parameters of the TCD, impact energy cannot be used quantitatively to predict failure in components, but only as a relative ranking of materials. However, many workers have attempted to use impact results to estimate fracture mechanics parameters, and this can be done with reasonable success, at least for fairly brittle materials. Thus, for example, Williams and co-workers (Plati and Williams, 1975; Williams, 1984) describe a technique for obtaining G_c values from notched impact tests which involves testing at a number of different notch depths. They applied the PM to predict the effect of root radius, using exactly the same theory as described above, which gives, for the measured value for a notched specimen, G_{cm}, compared to the sharp-crack value G_c:

$$\frac{G_{cm}}{G_c} = \frac{\left(1+\frac{\rho}{L}\right)^3}{\left(1+\frac{2\rho}{L}\right)^2} \tag{6.5}$$

Note that this becomes identical to Eq. (6.2) if all terms are squared. They collected a considerable amount of data on the effect of both temperature and notch radius on the impact performance of a range of polymeric materials. Our analysis of the tests of Inberg and Gaymans (in Sections 6.3 and 6.4) suggests that data obtained in conventional tensile tests carried out at speeds approaching those of a typical impact test can also be interpreted using the TCD. In general, then, we can expect that the TCD should be applicable at any temperature and strain rate, provided the failures occur by brittle fracture under small-scale yielding conditions.

6.6 Discussion

In this chapter, we have seen that the TCD can be applied to polymeric materials; these materials, whilst they have low toughness and often display classic brittle behaviour, also show considerable non-linear and plastic deformation before failure. This added complication is surely the underlying reason for the fact that we can no longer calculate L using σ_u as the characterising stress, as we did for ceramics in Chapter 5, but rather must find a new critical stress, σ_o, and corresponding L value. The plain specimen strength is no longer a useful guide to the behaviour of notched specimens, and this recalls a point that was discussed in the previous chapter with regard to samples that displayed low apparent σ_u values due to inherent defects. The plain-specimen tensile strength can be influenced by a range of factors which do not affect the notched strength, such as inherent defects, plastic deformation (including necking instability which is affected by the work-hardening rate) and other damage modes such as large-scale crazing. One might come to the conclusion that the tensile test, though it is the basic test for characterising all materials, is in fact not particularly useful when it comes to predicting behaviour under real service conditions, that is behaviour in the presence of stress concentrations. The failure point in the tensile test is a very particular one, unique to the type of loading, and real components are only rarely loaded in pure tension. One can of course think of important exceptions, such as ropes, but in general the designer must accept the presence of bending and torsion in almost all components. The moral of the story, then, at least as far as the prediction of brittle fracture is concerned, is to ignore the measured tensile strength and to use the TCD with parameters derived from the testing of notched specimens.

Another aspect of the behaviour of polymers, which we did not have to face with ceramics, is the existence of features which have no effect on strength. We found two different types of such features: those notches and cracks which were non-damaging as a result of their small size, and those which were non-damaging due to their low stress-concentration factors. In both cases, the phenomenon was predicted by the TCD, arising due to the difference between σ_o and σ_u. For these non-damaging features the specimen strength was found to be simply equal to σ_u, so the effect from a predictive point of view is that we have two different predictions for σ_f, one being the TCD and other being $\sigma_f = \sigma_u$. These two predictions interact in a very simple way: the correct prediction is simply the lower of the two. This simple interaction is not one which could have been predicted, in fact it is rather surprising from a fracture mechanics point of view. Its success probably relies on the fact that provided brittle fracture occurs, then the zone of plastic (or other non-linear) deformation around the notch remains small. In fact this was directly observed by some workers (Nisitani and Hyakutake, 1985), who found small plastic zones in all their brittle fractures, even for conditions approaching the brittle–ductile transition. This means that the all-important small-scale yielding criterion is maintained. This criterion (see Chapter 1) is one of the essential prerequisites for LEFM, and we would also expect it to be a necessary condition for the use of the TCD, because without it the elastic stresses near the notch cannot be relied on as a means of characterising behaviour. This issue will be discussed at more length in Chapter 13.

Table 6.3 provides a list of parameter values for all the materials considered in this chapter. Some of the L values are of the same order of magnitude as those found in engineering ceramics, whilst others are significantly larger. In the case of the ceramics, it was suggested that the value of L might be related to the size of microstructural features such as grains; however, we found that there was no simple relationship, L varying between 1 and 10 times the grain size. Many polymers, including most of the ones which we have been considering here, have no such microstructure, being amorphous, so what could the value of L correspond to? One possible candidate is the craze. Crazes tend to form at a fixed size which is certainly of the same order of magnitude as L. For example, in PMMA the typical craze length is 70–100 μm, a value that seems to stay more or less constant and which is identical to the values of L which we determined. Stress values determined at $L/2$, or averaged over $2L$, may thus be characteristic of the amount of stress being applied to a craze which, if it exceeds a critical value, will cause the craze to propagate. In PC, on the other hand, where there is significant plastic deformation as well as crazing, the observed craze lengths and the plastic zone sizes at failure (Tsuji et al., 1999) were considerably larger than L, so no obvious connection can be made in that case.

Another argument is that the value of L relates not to the size of any inherent feature in the material, but rather to the size of the process zone ahead of the notch, in which any non-linear processes such as plasticity and damage, occur. This is a point which we will return to in more detail in Chapter 13 when we have the benefit of considering all the experimental data covering different materials and failure processes. We can note, as we did previously in the case of ceramics, that a number of workers have used process-zone models of various kinds, from a simple consideration of process zone size (Tsuji et al., 1999) to the attribution of complex stress-deformation characteristics for the material in the process zone using a cohesive zone model (Elices et al., 2002; Gomez and Elices, 2003; Gomez et al., 2000).

Table 6.3. Mechanical property values for polymers

Material	L (μm)	σ_u(MPa)	σ_o(MPa)	K_c(MPa.m$^{1/2}$)	Reference
Epoxy	2		340	(0.85)	(Kinloch et al., 1983)
Epoxy (rubber modified)	20		200	(1.6)	(Kinloch et al., 1983)
Epoxy (various)	0.24 −1.48		495 −340	(0.43 −0.73)	(Kinloch and Williams, 1980)
PC	61	70.2	250	3.47	(Tsuji et al., 1999)
PC	76.5	68.5	200	(3.1)	(Nisitani and Hyakutake, 1985)
PC (high strain rate)	46	(73.5)	250	3.0	(Inberg and Gaymans, 2002a)
PC/ABS (high strain rate)	176	(69)	250	5.9	(Inberg and Gaymans, 2002a)
PMMA	60		136	(1.87)	(Gomez et al., 2000)
PMMA	107	71.5	146	2.23	(Taylor et al., 2004)
PMMA bone cement	154	52	104	1.6	(Taylor et al., 2004)
PS	420	41.9	57.6	1.8	Unpublished work (D.Lavin)
PVC Foam (4 different densities)	534 −802	2.51 −9.38	3.51 −12.5	0.1 −0.42	(Grenestedt et al., 1996)

Note: Brackets indicate approximate or estimated values.

The research literature on polymers shows work in the 1980s on the use of theories similar to the TCD, including some implementations of the PM and LM which are identical to our approach, going back to the work of Williams, Kinloch and colleagues as described above. Given that this work was done over 20 years ago, I was surprised to find relatively little sign of it in more recent publications in this field. As mentioned above, almost all recent references to TCD-like theories (e.g. Carpinteri and Pugno, 2005; Grenestedt et al., 1996; Seweryn and Lukaszewicz, 2002) were concerned with predicting the behaviour of sharp V-shaped notches with (assumed) zero root radius. The interest in these notches, apart from predicting the effect of notch angle, was to study their behaviour under applied multiaxial loading. This is certainly a very interesting subject (and one which will be discussed later in Chapter 11) but it is surprising that relatively little work has been done applying the TCD to notches of finite root radius, with almost no consideration of the very blunt and very small features which can be non-damaging. Thus, whilst the background work has certainly been done as regards the application of the TCD to the failure of polymers, there is a lot more useful research which could be undertaken to allow us to understand fully their behaviour for a wide range of stress concentration features. Such work will certainly be rewarded, given the strong need for a general procedure that can be used in industrial design of load-bearing, polymeric components.

References

Berry, J.P. (1961a) Fracture processes in polymeric materials I. The surface energy of poly(methyl methacrylate). *Journal of Polymer Science* **L**, 107–115.

Berry, J.P. (1961b) Fracture processes in polymeric materials II. The tensile strength of polystyrene. *Journal of Polymer Science* **L**, 313–321.

Carpinteri, A. and Pugno, N. (2005) Fracture instability and limit strength condition in structures with re-entrant corners. *Engineering Fracture Mechanics* **72**, 1254–1267.

Culleton, T., Prendergast, P.J., and Taylor, D. (1993) Fatigue failure in the cement mantle of an artificial hip joint. *Clinical Materials* **12**, 95–102.

Elices, M., Guinea, G.V., Gomez, F.J., and Planas, J. (2002) The cohesive zone model: advantages, limitations and challenges. *Engineering Fracture Mechanics* **69**, 137–163.

Fleck, N.A., Kang, K.J., and Ashby, M.F. (1994) The cyclic properties of engineering materials. *Acta Metall.Mater.* **42**, 365–381.

Gomez, F.J. and Elices, M. (2003) Fracture of components with V-shaped notches. *Engineering Fracture Mechanics* **70**, 1913–1927.

Gomez, F.J., Elices, M., and Valiente, A. (2000) Cracking in PMMA containing U-shaped notches. *Fatigue and Fracture of Engineering Materials and Structures* **23**, 795–803.

Grenestedt, J.L., Hallestrom, S., and Kuttenkeuler, J. (1996) On cracks emanating from wedges in expanded PVC foam. *Engineering Fracture Mechanics* **54**, 445–456.

Inberg, J.P.F. and Gaymans, R.J. (2002a) Polycarbonate and co-continuous polycarbonate/ABS blends: influence of notch radius. *Polymer* **43**, 4197–4205.

Inberg, J.P.F. and Gaymans, R.J. (2002b) Polycarbonate and co-continuous polycarbonate/ABS blends: influence of specimen thickness. *Polymer* **43**, 3767–3777.

Inberg, J.P.F., Takens, A., and Gaymans, R.J. (2002) Strain rate effects in polycarbonate and polycarbonate/ABS blends. *Polymer* **43**, 2795–2802.

Kinloch, A.J. and Williams, J.G. (1980) Crack blunting mechanisms in polymers. *Journal of Materials Science* **15**, 987–996.

Kinloch, A.J. and Young, R.J. (1983) *Fracture behaviour of polymers*. Applied Science Publishers, London.

Kinloch, A.J., Shaw, S.J., and Hunston, D.L. (1982) Crack propagation in rubber-toughened epoxy. In *International Conference on Yield, Deformation and Fracture, Cambridge* pp. 29.1–29.6. Plastics and Rubber Institute, London.

Kinloch, A.J., Shaw, S.J., and Hunston, D.L. (1983) Deformation and fracture behaviour of a rubber-toughened epoxy: 2. Failure criteria. *Polymer* **24**, 1355–1363.

Leguillon, D. (2002) Strength or toughness? A criterion for crack onset at a notch. *European Journal of Mechanics A/Solids* **21**, 61–72.

Nishitani, H. and Hyakutake, H. (1985) Condition for determining the static yield and fracture of a polycarbonate plate specimen with notches. *Engineering Fracture Mechanics* **22**, 359–368.

Plati, E. and Williams, J.G. (1975) Effect of temperature on the impact fracture toughness of polymers. *Polymer* **16**, 915–920.

Seweryn, A. and Lukaszewicz, A. (2002) Verification of brittle fracture criteria for elements with V-shaped notches. *Engineering Fracture Mechanics* **69**, 1487–1510.

Taylor, D., Merlo, M., Pegley, R., and Cavatorta, M.P. (2004) The effect of stress concentrations on the fracture strength of polymethylmethacrylate. *Materials Science and Engineering A* **382**, 288–294.

Tsuji, K., Iwase, K., and Ando, K. (1999) An investigation into the location of crack initiation sites in alumina, polycarbonate and mild steel. *Fatigue and Fracture of Engineering Materials and Structures* **22**, 509–517.
Williams, J.G. (1984) *Fracture mechanics of polymers*. Ellis Horwood, Chichester.
Zheng, X.L., Wang, H., and Yan, J.H. (2003) Notch strength and notch sensitivity of polymethyl methacrylate glasses. *Materials Science and Engineering A*.

CHAPTER 7

Metals

Brittle Fracture in Metallic Materials

7.1 Introduction

During the past 50 years an enormous amount of research has been carried out on the study of fracture processes in metals; great advances have been made in this field, especially in the understanding of toughness, the relationship between toughness and strength, and the mechanisms by which fracture occurs in metallic materials. Today's transport and energy industries would be impossible without the results of this research, applied to the design of pipelines, reactor pressure vessels, jet engines and so on. Nevertheless, brittle fracture is still an ever-present danger, even in metals such as aluminium which are often thought of as essentially ductile (Colour Plate 3).

Before we proceed, then, we should clarify the definition of 'brittle', as it is used in this book. A brittle fracture is defined as any fracture that occurs by the rapid, unstable extension of a crack. The use of the terms 'brittle' and 'ductile' can be confusing when talking about metals, because they are often used in a different sense: to describe microscopic mechanisms of crack extension. In metals, crack growth can occur by one of the three micro-mechanisms: cleavage, void growth and shear. Cleavage, which occurs in steels at low temperatures, and in some other metals, involves crack growth by the separation of atomic planes. It is usually initiated by cracking in a brittle microstructural feature such as a precipitate or inclusion. The propagation of this microscopic crack is primarily controlled by local tensile stress. This mechanism is frequently referred to as brittle, though interestingly it requires plastic deformation before it can commence, so cleavage always begins within the plastic zone. The other two mechanisms of crack growth are referred to as ductile mechanisms. Void growth involves the initiation of cavities in regions of intense local plastic strain. These cavities, which are ellipsoidal in shape, are initiated at microstructural features such as inclusions: their growth is dependent on both plastic strain and triaxiality. Growth continues until either the voids meet or else the remaining material between them fails by some other means, usually shear. Shear failure involves intense dislocation motion confined to narrow bands, typically at 45° to the tensile axis. It is common in thin sections and in materials prone to strain localisation: it is controlled by effective stress parameters such as the Von Mises stress. In this book, these micro-mechanisms will be referred to by the terms 'cleavage', 'void-growth' and 'shear', or, where necessary for clarity, as the 'brittle micro-mechanism' and the 'ductile micro-mechanisms'.

In reading the research literature on this subject over the last 50 years, one is struck by the impression of a goal which always seems to be just within reach, yet which keeps receding as time goes on. Irwin, writing a review article in 1964 which summarised much of the work of the previous two decades, felt able to make the following confident statement:

> . . . linear elastic fracture mechanics already provides a rather complete set of mathematical tools. Additional experimental observations rather than additional methods of analysis are now the primary need for practical applications (Irwin, 1964)

Since that time, we have found that a number of phenomena, most of which were in fact known to Irwin, have turned out to be much more difficult to understand and quantify that could ever have been expected 40 years ago. Principal among these are the slow, stable crack extension which frequently precedes unstable failure, and the effect of constraint. These issues were mentioned previously in Chapter 1, and will be discussed more fully in the context of recent research later in this chapter. They have come to dominate work in this field because, in modern, high-quality alloys, failure is often preceded by considerable amounts of plastic deformation (thus the small-scale yielding criterion no longer applies) and generally involves a period of stable crack growth under conditions of complex 3D stress. That is to say, the LEFM conditions envisaged by Irwin are now the exception rather than the rule in engineering structures made from metallic materials. The difference between brittle and ductile failures, even at the macroscopic level, becomes unclear given that, in a specimen containing a notch, failure may occur either close to the notch or in the centre of the cross section as a result of high local stresses developed under conditions of constraint. The latter mode of failure is typical of the behaviour of a simple tensile specimen after the formation of a neck. In some cases, a stress concentration feature may have no effect in reducing strength, but on the other hand even a classic ductile failure may involve the initiation and growth of crack-like damage; thus there is a blurring of the distinction between brittle and ductile failure modes.

It is certainly not my intention to solve all these problems here. This chapter has two specific aims:

(1) To demonstrate that the TCD can be applied to predict fracture in metals in cases where constrained yielding occurs. We will see that in situations where behaviour conforms to the criteria of LEFM, then notches and cracks can validly be studied using the TCD.

(2) Taking a wider view of the subject, to show that the use of a material-dependant distance parameter is an essential ingredient of any theoretical model of fracture. Thus, whilst the particular linear, elastic form of the TCD which is used in

this book is invalid in many cases, nevertheless some form of critical distance approach will always be needed to recognise the existence of microstructure and other size-dependant phenomena.

7.2 Predicting Brittle Fracture Using the TCD

In this section, we will proceed very much in the same way as we did in the previous two chapters, where we discussed ceramics and polymers. By examining the experimental data and making predictions using simple analytical solutions or FE models, we shall show that the TCD is able to predict the onset of brittle fracture from notches in various metallic materials. In addition to the effects of notch root radius and K_t (which turn out to be very similar in metals and non-metals), attention will be paid to the effect of constraint and to the role of material microstructure.

7.2.1 The effect of notch root radius

Figure 7.1 shows a typical set of experimental data recording the effect of notch root radius on the measured fracture toughness of steel. These results are due to Wilshaw et al., who tested a mild steel at a temperature of $-196\,°C$; failure occurred by the cleavage mechanism (Wilshaw et al., 1968). Notched bar specimens were used, of dimensions $10 \times 10 \times 60\,mm$, containing a 2 mm-deep notch with an included angle of $45\,°$, loaded in three-point bending. We have already seen graphs similar to this in

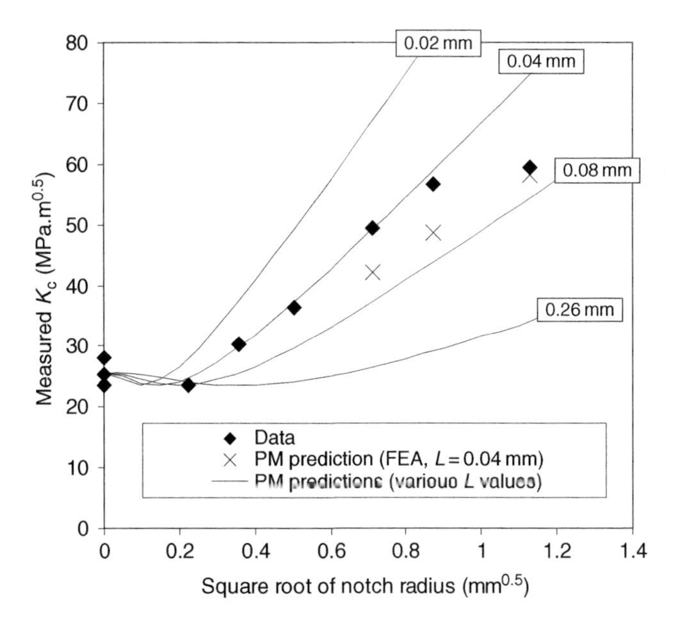

Fig. 7.1. Experimental data due to Wilshaw et al. Lines show predictions using the PM (Eq. 7.2) with various values of L. Also shown are predictions at $L = 0.04\,mm$ using FEA for more accurate stress analysis.

Chapters 5 and 6; data points at $\rho = 0$ record the fracture toughness K_c, equal to the plain-strain toughness if plain-strain conditions apply, as in fact they do in this case. There is a critical root radius, ρ_c, beyond which the measured value of toughness increases, showing an approximately linear relationship with the square root of the radius.

We proceed in the same manner as described in previous chapters (see Sections 5.2.2 and 6.2.1), making use of an approximate prediction of the stress field near the notch root (Creager and Paris, 1967), for notches of length a and root radius ρ:

$$\sigma(r) = \frac{K}{\sqrt{\pi}} \frac{2(r+\rho)}{(2r+\rho)^{3/2}} \tag{7.1}$$

Using this equation with the PM, we obtain a value for the measured K_c value of a notched specimen, K_{cm}, in terms of the critical distance, L:

$$K_{cm} = K_c \frac{(1+\rho/L)^{3/2}}{(1+2\rho/L)} \tag{7.2}$$

The value of K_c for cracks in this material was 25.7 MPa$(m)^{1/2}$ and the yield strength was 829 MPa; the UTS (σ_u) was not given in the paper, but comparing it with similar materials we can estimate a value of 900 MPa.

Figure 7.1 shows prediction lines, choosing various values of L. The value of L, which we would calculate using σ_u, which previously we termed L_u, is 0.26 mm. Recall that L_u is defined as:

$$L_u = \frac{1}{\pi}\left(\frac{K_c}{\sigma_u}\right)^2 \tag{7.3}$$

Clearly this value of L gives very poor predictions of the data: better predictions can be made using a much smaller value, of the order of 0.04 mm. Even for this L value there is some deviation at the larger values of ρ, but this error is due to inaccuracies in Eq. (7.1), which is valid only when $\rho << a$ and assumes infinite specimen dimensions. Calculations made using FEA to obtain the stress field give, with the same value $L = 0.04$ mm, significantly lower predictions for the blunter notches. When accurate stress analysis is used, the best fit to the data is obtained for an L value of 0.035 mm as shown in Fig. 7.2. It is interesting to note that this is exactly equal to the measured grain size of the material. Recall that L is linked to a parameter which we call the inherent strength of the material, σ_o, as follows:

$$L = \frac{1}{\pi}\left(\frac{K_c}{\sigma_o}\right)^2 \tag{7.4}$$

The value of σ_o in this case is 2447 MPa, which is higher than σ_u by a factor of approximately 2.7, and higher than the yield strength of the material by 2.95. Clearly the same effect is occurring here as we found in polymers in Chapter 6 (though not in ceramics in Chapter 5): the value of σ_u cannot be used in making the TCD prediction because it describes the behaviour of plain specimens, which fail by a different mechanism, involving extensive plastic deformation throughout the specimen before failure.

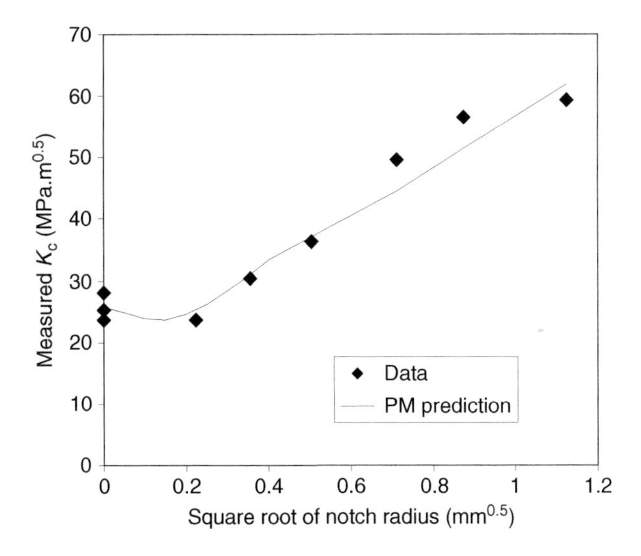

Fig. 7.2. The same data as in Fig. 7.1, with PM predictions at the optimum L value of 0.035 mm (using accurate stress analysis throughout).

We can also make predictions using the LM, as we did before; the equivalent expression for measured toughness is

$$K_{cm} = K_c(\rho/4L + 1)^{1/2} \tag{7.5}$$

Whilst both the PM and the LM can be used with reasonable accuracy, it was found that the PM was somewhat better for describing the data of Fig. 7.2 and indeed other data where the micro-mechanism is cleavage failure. In all this data there is a very clearly defined value of ρ_c, with almost no change in K_c between $\rho = 0$ and $\rho = \rho_c$. Indeed in some cases there is possibly a sign of the slight decrease in K_c around ρ_c which the PM would predict.

On the other hand, some results, typified by Fig. 7.3, show a monotonically increasing curve with no clearly defined value of ρ_c. This example comes from tests on a dispersion-strengthened aluminium alloy, DISPAL-2, tested at four different temperatures (Srinivas and Kamat, 2000). Here the LM modelled the data very accurately. Crack propagation in this material occurred by the ductile micro-mechanism of void growth, so the difference in behaviour may possibly be related to the mechanism of failure. The optimum value of L was constant at 0.045 mm for the three lower temperatures, rising to 0.075 mm at 350 °C. The corresponding σ_o values were again of the order of 3 times the relevant yield strength. Plane-strain conditions were maintained throughout.

These results, and others which will be presented below, clearly show that the TCD can be used to describe the effect of notch root radius when brittle fracture occurs, whether the crack extension mode is cleavage or a ductile micro-mechanism. In these cases, however, strict conformance to LEFM conditions was maintained, in respect of

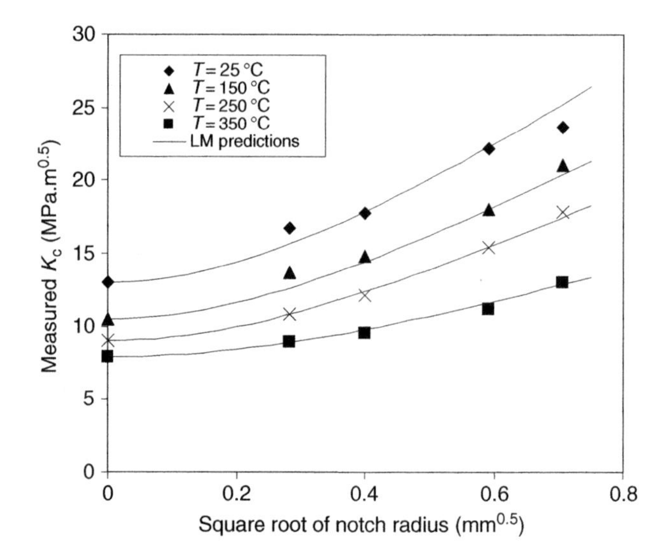

Fig. 7.3. Data from Srinivas and Kamat on an aluminium alloy tested at four different temperatures. Predictions using the LM.

contained yielding and plane strain. We shall now see what happens when we depart from these conditions and examine cases of reduced constraint.

7.2.2 The effect of constraint

It is well known that the value of K_c depends on the level of constraint. This issue was briefly discussed in Chapter 1 with respect to the difference between plane-stress and plane-strain conditions, that is the so-called 'out-of-plane constraint'; it was also mentioned that 'in-plane constraint' effects occur due to changes in the stresses parallel to the crack growth direction. These effects will be considered in more detail later on in this book, especially in the chapter on multiaxial loading. Suffice it to say that the effect of constraint is a major topic of current research. To date, there is no agreed method for predicting the effect of constraint on fracture toughness, and even the quantification of the level of constraint in a given situation is not a trivial matter.

The present section is concerned with investigating how the TCD might be used in cases of varying constraint. If the TCD can be used at all, it is clear that some changes to the material constants will occur, for the same reason that changes occur in the value of the conventional fracture toughness. In what follows (as in earlier chapters), we shall use the notation K_c to refer to *any* fracture toughness value, independent of the level of constraint, that is we do *not* use the convention in which K_c refers solely to the plane-stress toughness and K_{IC} to the plain-strain toughness. To study this problem we will take the relatively tractable case of out-of-plane constraint for which, thanks to extensive experimental work, it is possible to define the level of constraint as a function of applied stress conditions with some confidence.

The specimen dimensions required to ensure plane-strain conditions are specified by various national and international standards (e.g. British Standards Institute London, 1991). A typical requirement is that the specimen thickness B shall be larger than some critical value B_c, a function of the plane strain K_c and the yield strength σ_y:

$$B_c = 2.5(K_c/\sigma_y)^2 \tag{7.6}$$

The same restriction applies to other dimensions: the crack length a and remaining ligament width (W-a). Rearranging, this equation gives us a value for K_c which we will refer to as the 'plane-strain limit':

$$K_c \; [\textit{plane-strain limit}] = \sigma_y(B/2.5)^{1/2} \tag{7.7}$$

This condition is designed to be a conservative one, so we can say that if K_c is less than the value given by Eq. (7.7), then we certainly have conditions of plane strain, but even in cases where K_c is somewhat larger than this value, plane strain may still exist.

Constraint is reduced through the specimen thickness by the spread of plasticity. As thickness is decreased (or applied load increased) the plane-stress regions, which always occur at the two surfaces, occupy an increasing fraction of the thickness, causing K_c to increase. Many workers have attempted to estimate the point at which full plane-stress conditions occur, using either analytical models or experimental observations (e.g. Ando et al., 1992; Irwin, 1964; Knott, 1973). Irwin (1964) estimated the plane-stress plastic zone size r_y as:

$$r_y = (1/\pi)(K/\sigma_y)^2 \tag{7.8}$$

He noted from experimental results that when $r_y = B$ specimens showed 50% or more of slanted fracture, which is associated with plane stress, and that this increased to almost 100% if $r_y = 2B$. Knott (1973) pointed out that the measurement of slant fracture will tend to underestimate the amount of plane stress, since some plane-stress fracture will produce flat surfaces. Given this, we will use the condition $r_y = B$ to indicate the 'plane-stress onset', that is to indicate a value of K_c above which plane-stress conditions will begin to dominate:

$$K_c \; [\textit{plane-stress onset}] = \sigma_y(\pi B)^{1/2} \tag{7.9}$$

To apply this condition to notches, we note that Eq. (7.8) has the same general form as Eq. (7.4), so we can use a variation of the PM in which the critical stress is σ_y and the critical distance is $B/2$. This will be an exact prediction of the size of the plane-stress plastic zone for a crack, and an approximate prediction in the case of a notch. Tsuji et al. used a slightly different approach based on matching areas under the stress/distance curves for elastic and plastic conditions (Tsuji et al., 1999). Their method is probably more accurate than the one used here but we found that the two methods gave very similar predictions (within 10%): in any case the aim here is only to indicate the approximate value of K at which plane-stress conditions emerge.

For the data of Wilshaw et al. shown above, all fractures occurred at K values below the plane-strain limit (Eq. 7.7). Figure 7.4 shows further results on low-temperature

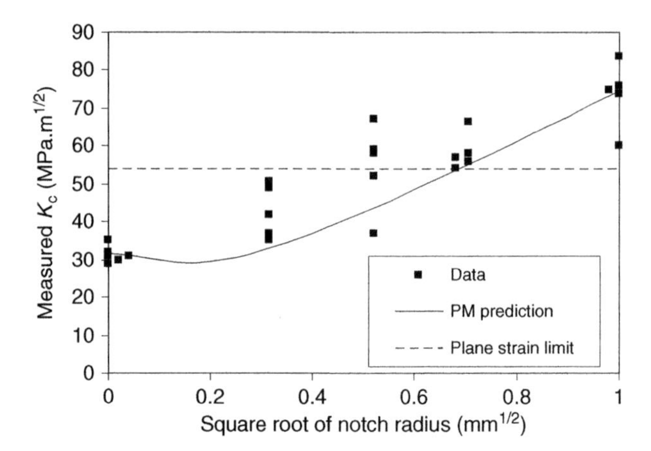

Fig. 7.4. Data from Tsuji et al.: predictions using the PM. The dashed line indicates the plain-strain limit (Eq. 7.7). The plane stress onset (Eq. 7.9) occurs at a value of K_c which is higher than any of the data points.

cleavage fracture of steel, in this case from Tsuji et al. (1999). There is more scatter in this data, but a PM prediction also fits reasonably well, using an L value of 0.05 mm. All data points lie below the plane-stress onset value (not shown). The plane-strain limit is shown on the graph: it goes through the middle of the data but, as noted above, this limit is a conservative one; all fractures almost certainly occurred under plane-strain conditions. Figure 7.5 shows data on a steel which was similar to that used by Tsuji et al., but tested in the form of thinner specimens (Yokobori and Konosu, 1977).

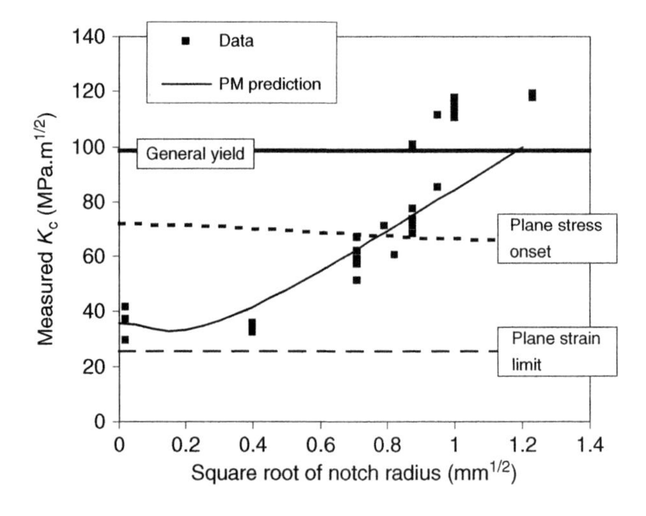

Fig. 7.5. Data from Yokobori and Konusu. Lines indicate the estimated limit of plane strain, the onset of plane stress and the onset of general yield in these specimens. The PM predictions match well to the experimental data for all points below the plane stress onset.

If we attempt to use the TCD here, we find that there is no single value of L which will fit this data, using either the PM or the LM. However, the data for root radius values up to about 1 mm can be predicted quite well using the PM with a value of L which is identical to that used for the Tsuji data (0.05 mm). Using the approach described above we can draw a line on this graph corresponding to the onset of plane stress: the line is almost horizontal but does incline slightly as ρ increases. Note that the experimental data points begin to deviate from the PM prediction line just at the point of plane-stress onset, that is the PM prediction works well for all data up to the point at which constraint is lost. Also shown on the graph is a line corresponding to general yield in these specimens, indicating that the failures in the blunter notches occurred under conditions of full plasticity, though the micro-mechanism of failure in these cases was still cleavage.

It is interesting to note that around the transition point ($(\rho)^{1/2}$ values of 0.9 mm$^{1/2}$ and 1 mm$^{1/2}$) there is more scatter in the data than elsewhere, perhaps indicating a change in fracture mechanism with some specimens failing under plane-strain conditions and others being affected by reduced constraint and therefore failing at higher stress levels, after general yielding has occurred.

Figure 7.6 shows a similar situation for a different material: in this case a high-strength steel tested at room temperature, which had a K_c value of 29.6 MPa(m)$^{1/2}$ (Irwin, 1964). Again there was no single value of L which could predict all the data: a very small value of 2.3 μm was successful at low notch radii and the data shifted to values above the prediction line in the region between the plane-strain and the plane-stress limit

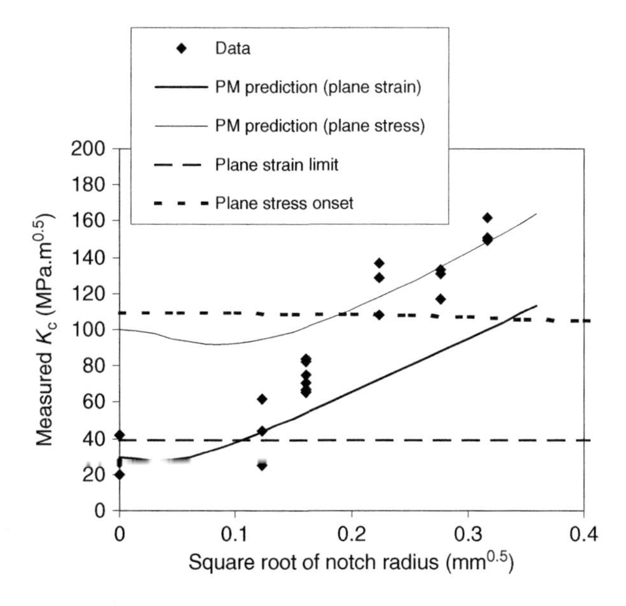

Fig. 7.6. Data and predictions for a high-strength steel (Irwin, 1964) showing similar behaviour to the data of Fig. 7.5. In this case a tentative plane-stress prediction has been included as well, which fits the data at higher root radii.

lines. This time a second prediction line has been drawn, also using the PM, which passes through the data points for the blunter notches and may represent plane-stress conditions. This prediction is a very tentative one, since we do not know the value of K_c for plane stress so it was necessary to choose values for both K_c and L. The resulting values were $K_c = 100\,\text{MPa(m)}^{1/2}$, $L = 0.015\,\text{mm}$; this value of K_c is plausible given that the plane-stress toughness is typically three times higher than in plane strain (Knott, 1973). The very small plane-strain value of L probably reflects the fact that the relevant microstructural parameter in this quenched and tempered steel will be the lath width, rather than the grain size. The plane-strain value of σ_o was very high, at 11,010 MPa, which is 6.9 times the yield strength ($\sigma_y = 1590\,\text{MPa}$), showing that there is no fixed relationship between σ_o and σ_y in different materials. The plane-stress value of σ_o was even higher, at 14,570 MPa ($9.1\sigma_y$). Finally, Fig. 7.7 shows data obtained under fully plane-stress conditions, using thin specimens of aluminium alloy 7075-T6 which had a yield strength of 498 MPa (Mulherin et al., 1963). Good predictions were obtained using the PM with a K_c value of $77\,\text{MPa(m)}^{1/2}$ and an L value of 0.07 mm, giving $\sigma_o = 5190\,\text{MPa}$, which is $10.4\sigma_y$.

This section has shown that, not surprisingly, TCD predictions using parameters obtained under plane-strain conditions are not applicable under conditions of reduced constraint. The data presented here highlight a particular problem in the prediction of notch behaviour: as the notch root radius increases, necessitating higher applied loads to failure, the level of constraint can reduce as plastic zones become larger. Thus, for the same sheet thickness, a crack may be in plane strain but a notch of the same length may experience plane stress, or intermediate constraint conditions. In some cases, such as the cleavage-fracture data reported in Fig. 7.5, TCD predictions seem to be accurate up to the point at which plane-stress conditions are expected to dominate – that is the plane-strain TCD analysis was valid also in the region of intermediate constraint – and the onset of plane stress heralded a change of behaviour in which general yield occurred before fracture. From an engineering point of view, one would regard general yield as an absolute limit of the load-carrying potential of a structure anyway, so there is little

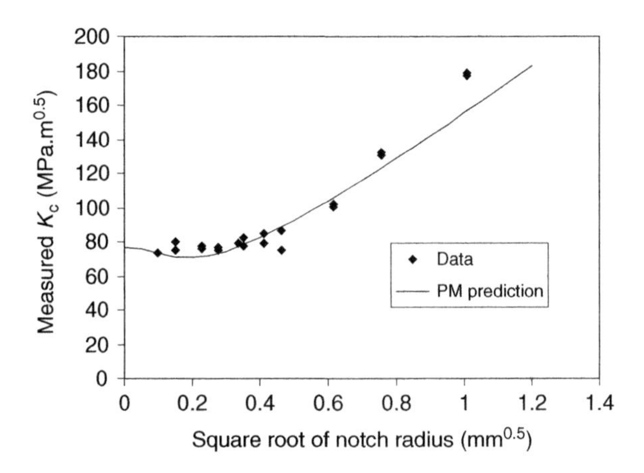

Fig. 7.7. Data and predictions for an aluminium alloy (Mulherin et al., 1963) tested under conditions of plane stress.

practical value in being able to predict failures that occur after general yield. The success of the plane-strain TCD in the intermediate region is probably due to the fact that, in this material, brittle failure is being initiated in the centre of the specimen thickness where some plane-strain conditions still remain.

It has also been shown, in respect of two sets of data presented above, that the TCD may be able to describe results obtained under conditions of pure plane stress. However, this conclusion should be treated with some caution. Fracture under plane-stress conditions is notoriously more difficult to predict. Pre-cracked specimens invariably show some stable crack extension prior to failure, so that the actual length of the crack at the point of unstable fracture is always greater than its original length. The TCD, like LEFM, cannot predict this effect simply by using the initial conditions: this problem has led to the development of R-curve analysis to predict the extent of stable crack growth and the conditions for instability. However, it has been observed that this stable crack growth does not normally occur for notches with root radii greater than ρ_c; such notches usually fail unstably as soon as a crack initiates (Irwin, 1964). From this we conclude that it may be possible, given enough experimental data, to deduce values of the TCD parameters L and σ_o as a function of the operative level of constraint.

7.2.3 *The role of microstructure*

Can the values of L deduced above be related in any way to the sizes of microstructural features? It was already noted in the case of the cleavage-fracture data of Wilshaw et al. (Fig. 7.2) that the value of L was exactly equal to the grain size of the material: 35 μm. Yokobori and Konusu (1977) carried out tests on a similar material, heat treated to give a range of grain sizes: the data already presented above in Fig. 7.5 was for a grain size of 36 μm.

Figure 7.8 shows data for their largest grain size: 198 μm. There was little change in the value of K_c for the cracked specimens, but for finite root radii the larger grain size material was significantly weaker. The point of predicted plane-stress onset was only just reached at the largest radii tested, so it is not clear what is happening when constraint is lost. The value of L for the large grain size material was 240 μm, accurately reflecting the increased grain size. Figure 7.9 shows the value of L as a function of grain size for all the data of Yokobori and Konusu: there is a clear relationship between the two, given by $L = 1.2d$. This direct link to grain size is likely in the case of cleavage fracture because it is well known that grain size plays a strong role in determining toughness: this has been developed in micro-mechanical models such as the RKR model and its successors (see Discussion below).

Figure 7.10 shows data on AISI 4340 steel (Ritchie and Horn, 1978; Ritchie et al., 1976) in which two different grain sizes were achieved by the use of different austenitising temperatures. Again L is larger for the material with the larger grain size: the result is that this material is superior at $\rho = 0$ but soon becomes weaker than the fine-grain material as the notch radius increases. This explains the observation reported by these workers that the large grain size material had inferior Charpy impact energy, since Charpy specimens have a root radius of 0.25 mm. Interestingly the mechanism of failure changed from quasi-cleavage in the small-grain material to intergranular fracture in the large-grain

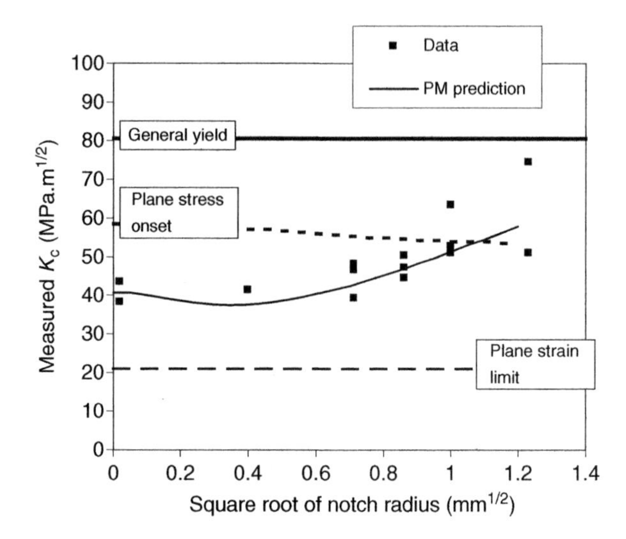

Fig. 7.8. Further data from Yokobori and Konusu (1977), for the same material as shown in Fig. 7.5, heat-treated to give a larger grain size.

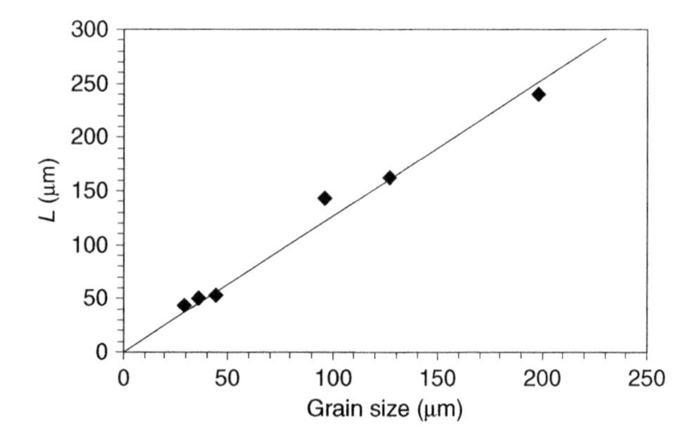

Fig. 7.9. The relationship between L and the grain size d for all six grain sizes tested by Yokobori and Konusu. The line corresponds to $L = 1.2d$.

material (mixed with fibrous rupture in both cases). This change in mechanism accords with the change in L value which we calculated: in the small grain material an L value of 6 μm reflects the fine bainitic structure, whilst an L value of 120 μm, of the same order of magnitude as the grain size, occurred in the large grain material. The ratio σ_o/σ_u was also very different: 3.7 in the small-grain material but only 1.3 in the large-grain material.

Microstructural parameters have also been shown to play a role when ductile micromechanisms are involved. For example, the spacing of inclusions is of obvious

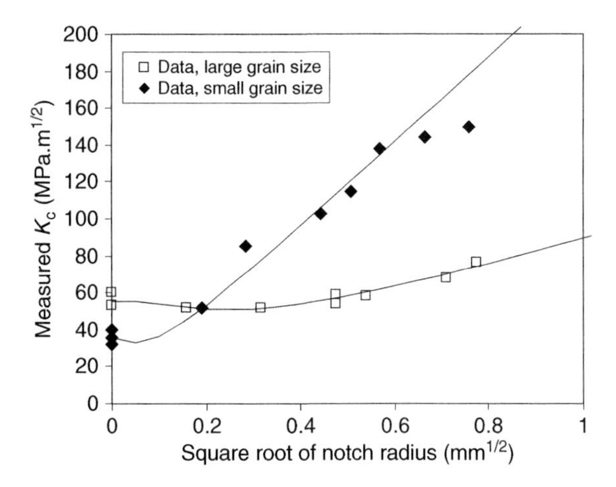

Fig. 7.10. Data from Ritchie et al., 1976 illustrating how the effect of grain size alters with increasing root radius. Lines indicate PM predictions.

importance in the void growth mechanism when these inclusions act as the initiation points for voids: widely spaced inclusions will give rise to greater toughness because more void growth will be needed before failure. Many workers have incorporated inclusion spacing into theoretical models of this process. For example, Firrao and co-workers (Doglione and Firrao, 2000; Roberti et al., 1981) demonstrated that the critical root radius ρ_c was equal to the spacing of inclusions, s, and developed a simple relationship to predict the increase in toughness (expressed in terms of the critical J integral, J_c) for $\rho > \rho_c$:

$$J_{c(notch)}/J_{c(crack)} = \rho/s \tag{7.10}$$

The fact that there is a relationship between s and ρ_c implies that there will necessarily be a relationship between s and L.

7.2.4 *Blunt notches and non-damaging notches*

In previous chapters we saw that, for large blunt notches, in which the stress gradient over a distance L from the notch is small, the TCD prediction reduces to a simpler result: the applied stress to cause fracture, σ_f, is simply related to σ_o and the notch stress-concentration factor K_t, thus:

$$\sigma_f = \sigma_o/K_t \tag{7.11}$$

We saw, in the case of polymers for which $\sigma_o/\sigma_u > 1$, that this led to the existence of 'non-damaging notches' – notches which had $K_t < \sigma_o/\sigma_u$ and therefore were predicted to have no effect in reducing strength. Exactly the same effect can be expected in metals. Figures 7.11 and 7.12 (Zheng, 1989) illustrate two different types of behaviour. Figure 7.11 shows a plot of fracture stress σ_f as a function of K_t for notched specimens of a quenched and aged Ti-2.5Al-16V alloy tested at a low temperature. This material

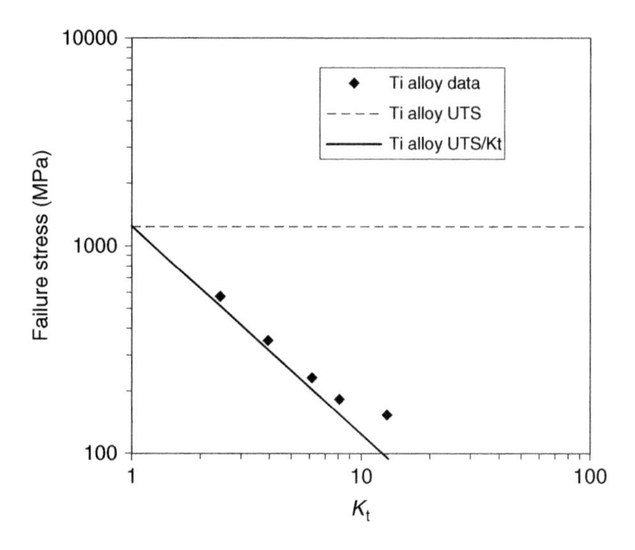

Fig. 7.11. Failure stress σ_f as a function of K_t for large notches in a brittle Ti alloy (Zheng, 1989). The failure stress is accurately predicted by dividing the UTS (σ_u) by K_t, for all except the highest K_t values.

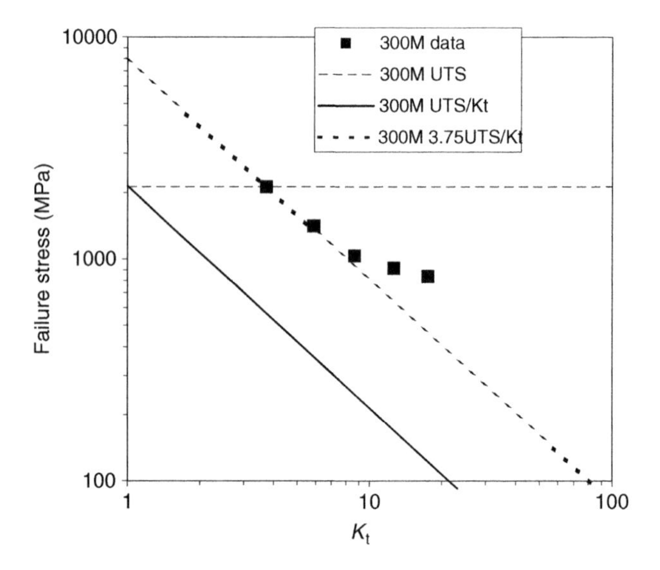

Fig. 7.12. A plot similar to Fig. 7.11, for 300M steel (Zheng et al., 1989). The prediction line now shifts to $3.75\sigma_u/K_t$. Notches with $K_t < 3.75$ will be non-damaging in this material.

is clearly very brittle, in fact it displays the classic brittle behaviour that we would expect of a ceramic: Eq. (7.11) applies and σ_o is equal to σ_u, the plain-specimen tensile strength. On the other hand, the quenched and tempered 300M steel shows data which, whilst it largely conforms to Eq. (7.11), has a value of $\sigma_o = 3.75\sigma_u$. We would predict that a notch with $K_t = 3.75$ or less in this material will have no effect on σ_f, as the graph shows. For the highest K_t values tested in both these materials, the data points separate from the prediction line based on Eq. (7.11). This is to be expected as the high-K_t notches have higher stress gradients: a TCD analysis would predict this effect.

7.3 Discussion

7.3.1 Applicability of the TCD

We have shown above that the TCD is capable of predicting brittle fracture in metallic materials, at least in certain circumstances. This conclusion has been tested, as it was in the case of ceramic and polymeric materials in the previous two chapters, by a direct comparison of TCD predictions with experimental data. This exercise has provided ample evidence that the theory is applicable in situations of fully constrained yielding, that is cases in which the plastic zone is smaller than any specimen dimensions, and therefore plane-strain conditions prevail. We also saw some evidence to suggest that the TCD may be useful in cases of plane stress, but further investigation is clearly needed here.

It has been noted by a number of workers that the measured toughness increases approximately as the square root of notch radius, for radii above the critical value; this result can be predicted using the present approach, because as ρ increases and becomes much larger than L, so, in Eq. (7.2), K_{cm} becomes proportional to $(\rho)^{1/2}$: another way of saying this is that K_{cm} becomes proportional to the stress at the notch root. However, whilst this relationship is a useful approximate one it is not exact; in general, the result will depend not only on root radius but also on notch length and specimen dimensions. The TCD can still be used but FEA will be needed to provide an accurate description of the stress field, as we saw above.

The values of L found in this work are of the same order of magnitude as microstructural features such as grains or bainite laths or the spacing of inclusions: this result is to be expected since the underlying reason for the deviation from LEFM behaviour is that physical quantities (crack length, notch radius) become similar in magnitude to these microstructural features. This gives a clue to the operative failure mechanisms and the role played by microstructure, and may pave the way to a more mechanistic form of the TCD.

On the other hand, the strength parameter σ_o is unlikely to have any physical meaning. The values found for this parameter for cleavage in steels are considerably higher than measured values of the cleavage fracture stress, which is typically of the order of 1000 MPa (e.g. Ritchie et al., 1973; Wilshaw et al., 1968). Some authors working with polymers have suggested that, being approximately three times larger than the yield strength, σ_o may be related to the peak stress value ahead of a crack or notch in plane strain (Kinloch and Williams, 1980). However, the peak stress occurs at a distance different from $L/2$ and its magnitude is a feature of the elastic/plastic stress distribution. In any case, we have seen in the examples above that the ratio σ_o/σ_y varies widely from material to material. Figure 7.13 shows some recent data obtained by my colleague Luca

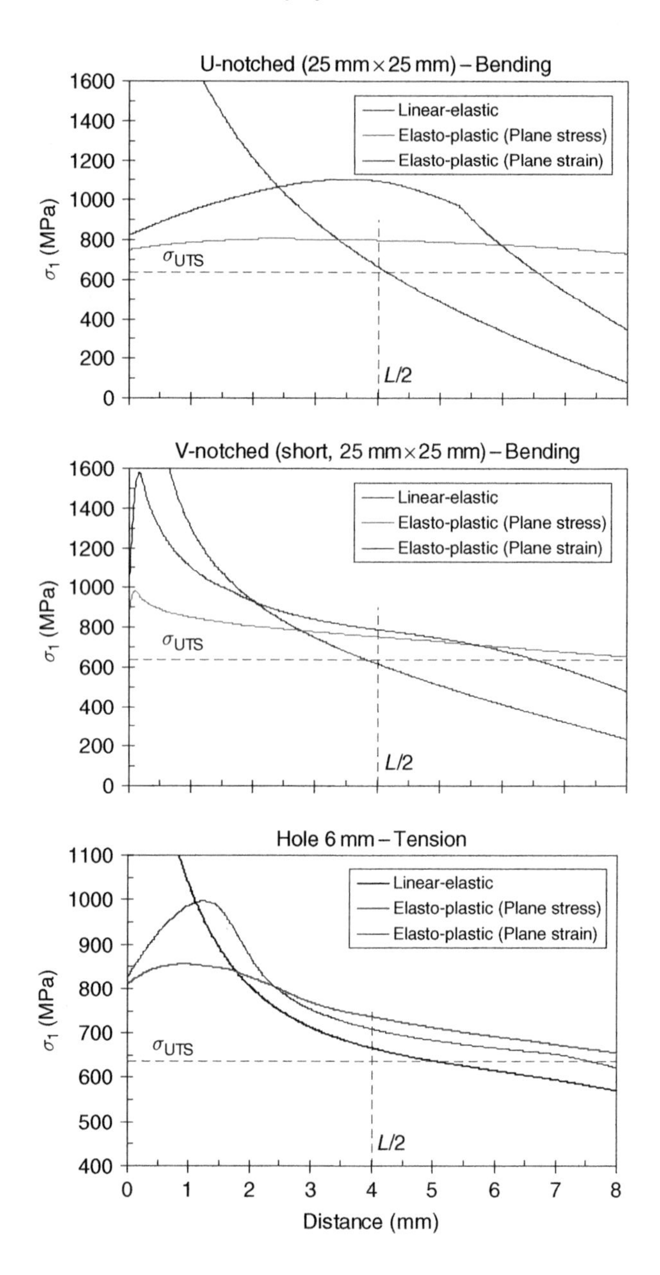

Fig. 7.13. Data on En3b steel courtesy of L. Susmel. Stress–distance curves at failure for V and U-shaped notches and holes. $L/2$ calculated using the UTS and plane strain K_c. Three other notch types were also tested, with similar results.

Susmel, who tested the plain carbon steel En3b at room temperature, where it failed by a ductile micro-mechanism. As the stress–distance curves show, this material clearly conforms to the TCD with $\sigma_o = \sigma_u$ and a value of L calculated using σ_u and the plane strain K_c. Also shown are the elastic plastic stress–distance curves, indicating that these would give less accurate predictions if used with the TCD.

7.3.2 *Other theoretical models*

To my knowledge the TCD as we are using it here has not been used in the past to assess notches in metallic materials. The only exception to this is some very recent work on V-shaped notches of zero root radius (Seweryn, 1994; Strandberg, 2002) which, being directed towards multiaxial loading, will be discussed in a later chapter. In the early decades of fracture mechanics, critical distance theories based on local stress or strain were suggested (e.g. Neuber, 1958; McClintock, 1958), and Irwin certainly used Neuber's fictitious radius approach (Irwin, 1964) which is itself derived from the LM. However, these early attempts were not developed into a fully-fledged approach, the reason being that workers in this field perceived other priorities. In particular, it was appreciated at an early stage that crack propagation in many engineering alloys did not conform to LEFM conditions: their high toughness values (or low yield strengths) gave rise to large amounts of plasticity before failure. Therefore the majority of effort was directed into developing forms of fracture mechanics which would apply under these conditions: the field of Elastic Plastic Fracture Mechanics (EPFM) was born. The major preoccupations in this area were the development of new parameters to characterise toughness: the COD and J integral, and the attempt to understand failure when preceded by periods of stable crack extension. The latter effort gave rise to the concept of R-curves.

Whilst these developments were occurring in the field of continuum mechanics, other workers were developing models which addressed the actual mechanisms of failure in metallic materials (e.g. Ritchie et al., 1973; Yokobori et al., 1976). It is in the realm of these micro-mechanical models that we see the introduction of material length constants and, in some cases, theories very similar to the TCD. The most obviously similar theory, and one which marked a turning point in the understanding of cleavage fracture in steels, was the model developed by Ritchie, Knott and Rice – the so-called 'RKR model' (Ritchie et al., 1973), which has been mentioned previously (Section 4.3). In this model the mechanism for cleavage fracture was envisaged to be the initiation and subsequent propagation of a small crack, formed by the fracture of a carbide particle. The initial cracking of the carbide requires plastic strain and so can only occur within the plastic zone, and it occurs relatively easily. The critical stage was envisaged to be the propagation of this crack into the surrounding material, an event which can be modelled as a classic Griffith brittle fracture process, depending only on crack size and local tensile stress. A critical distance comes into the model because these carbides reside in grain boundaries. It was shown that accurate predictions of fracture toughness could be made using the tensile stress at a distance from the crack tip equal to twice the average grain diameter.

This theory is clearly very similar to our TCD, but differs from it in two important ways. First, the RKR model is essentially a micro-mechanical one: it starts from a presumed mechanism of failure and derives material constants which have a real physical meaning – the tensile stress needed to propagate a micro-crack and the grain size of the material. In the TCD, on the other hand, we arrive at our material constants of length and stress in a different way. Secondly, the RKR model uses the actual stress field, that is the elastic/plastic stress field ahead of the crack, whilst the TCD uses the elastic stress field. The justification for using this approach (of which more will be said in Chapter 13) lies in the fact that the TCD is a linear, elastic, continuum theory, and thus holds true to the philosophy of LEFM. Micro-mechanical models provide great insight into structure/property relationships but they are difficult to use in practice because, to be faithful to their origins, they must contain all the complexities of the real situation. In this case that means an elastic/plastic stress field and a real microstructure with all the complexities of grain boundaries, secondary phases, residual stress and so on.

It is interesting to chart the development of cleavage fracture theories which sprang from the evolution of the RKR model. One obvious improvement was to introduce stochastic parameters, recognising the fact that microstructural distances such as carbide particle size and grain size are not constants but can be described statistically (Lin et al., 1986). Improvements in numerical analysis, especially in the development of large FE models, allowed researchers to simulate the entire plastic zone region in detail. Now any point in the plastic zone could be considered as a potential source of cracking, and the overall probability of failure could be computed. This gave rise to models such as the so-called 'Local Approach' (Beremin, 1983). Beremin's model has since been used and modified by many other workers (e.g. Faleskog et al., 2004; Moltubakk et al., 1999): in some of these models there is really little trace of the micro-mechanical concepts of the original RKR model. On the other hand, we find researchers who have retained and developed the mechanistic approach, considering, for example, different types of fracture origin in addition to the cracked carbide (Mantyla et al., 1999), and including more detail about the various stages of the process, such as crack initiation and growth to the first grain boundary (Chen et al., 2003; Moya et al., 2004). Some good examples of these models, and of fractographic studies which support them, can be found in the proceedings of a recent symposium in honour of John Knott, one of the original authors of the RKR model (Soboyejo et al., 2002). These more mechanistic models invariably use one or more material length parameters, such as grain size. This is obviously necessary; what is less obvious, however, is that in recent years a critical length scale has also become the norm in the Beremin-type models. It was found that, in order to achieve reasonable predictions, parameters could not be considered on a point-by-point basis, but had to be averaged over a certain volume, V_0. Values of V_0 found by trial and error tend to be of the same order of magnitude as the grain size (Faleskog et al., 2004; Mirzaee et al., 2004; Yahya et al., 1998). Thus the 'Local Approach' becomes (in terms of the definitions in Chapter 4) a 'non-local approach', since information from the surroundings is used when making a calculation at a particular point.

In parallel to this work on cleavage fracture, similar developments can be traced in the prediction of crack propagation by the void growth mechanism. In this case the original models were those of Rice and Tracey, who considered the growth of a single void, and Gurson, who modelled a series of regularly spaced voids (see, for example, Pardoen et al., 1998). The key feature here, and one which immediately leads to the use of a critical distance parameter, is the origin of the voids, which invariably initiate at microstructural features, usually inclusions. Thus the spacing of inclusions, s, is an almost essential feature in any model of void growth. In recent years, these models have developed great sophistication, being used in conjunction with numerical analysis and in some cases merged into models of the process-zone type (Dos Santos and Ruggieri, 2003). Many workers use a two-parameter approach, the parameters most commonly chosen being plastic strain and triaxiality (expressed as the ratio between the mean of the three principal stresses and the Von Mises effective stress). Failure loci have been developed using these two parameters (Mackenzie et al., 1977; Schluter et al., 1996).

As notch radius, or material toughness, increases, a point can be reached where failure is no longer initiated in the vicinity of the notch root but rather moves to the centre of the specimen, producing a ductile fracture by initiation and growth of damage of the void-linkage type (Spencer et al., 2002; Spencer et al., 2002; Geni and Kikuchi, 1999). This type of failure is very specimen-specific, being more likely to occur in deeply notched tensile specimens in which the degree of triaxiality in the centre can be very large. These kinds of failures are relatively rare in industrial components, for which non-uniform loading such as bending and torsion will tend to favour notch-initiated failures. TCD-like theories have been used also in this context. For example, Schulter et al. used a critical distance approach in conjunction with a two-parameter method, to predict the behaviour of a structural steel (Schluter et al., 1996). The critical distance was taken to be the spacing of void-nucleating inclusions. Likewise, models of shear failure often include a critical microstructural distance (Biel-Golaska, 1998).

Process zone theories such as the cohesive zone model are more often applied to brittle and quasi-brittle materials but have also been applied to metals (e.g. Elices et al., 2002). In principle, these offer the possibility of a computer simulation in which the onset and growth of a crack can be predicted and followed. In practice, there are still some serious computational problems to be solved (de Borst and Remmers, 2004). In the realm of continuum mechanics models which do not use any material length constants, we find the work of Zheng, who has modified the local strain approach originally suggested by Neuber (Zheng, 1989); this kind of model is still frequently used in the area of LCF, but not so often to predict brittle fracture, though here Zheng has made some very successful predictions. Also the NSIF approach of Pluvinage (previously discussed in Chapter 4) has been applied to predict brittle fracture, though this seems to necessitate some redefinition of parameters such as the critical stress.

These last two theories notwithstanding, the overwhelming feeling among researchers at the present time is that, whatever kind of theoretical model is used to predict fracture, some form of material length scale, that is some kind of critical distance parameter, is an essential feature. In some cases, this length is associated with a particular feature of the microstructure (e.g. grain size) but more often it is simply found by trial and error, that is by using an empirical approach, just as we have done here.

This chapter concludes with a table listing the relevant parameters for various materials analysed above (Table 7.1). In the following chapter, we shall complete our survey of monotonic fracture in different types of materials by considering composites.

Table 7.1. Mechanical property values for metals

Material	L (μm)	σ_y (MPa)	σ_u (MPa)	σ_o (MPa)	K_c (MPa.m$^{1/2}$)	Reference
Mild steel (−170 °C)	50	700	810	2538	31.8	(Tsuji et al., 1999)
Mild steel (−196 °C)	50 −240	718 −585		2872 −1485	36 −40.8	(Yokobori and Konosu, 1977)
Various grain sizes						
Mild steel (−196 °C)	35	829	(900)	2298	25.7	(Kinloch and Williams, 1980)
Plain carbon steel En3b	8010	606	638.5	638.5	101.3	L. Susmel (unpublished work)
Alloy steel AISI4340	6	1593	2217	8291	36	(Ritchie et al., 1976)
Small grain size						
Alloy steel AISI4340	120	1593	2193	2858	55.5	(Ritchie et al., 1976)
Large grain size						
High strength steel H-11	2.3	1589		11011	29.6	(Irwin, 1964)
High strength steel H-11.	(15)	1589		(14566)	(100)	(Irwin, 1964)
Plane Stress						
Aluminium alloy DISPAL	45 −75	320 −161	360 −173	1240 −630	13 −7.9	(Srinivas and Kamat, 2000)
Various temps from room temp to 350 °C						
Aluminium alloy 7075-T6	70	498		5190	77	(Mulherin et al., 1963)
Plane Stress						

Note: Brackets indicate approximate or estimated values; data are for plane strain at room temperature unless otherwise stated.

References

Ando, K., Mogami, K., and Tuji, K. (1992) Probabilistic aspect of cleavage crack initiation sites and fracture toughness. *Fatigue and Fracture of Engineering Materials and Structures* **15**, 1171–1184.

Beremin, F.M. (1983) A local criterion for cleavage fracture of a nuclear pressure vessel steel. *Metallurgical Transactions A* **14A**, 2277–2287.

Biel-Golaska, M. (1998) Analysis of cast steel fracture mechanisms for different states of stress. *Fatigue and Fracture of Engineering Materials and Structures* **21**, 965–975.

British Standards Institute London (1991) *BS 7448-1:1991 Fracture mechanics toughness tests – part 1: Method for determination of Kic, critical CTOD and critical J values of metallic materials.*

Chen, J.H., Wang, Q., Wang, G.Z., and Li, Z. (2003) Fracture behaviour at crack tip – A new framework for cleavage mechanism of steel. *Acta Materialia* **51**, 1841–1855.

Creager, M. and Paris, P.C. (1967) Elastic field equations for blunt cracks with reference to stress corrosion cracking. *International Journal of Fracture Mechanics* **3**, 247–252.

de Borst, R. and Remmers, J.C. (2004) Computational aspects of cohesive-zone models. In *The 15th European Conference of Fracture. Advanced Fracture Mechanics for Life and Safety Assessments* (Edited by Nilsson, F.) KTH, Stockholm.

Doglione, R. and Firrao, D. (2000) Inclusions effect on the notch behaviour of a low-alloy tempered steel. In *Notch Effects in Fatigue and Fracture* (Edited by Pluvinage, G. and Gjonaj, M.) pp. 39–50. Kluwer Academic Publishers, Dordrecht.

Dos Santos, F.F. and Ruggieri, C. (2003) Micromechanics modelling of ductile fracture in tensile specimens using computational cells. *Fatigue and Fracture of Engineering Materials and Structures* **26**, 173–181.

Elices, M., Guinea, G.V., Gomez, F.J., and Planas, J. (2002) The cohesive zone model: advantages, limitations and challenges. *Engineering Fracture Mechanics* **69**, 137–163.

Faleskog, J., Kroon, M., and Oberg, H. (2004) A probabilistic model for cleavage fracture with a length scale – Parameter estimation and predictions of stationary crack experiments. *Engineering Fracture Mechanics* **71**, 57–79.

Geni, M. and Kikuchi, M. (1999) Void configuration under constrained deformation in ductile matrix materials. *Computational Materials Science* **16**, 391–403.

Irwin, G.R. (1964) Structural aspects of brittle fracture. *Applied Materials Research* **3**, 65–81.

Kinloch, A.J. and Williams, J.G. (1980) Crack blunting mechanisms in polymers. *Journal of Materials Science* **15**, 987–996.

Knott, J.F. (1973) *Fundamentals of fracture mechanics.* Butterworths, London.

Lin, T., Evans, A.G., and Ritchie, R.O. (1986) A statistical model of brittle fracture by transgranular cleavage. *Journal of the Mechanics and Physics of Solids* **34**, 477–497.

Mackenzie, A.C., Hancock, J.W., and Brown, D.K. (1977) On the influence of state of stress on ductile fracture initiation in high strength steels. *Engineering Fracture Mechanics* **9**, 167–177.

Mantyla, M., Rossol, A., Nedbal, I., Prioul, C., and Marini, B. (1999) Fractographic observations of cleavage fracture initiation in a bainitic A508 steel. *Journal of Nuclear Materials* **264**, 257–262.

McClintock, F.A. (1958) Ductile fracture instability in shear. *Journal of Applied Mechanics* **25**, 582–588.

Mirzaee, A., Hidadi-Moud, S., Truman, C.E., and Smith, D.J. (2004) Application of the local approach to predict brittle fracture following local compression. In *The 15th European Conference of Fracture – Advanced Fracture Mechanics for Life and Safety Assessments* (Edited by Nilsson, F.) KTH, Stockholm.

Moltubakk, T., Thaulow, C., and Zhang, Z.L. (1999) Application of local approach to inhomogeneous welds. Influence of crack position and strength mismatch. *Engineering Fracture Mechanics* **62**, 445–462.

Moya, C., Martin-Meizoso, A., and Ocana, I. (2004) Micromechanisms of cleavage fracture in HAZ of C-Mn steel welds. In *The 15th European Conference of Fracture. Advanced Fracture Mechanics for Life and Safety Assessments* (Edited by Nilsson, F.) KTH, Stockholm.

Mulherin, J.H., Armiento, D.F., and Marcus, H. (1963) Fracture characteristics of high strength aluminium alloys using specimens with variable notch root radii. In *ASME conference paper 63-WA-306* ASME (USA).

Neuber, H. (1958) *Theory of notch stresses: Principles for exact calculation of strength with reference to structural form and material.* Springer Verlag, Berlin.

Pardoen, T., Doghri, I., and Delannay, F. (1998) Experimental and numerical comparison of void growth models and void coalescence criteria for the prediction of ductile fracture in copper bars. *Acta Materialia* **46**, 541–552.

Ritchie, R.O., Francis, B., and Server, W.L. (1976) Evaluation of toughness in AISI 4340 steel austenitised at low and high temperatures. *Metallurgical Transactions* **7A**, 831–838.

Ritchie, R.O. and Horn, R.M. (1978) Further considerations on the inconsistency in toughness evaluation of AISI4340 steel austenitised at increasing temperatures. *Metallurgical Transactions A* **9A**, 331–339.

Ritchie, R.O., Knott, J.F., and Rice, J.R. (1973) On the relationship between critical tensile stress and fracture toughness in mild steel. *Journal of the Mechanics and Physics of Solids* **21**, 395–410.

Roberti, R., Silva, G., Firrao, D., and DeBenedetti, B. (1981) Influence of notch root radius on ductile rupture fracture toughness evaluation with Charpy-V type specimens. *International Journal of Fatigue* **3**, 133–141.

Schluter, N., Grimpe, F., Bleck, W., and Dahl, W. (1996) Modelling of the damage in ductile steels. *Computational Materials Science* **7**, 27–33.

Seweryn, A. (1994) Brittle fracture criterion for structures with sharp notches. *Engineering Fracture Mechanics* **47**, 673–681.

Soboyejo, W.O., Lewandowski, J.J., and Ritchie, R.O. (2002) *Mechanisms and mechanics of fracture*. TMS, Warrendale, Pennsylvania, USA.

Spencer, K., Corbin, S.F., and Lloyd, D.J. (2002) Notch fracture behaviour of 5754 automotive aluminium alloys. *Materials Science and Engineering A* **332**, 81–90.

Srinivas, M. and Kamat, S.V. (2000) Influence of temperature and notch root radius on the fracture toughness of a dispersion-strengthened aluminium alloy. *Fatigue and Fracture of Engineering Materials and Structures* **23**, 181–183.

Strandberg, M. (2002) Fracture at V-notches with contained plasticity. *Engineering Fracture Mechanics* **69**, 403–415.

Tsuji, K., Iwase, K., and Ando, K. (1999) An investigation into the location of crack initiation sites in alumina, polycarbonate and mild steel. *Fatigue and Fracture of Engineering Materials and Structures* **22**, 509–517.

Wilshaw, T.R., Rau, C.A., and Tetelman, A.S. (1968) A general model to predict the elastic-plastic stress distribution and fracture strength of notched bars in plane strain bending. *Engineering Fracture Mechanics* **1**, 191–211.

Yahya, O.M.L., Borit, F., Piques, R., and Pineau, A. (1998) Statistical modelling of intergranular brittle fracture in a low alloy steel. *Fatigue and Fracture of Engineering Materials and Structures* **21**, 1485–1502.

Yokobori, T., Kamei, A., and Konosu, S. (1976) A criterion for low-stress brittle fracture of notched specimens based on combined micro- and macro fracture mechanics – I. *Engineering Fracture Mechanics* **8**, 397–409.

Yokobori, T. and Konosu, S. (1977) Effects of ferrite grain size, notch acuity and notch length on brittle fracture stress of notched specimens of low carbon steel. *Engineering Fracture Mechanics* **9**, 839–847.

Zheng, X.L. (1989) On an unified model for predicting notch strength and fracture toughness of metals. *Engineering Fracture Mechanics* **33**, 685–695.

CHAPTER 8

Composites

Brittle Fracture in Fibre Composite Materials

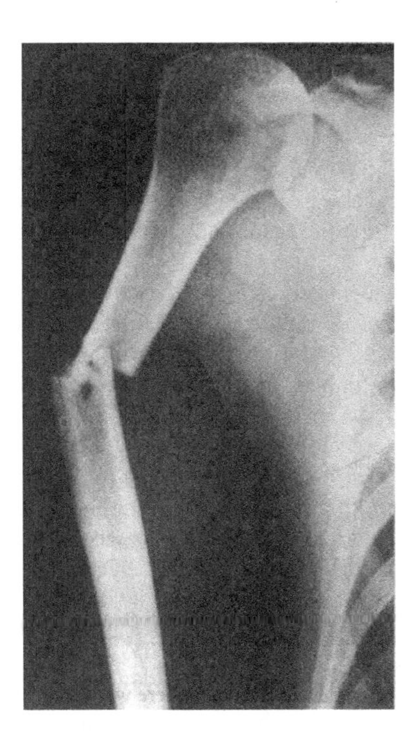

Fig. 8.1. Natural composites, such as bone, display fracture behaviour which is very similar to that of engineering fibre composites.

8.1 Introduction

The term 'composites' covers a wide range of materials; indeed, if interpreted in its broadest sense to mean any material containing two or more constituents, then one can argue that all engineering materials are in fact composites. This chapter is concerned mainly with industrial fibre composites, that is man-made materials in which fibres of a stiff, strong but usually brittle material are added to reinforce a weaker, more elastic matrix. Composites using polymer matrices, especially epoxy resins reinforced with long fibres of glass or carbon, became popular in the 1960s and are rapidly expanding their range of applications, replacing metals in components from golf clubs to aircraft wings. These materials can now be made with strength and toughness values which are very comparable to those of metals, and have the added advantage of low density. In recent decades, composite materials with metallic and ceramic matrices have also been developed and are finding important applications. Composites employing short, discontinuous fibres generally have inferior properties but are easier to make, for example by injection moulding. Many natural materials such as bone (Fig. 8.1) and wood also have composite structures: the techniques which have been developed to study industrial fibre composites can usefully be applied to understand the fracture behaviour of these materials.

Optimal properties for composites used in load-bearing components are usually achieved by using long, essentially continuous fibres in carefully oriented patterns such as laminates. High volume fractions of fibres are used, with fibre orientations chosen to coincide with principal stress directions in the component. The general rule is that an increase in fibre volume fraction tends to increase all three of the principal mechanical properties: stiffness (E), strength (σ_u) and toughness (K_c); this is very different from the situation that applies in most other classes of materials, where some type of 'trade-off' usually occurs between material properties, especially between strength and toughness.

The fracture mode of these materials is almost always brittle, that is there is very little plastic deformation before failure, though there may be significant non-linearity in the stress–strain curve due to the build-up of damage prior to failure. This damage can take many forms: fibre failure, matrix/fibre interface failure, delamination and so on. Large damage zones form ahead of notches and other stress concentrators; impact by foreign bodies can also create significant regions of non-critical damage. This ability to sustain damage without catastrophic failure is an important advantage of composites which sets them apart from other materials which fail by brittle fracture modes, such as ceramics. The scale on which this sub-critical damage occurs is large – of the order of millimetres or even centimetres – and so can be detected by non-destructive inspection methods such as ultrasonics. It is also a significant source of toughening, since the formation of damage requires energy and tends to reduce local stress concentrations. Other toughening mechanisms also operate, such as crack deflection, and crack bridging by intact fibres. The description of failure mechanisms and the development of theoretical models of a mechanistic nature has been, and continues to be, a very active area of research. It has proven to be very difficult to develop these models owing to the number and variety of different mechanisms of damage and toughening and their interdependence. More will be said concerning these mechanistic models at a later stage in this chapter.

8.2 Early Work on the TCD: Whitney and Nuismer

In researching this topic, I was surprised to discover that the TCD is well known and frequently used to predict failure in composite materials. The concept was first suggested in the 1970s and seems to have developed quite independently, with little reference to parallel developments in the fields of polymers and metals. However, whilst the TCD failed to become popular in these other fields and has largely fallen out of use until recent years, its use in the field of composite materials has grown steadily, to the point where it is now commonly employed not only in academic research but also in many practical applications for predicting failure in engineering components.

The use of the TCD in composite materials can be traced to the work of Whitney and Nuismer. Their original publication (Whitney and Nuismer, 1974) is still the fundamental reference on this subject: in a recent literature search, I found over 200 citations to this paper in modern journals. The same concept, in a slightly more developed form, can be found in *Experimental Mechanics of Fibre Composite Materials* (Whitney et al., 1982), an excellent early textbook on this subject. These publications contain most of the same theory which we have already seen developed and applied to other materials in the preceding three chapters. Whitney and Nuismer suggested both the PM and the LM (which they called the Point Stress Criterion and Average Stress Criterion) and made predictions of the effect of both sharp (crack-like) notches and circular holes. The values of the critical distances (which they referred to as d_o and a_o for the PM and LM respectively) were determined empirically, from data on holes and notches of different sizes, but the theoretical link to fracture mechanics, through K_c, was also outlined, with the resulting conclusion that, in theory, d_o should be equal to $4a_o$, as we would expect since, using our terminology, $d_o = L/2$ and $a_o = 2L$. The value of the critical stress was assumed to be the plain-specimen tensile strength, σ_u.

Figures 8.2 and 8.3 show results and predictions from the original paper, applying the PM and LM to data on plates containing a central, circular hole, loaded in uniaxial tension. Varying the hole radius has a strong effect on the measured stress to failure: prediction lines are shown using various values of the critical distance (equivalent to our $L/2$ for the PM in Fig. 8.2 and $2L$ for the LM in Fig. 8.3). It can be seen that reasonably good predictions (with less than 10% error) are possible with a single value of the critical distance, though there is a slight tendency for the optimal value to increase with hole size, which is an effect which we will return to below. Figures 8.4 and 8.5, also taken from the original Whitney and Nuismer paper, show results and predictions for tensile specimens containing sharp notches. Again a single value of the critical distance, combined with a critical stress of σ_u, gives very good predictions. Note that the values of the critical distances here are relatively large, compared to values obtained for ceramics, metals and polymers in previous chapters.

Whitney and Nuismer suggested two possible mechanistic reasons for the success of this method. The first was that the critical distance might correspond to some zone of damage ahead of the notch: this idea would justify the use of a stress averaging method such as the LM. The second reason was that failure might be initiated from some pre-existing flaw in the material: if the size and location of flaws is imagined to have some

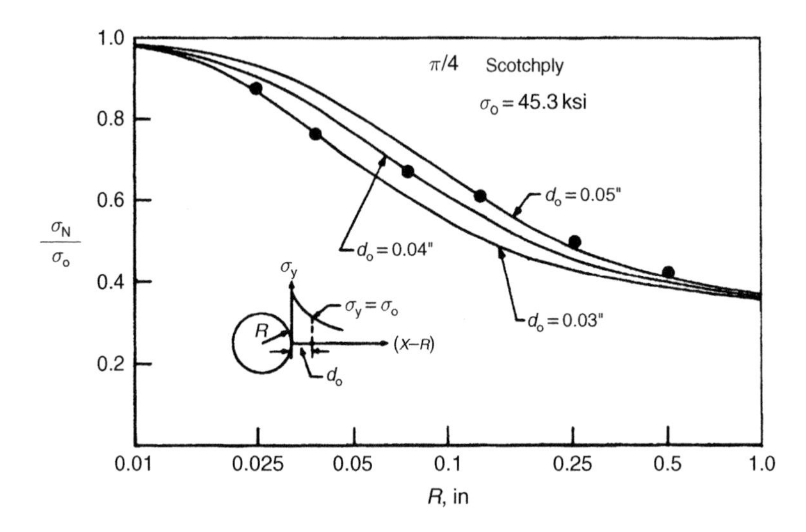

Fig. 8.2. Data from Whitney and Nuismer (1974); the effect of hole radius on fracture strength (normalised by the plain-specimen strength) in quasi-isotropic glass–epoxy laminate. The lines indicate predictions using the PM at three different values of the critical distance (d_0 here is equivalent to $L/2$).

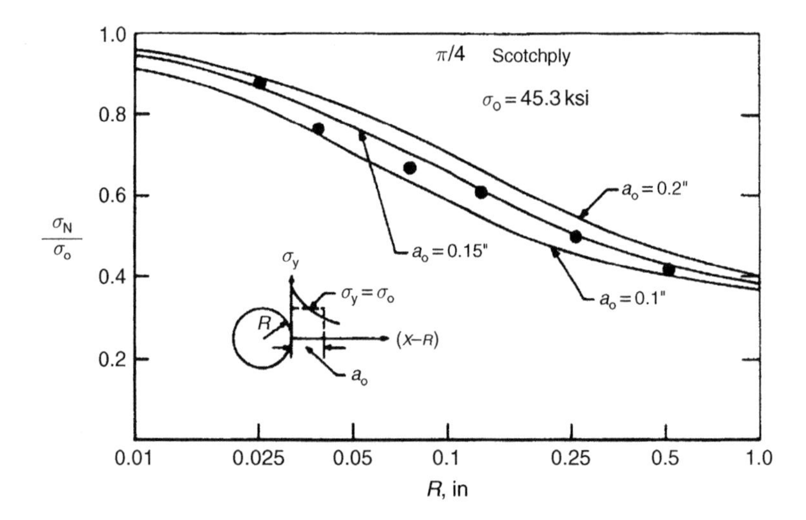

Fig. 8.3. Data as in Fig. 8.2: predictions using the LM (a_0 here is equivalent to $2L$).

statistical distribution, then this leads to a justification of the PM. These concepts, along with other possible explanations for the TCD, will be discussed in Chapter 13.

It is remarkable that this theory appeared all at once, emerging in its essentially complete form in Whitney and Nuismer's first paper. It is equally remarkable that there have been almost no further developments in this theory as applied to composite materials. Many

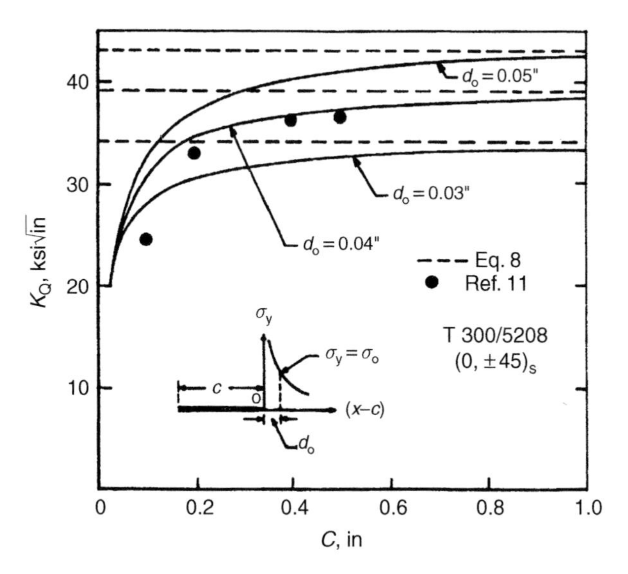

Fig. 8.4. Further data from Whitney and Nuismer (1974); measured toughness as a function of notch length for sharp notches in graphite–epoxy laminate. Prediction lines using the PM with various values of d_o.

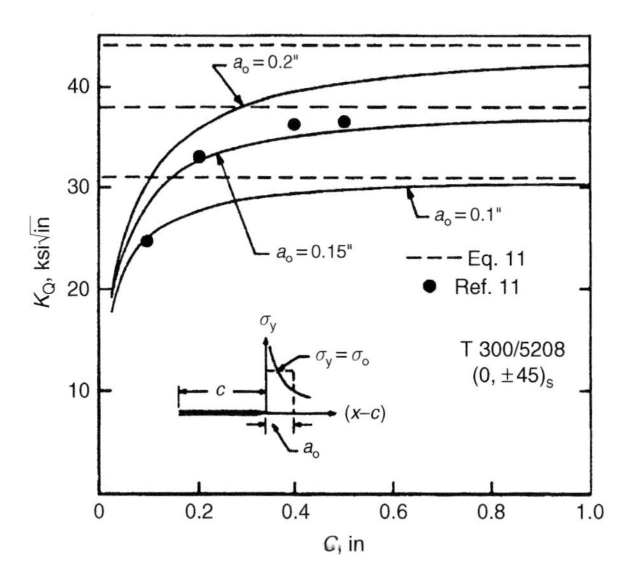

Fig. 8.5. Data as in Fig. 8.4: predictions using the LM with various values of a_o.

workers have been content to use the theory, with only minor modifications. Indeed, as we shall see below, even the experimental methods, which were confined to tensile testing of plates containing central holes and sharp notches, have set the tone for all subsequent work in this area. This may be justifiable because the method is so successful in predicting the fracture strength of many kinds of composites, but the effect has been that the dataset of experimental results, whilst being very large numerically, is limited to a narrow range of specimen types which do not include many of the stress concentration features encountered in real components.

This approach to the prediction of failure in fibre composites was very rapidly accepted by the research community, so much so that, a decade later, Awerbuch and Madhukar were able to present a comprehensive review demonstrating the accuracy of the TCD in predicting a large amount of experimental data (Awerbuch and Madhukar, 1985). They considered over 2800 test results in three types of composite: the commonly used graphite–epoxy, a newer graphite–polyimide material and a metal-matrix composite of boron fibres in aluminium. These were all continuous-fibre laminates but their orientation and laminate lay-up structure varied widely, from unidirectional materials loaded at various angles to the fibre direction through to laminates containing fibres at a wide range of angles giving quasi-isotropic behaviour.

The overwhelming conclusion was that the TCD was appropriate, giving accurate predictions of failure stress. The only cases for which difficulties were noted were some of the uniaxial graphite–polymer composites loaded at the extreme angles of 0° and 90° to the fibre direction. Both the PM and the LM were successful, with the LM giving slightly better accuracy overall. Values of the critical distance L were generally large, usually in the range 1–5 mm but sometimes as high as 15 mm. Considerable variation in the value of L occurred even within materials of the same type and laminate sequence, due to differences in fibre volume fraction and in the method of manufacture used. Wetherhold and Mahmoud also demonstrated that the Whitney and Nuismer approach was successful when applied to a large range of data on composites, including both continuous and discontinuous fibre materials (Wetherhold and Mahmoud, 1986). More recent work has substantiated the general applicability of this approach also for other types of composites such as woven and knitted-fibre materials (Belmonte et al., 2001; Khondker et al., 2004; Soriano and Almeida, 1999) and ceramic-matrix composites (Antti et al., 2004; McNulty et al., 2001). Other types of loading have been considered, including failure under compression (Khondker et al., 2004), shear (Pereira et al., 2004) and multiaxial loading (Tan, 1988). Fatigue failure can also be predicted (Huh and Hwang, 1999; McNulty et al., 2001): the application of the TCD in the field of fatigue will be discussed in more detail in the next chapter. Antti et al. (2004) used the Waddoups imaginary crack model (whose predictions are equivalent to the TCD – see below) to analyse the behaviour of a ceramic-matrix composite: interestingly as the temperature was raised, the L value changed from a large one typical of composite materials to a much smaller value typical of ceramics.

8.3 Does L Vary with Notch Size?

Whilst the great majority of results can be predicted using a constant value of L for a given material, it was noticed that, in some cases, the optimal value tended to vary, increasing with increased notch size. Figure 8.6 shows an example of this effect, using some of the data collected by Awerbuch and Madhukar (1985), in this case for sharp notches in a boron–aluminium laminate with fibre orientations of 0 and 45° with respect

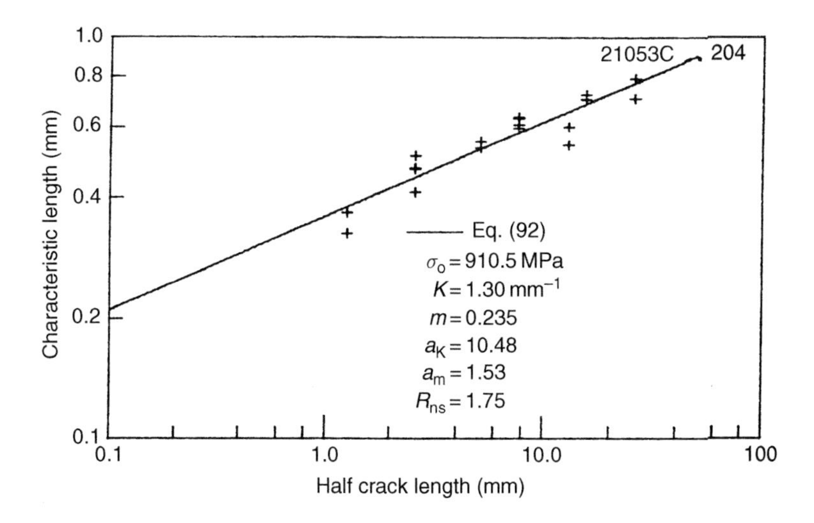

Fig. 8.6. Data from Awerbuch and Madhukar (1985) showing an increase in the characteristic length (equivalent to our *L*/2) in a boron–aluminium laminate.

to the loading axis. The critical distance clearly increases by a factor of 2 as notch length increases by a factor of 20. A number of workers have developed empirical laws to describe this effect. Karlak, who considered only data from circular holes (Karlak, 1977), proposed that the critical distance varies as the root radius of notch length, a:

$$L = C_1 a^{1/2} \quad (8.1)$$

Here C_1 is a constant. Pipes et al., who considered both holes and notches (Pipes et al., 1979), proposed a more general relationship using two constants, C_2 and m, as follows:

$$L = C_2 a^{m} \quad (8.2)$$

This has the advantage that it can also describe cases where there is no change in L, by setting $m = 0$. Conversely $m = 1$ corresponds to the case where notch size has no effect on fracture strength. These equations can give reasonably good descriptions of the data (for example, the results in Fig. 8.6 can be described using an m value of 0.235) but, being purely empirical, they throw no light on the reasons for the change in L with notch size. In recent years, other workers (e.g. Govindan Potti et al., 2000) have continued the trend of developing empirical equations of this kind.

Some insight into this issue can be gained by examining stress–distance curves at failure for various specimens. Figure 8.7 shows results from specimens of the glass-fibre/epoxy quasi-isotropic material (Whitney and Nuismer, 1974) shown in Fig. 8.2, containing holes of varying radius. In each case the stress–distance curve is plotted for applied loads corresponding to failure of the specimen. The plain strength σ_u is also shown, from which it can be seen that, whilst the point of intersection between the curves and the σ_u line (i.e. $L/2$) is approximately constant, there is a systematic change in which $L/2$ increases with radius by about a factor of 2.

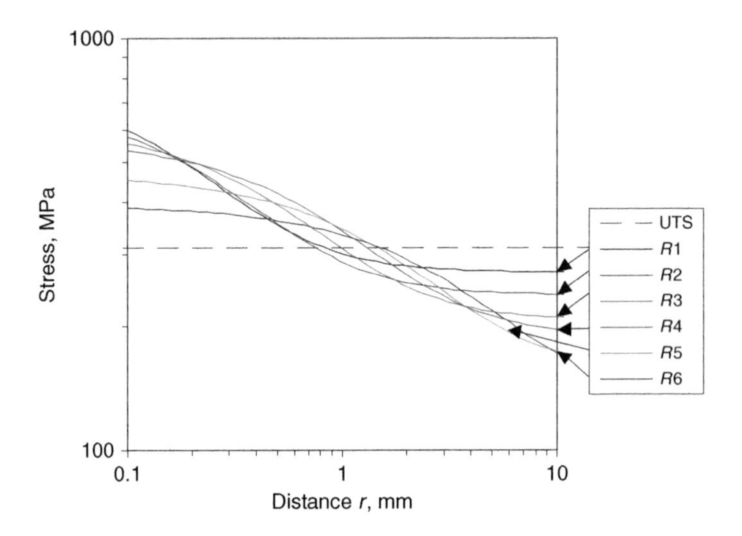

Fig. 8.7. Stress–distance curves at failure for data from Fig. 8.2. The lines labelled R1–R6 indicate increasing hole radius.

However, if we also include data from sharp notches in the same material (Fig. 8.8) we can see that the behaviour is not consistent: there is almost no change in $L/2$ with notch size (except for the smallest of the four notches); so in order to apply Eq. (8.2) we would have to use different constants for holes compared to those used for sharp notches. Finally, if we consider the stress axis, we realise that these apparently large changes in $L/2$ lead to very small changes in stress, due to the low slopes of the lines.

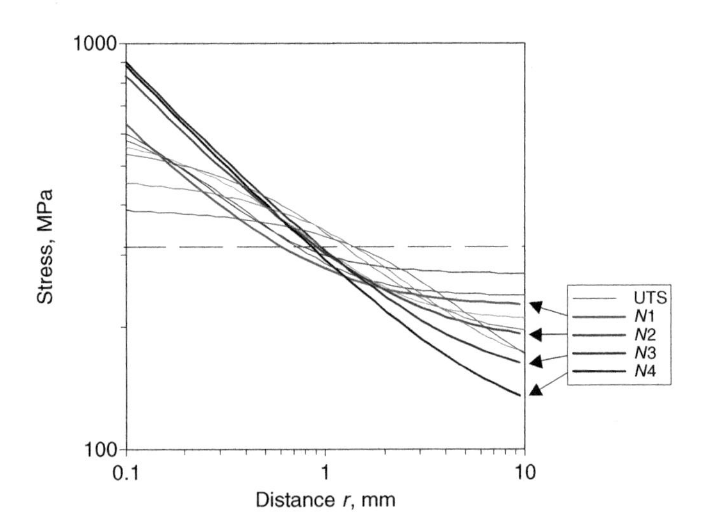

Fig. 8.8. As Fig. 8.7 but adding data on sharp notches for the same material. The lines labelled N1–N4 represent increasing notch length.

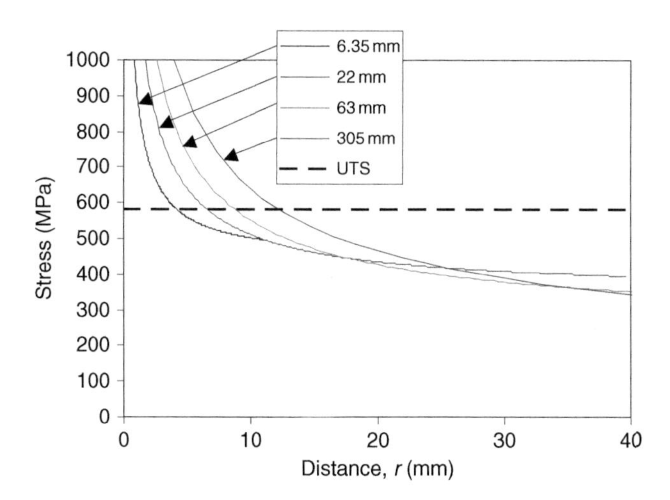

Fig. 8.9. Stress–distance curves at failure for data from Kennedy et al. (2002) for centre-cracked plates with four different crack lengths.

So the use of a single value of $L/2$ will give no more than 10% error in the prediction of fracture stress for all these notches and holes.

The largest variation in L which I was able to find is shown in Fig. 8.9, from a recent paper on an orthotropic graphite/epoxy material (Kennedy et al., 2002). In this paper the failure of TCD-like methods was commented on and used as a motivation to develop a more complex theoretical model. Here the value of $L/2$ at the intersection point varies by a factor of 3, from 4.2 to 12.2 mm. These results, which were obtained from sharp, central notches in plates loaded in tension, are convenient to analyse because the ratio between the notch radius and the half-width of the plate (a/W) was kept constant at 0.25 whilst the absolute size of the notch and specimen were varied through almost two orders of magnitude. Thus in this case we can rule out any complications arising from changing the a/W ratio. The value of L gives an approximate estimate of the size of the damage zone at failure, from which we can conclude that the larger specimens were failing under conditions that would be valid from an LEFM point of view (the damage zone being much smaller than either the crack length or the width of the remaining ligament W-a). However, as specimen size is reduced the damage zone occupies an increasing proportion of the total width until, for the smallest size tested, it probably covered the entire remaining ligament. In this case, then, the reduction of $L/2$ with notch size, leading to a relative weakening of the smaller notches, can be explained in terms of the increasing loss of linear-elastic conditions in the specimen as a whole. It is remarkable under these circumstances that the TCD, which is essentially a linear elastic theory, should continue to be even approximately accurate for the smaller specimen sizes. In fact all the results, from the smallest to the largest notch size, can be predicted using the TCD with a single value of $L/2$, with errors no larger than 13%.

Another possible reason for the apparent increase of L with notch size reported in some studies is the use of approximate methods of stress analysis. For tensile specimens of width $2W$ containing a symmetrical stress concentration (such as a central hole, radius a, a central crack, total length a or a pair of opposing edge cracks, length a) the stress–distance curve can be approximated by using the curve for the same feature in an infinite plate, multiplied by a correction factor.

For example, for a central hole one can use the Airy stress function for an infinite plate to describe the local stress, $\sigma(r)$ as a function of distance r and applied nominal stress σ:

$$\sigma(r) = \sigma\left(1 + \frac{1}{2}\left(\frac{a}{a+r}\right)^2 + \frac{3}{2}\left(\frac{a}{a+r}\right)^4\right) \tag{8.3}$$

One can correct for finite width by multiplying the stress by the following factor Y:

$$Y = \frac{2 + \left(1 - \frac{a}{W}\right)^3}{3\left(1 - \frac{a}{W}\right)} \tag{8.4}$$

This method was used in the original papers by Whitney and co-workers, the approach being to correct the measured fracture stress values so as to obtain the equivalent fracture stress for an infinite plate. The same approach has been used by many workers in this field ever since. However, this is an approximate method, which is valid only at relatively small distances from the notch root. As an example to illustrate the typical errors which can arise, Fig. 8.10 shows a plot of stress as a function of distance for a hole in a specimen which has $a/W = 0.375$. The stress is normalised by the maximum stress (at $r = 0$) and the distance is normalised by a. It can be seen that the accurate result (obtained from FEA) deviates considerably from the curve obtained using Eqs (8.3) and (8.4) above. Unacceptable errors (more than 10%) arise in this case if $r > 0.8a$; even greater errors can be expected for larger values of a/W. These conditions frequently arise in test specimens due to the relatively large values of L in these materials and the common practice of using quite large a/W ratios. It can easily be seen that this kind of

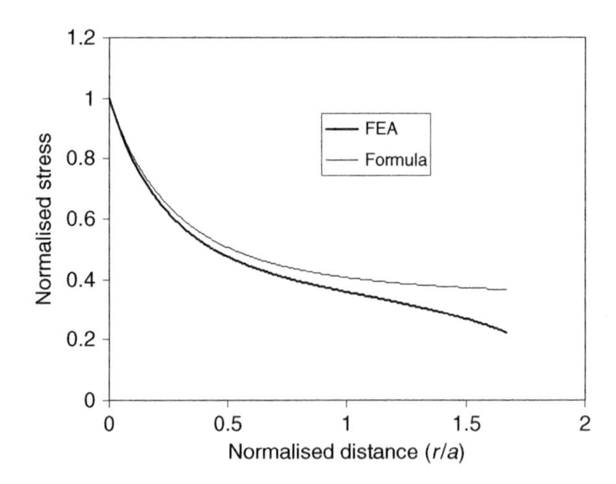

Fig. 8.10. Stress–distance curves calculated using the approximate formula (Eqs (8.3) and (8.4)) compared to an exact result using FEA. Central hole in a plate loaded in tension: $a/W = 0.375$.

error (in which the estimated stress at $L/2$ is greater than the actual stress) will lead to the effect in which $L/2$ appears to increase with increasing notch size, because if notch size increases, at constant W, the estimated stress will deviate more and more from the actual stress.

There is another effect which may contribute to the apparent increase of L with a, and this is the effect of constraint. The problem of out-of-plane constraint (to which we devoted considerable discussion in Chapter 7) is not usually given much consideration by researchers working on composite materials. At first sight this seems strange, considering that the relatively high K_c/σ_u ratios in these materials will mean that very large specimen thickness will be needed to ensure plane-strain conditions. In fact almost all specimens and components actually used will experience either plane-stress or mixed plane-stress/plane-strain conditions. Some workers have reported the kind of thickness effects which were noted for metals and polymers in previous chapters, whereby increasing thickness (causing more constraint) leads to increasing brittle behaviour. However, many workers have reported no effect and there are even reports of the opposite behaviour (see Awerbuch and Madhukar, 1985, p. 103). The relative insignificance of thickness effects may be due to the fact that yielding behaviour is usually unimportant in these materials. Some yielding of the matrix material can be expected (especially for metals and certain polymer matrices) but this is small compared to the effects of localised damage, the creation of which may be relatively unaffected by the degree of stress triaxiality. If any constraint effects do occur, then we can reason that they will cause an apparent increase in L with a. For example, if one tests a series of cracked specimens with decreasing crack size, then, for very short cracks, the applied K needed to cause failure will decrease, thus plane-strain conditions are more likely, with a concomitant decrease in L.

To conclude this section on the possible variation of critical distance with notch size, we can note that this may occur for various reasons, some connected to the mechanics of the situation (large damage zone size relative to specimen size; changing degree of constraint) and some due to inaccuracies in the methods of stress analysis used. The most important conclusion is that, even in cases where measurable changes do occur, a constant value of L can still be used, whilst maintaining acceptable levels of accuracy in the prediction of fracture stress.

8.4 Non-damaging Notches

In previous chapters (Chapters 6 and 7) we noted that non-damaging notches could occur, these being notches which had no effect on the strength of the specimen beyond that of reducing the net cross section. Thus, for these notches, the strength of the specimen (expressed in terms of the net section stress at failure) will be the same as that of a plain specimen. We saw that this arises in materials where the critical stress σ_o is different from the plain-specimen tensile strength σ_u. For large notches, the value of the ratio σ_o/σ_u gives the critical K_t factor below which the notch becomes non-damaging. Absolute size also plays a role: small notches and cracks less than approximately L in length can also be non-damaging if σ_o/σ_u is significantly greater than 1. Now in composite materials we have seen that the critical stress is equal to σ_u, so non-damaging notches would not be expected.

As an aside, some experimental data on composites seem to show convergence to a higher critical stress value. For example in Fig. 8.7, four of the six curves cross over at a single point, at which the stress is approximately 500 MPa and the distance is considerably smaller than the expected $L/2$ value. However, the curves corresponding to the other two holes do not cross at this point, and neither do the curves corresponding to the sharp notches (in Fig. 8.8). This emphasises the need to analyse data from a wide range of notch types, certainly including some very sharp notches as well as blunt notches or plain specimens. Many researchers working on composite materials have made the mistake of only using data from circular holes.

To return to the question of non-damaging notches, these can exist even when the critical stress is σ_u, in specimens with relatively large L and large a/W, as can be demonstrated using some recent data on composites containing discontinuous, randomly oriented fibres (Lindhagen and Berglund, 2000). Two different fibre lengths were used: short (2–9 mm) and long (about 25 mm), with two different matrix materials, an unsaturated polyester (UP) and a vinyl ester (VE). The mechanical properties of these four materials are shown in Table 8.1; the Young's modulus (E) and critical strain-energy release rate (G_c) were measured, allowing K_c and L to be calculated from first principles, using:

$$K_c = (G_c\, E)^{1/2} \tag{8.5}$$

$$L = \frac{1}{\pi}\left(\frac{K_c}{\sigma_u}\right)^2 \tag{8.6}$$

These materials showed particularly large L values, up to 22 mm. Interestingly, their fracture toughness values were almost identical, despite considerable differences in σ_u and L. The short-fibre materials had lower strength and built up larger damage zones before failure. Tensile specimens containing central holes were tested. Figure 8.11 shows the experimental data using a normalised plot of σ_f/σ_u versus a/L, along with predictions using the PM and LM. The data from all four materials could be analysed together in this way because the specimens all had the same ratio of a/W, equal to 0.375. It can be seen that both methods of prediction were accurate: errors were less than 20%, which was similar to the scatter in the individual test results. Interestingly, notches less than a certain size were non-damaging, failing at a net-section stress equal to the UTS (σ_u). This occurred for hole radii approximately equal to L, and was predicted by both the PM and the LM. The PM prediction crosses the UTS line and continues up to higher stress values: this is because in these specimens the stress at large distances from the notch (approaching the far side of the specimen) falls to values less than the applied nominal stress. The LM prediction, on the other hand, stops exactly at the UTS:

Table 8.1. Mechanical properties of the materials tested by Lindhagen and Berglund (2000)

Material	$K_c(\mathrm{MPa(m)})^{1/2}$	σ_u(MPa)	L (mm)
UP short fibres	28.4	107	22.4
UP long fibres	26.0	169	7.6
VE short fibres	26.8	109	19.5
VE long fibres	26.3	193	5.9

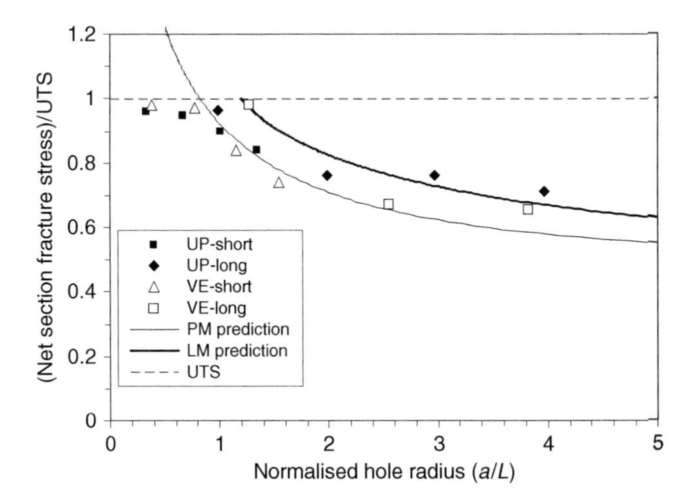

Fig. 8.11. Experimental data from Lindhagen and Berglund (2000), with predictions using the PM and LM. The fracture stress σ_f is normalised with respect to the UTS (σ_u) for each material, and the hole radius a is normalised with respect to L. Note the prediction (and actual existence) of non-damaging notches.

this corresponds to the point when stresses are being averaged over the entire remaining ligament $(W\text{-}a)$. The LM prediction becomes physically impossible at smaller values of a because one would be averaging over distances outside the specimen itself.

For hole sizes less than these crossover points, the fracture stress is equal to the plain specimen UTS. Thus we see that non-damaging notches can occur in situations where the remaining ligament of material is of a size similar to the critical distance. We did not encounter this situation before in the materials dealt with in previous chapters, because in most materials L is so small that this situation will only occur if the specimen or component is microscopic. We shall come across an example of this in a later chapter when discussing fatigue failure in some very small components. It is very encouraging to see that the TCD is, once again, able to make accurate predictions in these materials, even when the damage zone ahead of the notch is similar in size to the remaining portion of the specimen, implying non-LEFM conditions and highly non-linear deformation behaviour.

This discussion raises an interesting general question about notch sensitivity. Are composite materials notch sensitive? On the one hand, one can argue that the answer to this question is 'yes' because, with the exception of the non-damaging notches discussed above, almost any notch introduced into a composite material will have the effect of reducing strength. We can contrast this situation with that of metals and polymers for which many notches, some with quite high K_t factors, have no effect on strength at all beyond their effect in reducing the load-bearing cross section. On the other hand, the large L values of composite materials confer a greatly reduced notch sensitivity for many common sizes and shapes of notch. A circular hole a few millimetres in diameter will have a significant effect on a brittle polymer but a negligible effect in composite materials which have L values similar to the notch radius. In fact, it is difficult to find examples of composites containing holes so large that the full, theoretical K_t effect is actually realised in terms of strength reduction. Engineering ceramic materials, on the other hand, are truly notch sensitive in every sense of the word, having small L values and $\sigma_o/\sigma_u = 1$. We can see then that notch sensitivity is a complex concept. In the TCD, we have two parameters which both contribute to notch sensitivity: L and σ_o/σ_u. These issues, and especially the implications of the different parameter values in different material classes, will be discussed more fully in Chapter 13.

8.5 Practical Applications

Having been in common use for composite materials for over three decades, the TCD has found its way into the prediction of failure in engineering components made from these materials. In this respect, the research community has not been as useful to design engineers as it might have been. Most researchers, following the lead of Whitney and Nuismer and other early workers, have confined themselves to testing simple plate specimens in uniaxial tension, with a central circular hole or sharp notch, or sharp edge notches. It is understandable that flat plate specimens should be tested, since most of the important uses of composite materials are in the form of relatively thin sheets made by laminate construction. It is also understandable that circular holes should be tested, since sheets are often joined together using holes containing fasteners of various types: however, in this case the type of loading is very different from simple tension (see below). Some work has been done on tensile specimens containing eccentrically placed holes (Yao et al., 2003) but in this case the eccentricity had little effect. I was surprised to find almost no work on other types of stress concentration feature, such as corners, bends and changes in section, or on other types of loading such as bending and torsion. I would have thought that out-of-plane bending would be of particular concern for structures made from sheet material: this type of loading produces very different stress distributions around a hole or notch (see results for other materials in Chapter 6 and, later on, in Chapter 12).

Many practical failures associated with holes in composite materials are in fact bearing failures, due to the pressure of a bolt, pin or other fastener on the edge of the hole. An approach using the TCD in combination with the Yamada-Sun multiaxial failure law was developed some time ago (Chang et al., 1982) and has been recently applied by several different workers (Wu and Hahn, 1998; Wang et al., 1998; Xu et al., 2000) with good accuracy. The method includes different L values for tension and compression failure in the same material, leading to a locus of critical points around the circumference of the hole. The TCD was also used in a practical context to assess the strength of bolted joints in a bridge design (Zetterberg et al., 2001). Some other recent work illustrates nicely the very high stress gradients which arise in pin-loaded joints, emphasising the usefulness of the TCD in handling these kinds of problems (Aktas and Dirikolu, 2004).

Other practical problems for which the TCD has been successfully used include multiple elliptical holes (Xu et al., 2000), openings of various shapes in pressure vessels (Ahlstrom and Backlund, 1992), cutouts in wing spars of aircraft (Hollmann, 1991; Vellaichamy et al., 1990) and bonded joints (Engelstad and Actis, 2003). An interesting design problem with many applications is the size and placing of holes in beams intended to reduce weight without reducing strength: here again the TCD can form part of an overall analysis (Naik and Ganesh, 1993).

8.6 Other Theoretical Models

In addition to the Whitney and Nuismer (TCD) approach, a wide range of theoretical models has been developed to predict the failure of composites. We can find many of the same types of models as have been developed for other materials, such as imaginary crack and cohesive zone theories, which have already been described in outline in Chapter 3. In addition, other models have been developed which are aimed more specifically at the particular mechanisms of damage and failure known to occur in these materials.

A number of methods have been in use since the 1970s and 1980s and are well described in some of the publications mentioned above (Awerbuch and Madhukar, 1985; Whitney et al., 1982). Simple failure criteria based on maximum stress or strain or various combinations of elements from the 2D stress tensor include the well-known Tsai-Wu, Tsai-Hill and Yamada-Sun criteria. These may be used to predict the initial onset of damage at a notch but are of limited value in predicting final failure due to the changes which occur as damage builds up. An ICM was introduced by Waddoups a few years before Whitney and Nuismer's first paper (Waddoups et al., 1971): as we have seen previously, the ICM, used in conjunction with LEFM, is theoretically similar to the PM and LM, and indeed the Waddoups approach can give excellent predictions, at least in cases where K can be calculated for the resulting notch-plus-crack model. A model using the NSIF approach was developed a few years later (Mar and Lin, 1977) as was a strain-based fracture mechanics model using an imaginary crack (Poe and Sova, 1980), which was advocated for use with composites whose properties are not fibre-dominated.

To conclude this section on theoretical models, mention may be made of some more recent developments. Belmonte et al. developed a simple model based on the estimated size of the damage zone (Belmonte et al., 2001) and other models of the process-zone/cohesive-zone type have been constructed (e.g. Afaghi-Khatibi et al., 1996). Other workers have developed complex models of the damage-mechanics variety (Maa and Cheng, 2002; Wang et al., 2004). Whilst these models are capable of making good predictions, they are computationally much more complex than the TCD. It is interesting to note that several of the above workers also used the TCD and found that it gave satisfactory predictions – Maa and Cheng reported errors of about 10%, for instance – so it is difficult to justify using more complex models unless they shed some light on the underlying physical mechanisms involved, which damage-mechanics models generally do not do.

Many workers have recognised that the damage which develops ahead of the crack or notch in a composite has some unique features which require the development of more appropriate mechanistic models. Noteworthy among these are Reifsnider's critical element method and the combined fracture-mechanics/damage-mechanics model of Cowley and Beaumont (Cowley and Beaumont, 1997a; Cowley and Beaumont, 1997b; Reifsnider et al., 2000). It is very interesting to note that the TCD is an inherent feature in both of these models: in Reifsnider's model, stresses are averaged over a critical volume, whilst Cowley and Beaumont use the TCD to predict crack initiation. The TCD also forms an essential element in other theories as diverse as Zhang's prediction of creep crack growth (Zhang, 1999) and Leguillon's predictions of delamination stresses (Leguillon et al., 2001).

8.7 Fracture of Bone

Bone is a natural fibre-composite material, made up of very small fibres of collagen – a soft, polymeric material – and crystals of the hard, ceramic substance hydroxyapatite. These are arranged to form laminate structures in which fibre directions alternate very much as they do in industrial fibre composite materials. Bone displays considerable anisotropy, being about twice as strong in the longitudinal direction – parallel to the long axis of a load-bearing bone such as the femur or tibia – compared to the transverse direction.

We conducted some tests to investigate whether bone could be analysed using the TCD. Figs 8.12 and 8.13 show the stress–distance curves at the point of failure for plain specimens and specimens containing holes and sharp notches. For specimens loaded in the transverse direction (which therefore fail by cracks growing in the longitudinal direction) there is a common point of intersection, indicating that the PM is appropriate, with an L value of 1.36 mm. For the specimens loaded in the longitudinal direction there is some variation but the lines meet (within an error of 20%) at a distance of around 0.65 mm, implying an L value of 1.3 mm which is, interestingly, almost identical to that for the transverse specimens despite the considerable differences in strength and toughness in the two directions. These L values are also of the same order of magnitude as most of the industrial composite materials considered above.

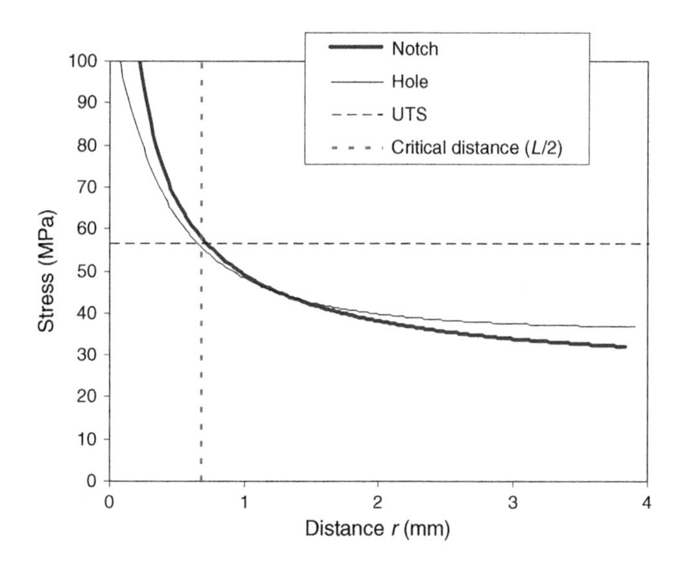

Fig. 8.12. Stress–distance curves at failure for bone specimens loaded in the transverse direction.

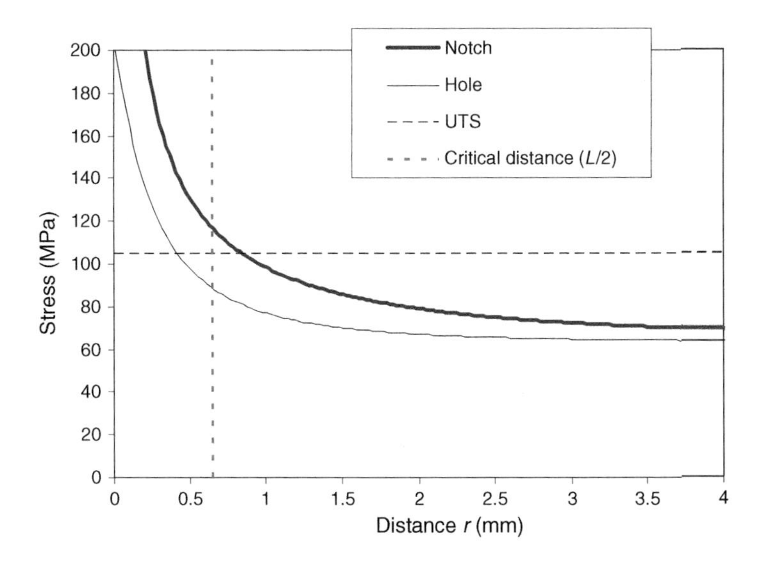

Fig. 8.13. As Fig. 8.12, loading in the longitudinal direction.

The mechanical properties of bone vary considerably, depending on the source of the material (different locations in the bone, different animals etc.): the bone which we tested, which was taken from the femurs of cows, had a rather low longitudinal tensile strength of 105 MPa: a value of 150 MPa would be more typical. It also had a rather high fracture toughness of $6.7\,\mathrm{MPa(m)}^{1/2}$. Whilst these values are certainly within normal limits (see Currey, 2002 for more data), they lead to a value of L which is probably rather larger than average for bone. Taking more typical values of strength (150 MPa) and toughness ($5\,\mathrm{MPa(m)}^{1/2}$), we would estimate a typical L value of 0.35 mm. Tests to measure the strengths of specimens containing sharp notches of various lengths (Lakes et al., 1990) found that cracks smaller than about 0.4 mm had no effect on strength, which accords with this value of L.

Two other publications on bone have used the ICM, with crack lengths (which should correspond in magnitude to L) of 0.34 and 1.82 mm respectively (Bonfield and Datta, 1976; Moyle and Gavens, 1986), so this is probably the relevant range of values for bone.

An important practical application here is the assessment of the reduction in strength that will occur if a hole is made in a bone during surgery, for example a drilled circular hole to receive a screw or take a sample for biopsy. Several workers have generated experimental data by testing whole bones which have had holes made in them, usually comparing the strength to that of the same bone in the opposite limb (the contralateral bone) which can be assumed to have a similar strength. Clark et al. tested rectangular holes and found no effect of corner radius below a value of 1.4 mm, which suggests an L value of the same order (Clark et al., 1977). Other workers came to similar conclusions regarding the impact strength of bones containing rectangular holes (Moholkar et al., 2002). Various workers have tested bones containing circular holes of different diameters, and an interesting finding is that the effect of the hole is generally much less when the bone is tested in torsion than when it is tested in tension. These data, and the resulting predictions, are discussed in detail in Chapter 11, as they have interesting consequences for the prediction of failure under multiaxial loading. Suffice it to say that the TCD can still be used, but only in conjunction with a multiaxial loading criterion.

8.8 Values of *L* for Composite Materials

The previous three chapters have concluded with a table listing values of *L* and other properties for the materials under consideration. To produce a complete list of all the composite materials mentioned in the literature would be impractical as it would run to several thousand entries. A brief statement will suffice to summarise the data. Classic engineering composites using continuous fibres and a polymer matrix in laminate structures, such as graphite–epoxy, exhibit a great variety of toughness and strength values but their *L* values tend to fall in a narrow range, typically 1–5 mm but occasionally reaching values as high as 15 mm (Awerbuch and Madhukar, 1985). The same holds true when these materials are prepared using woven fibres (e.g. Soriano and Almeida, 1999).

Metal matrix composites of long-fibre laminate structure, such as Boron–Aluminium, display a similar range of values (Awerbuch and Madhukar, 1985). Data for ceramic matrix composites are sparse: Antti et al. (2004) obtained values ranging from 2.7 mm at room temperature to 0.3 mm at 1100 °C.

Our analysis of bone (another laminate structure), though based on limited data, suggested *L* values in the range 0.35–1.8 mm. The few studies that have considered anything other than tensile loading, that is compression, shear and multiaxial stress states, suggest that *L* may vary somewhat with stress state whilst remaining in the same order of magnitude.

Discontinuous fibre composites can be expected to display larger *L* values, spread over a wider range: our analysis of just four materials from one source (Lindhagen and Berglund, 2000) revealed *L* values from 5.9 to 22.4 mm.

8.9 Concluding Remarks

We have seen in this chapter that the TCD is already widely used in the field of composite materials, both by the research community and by industrial designers. The use of a simple PM or LM with a critical stress equal to the plain-specimen strength can give accurate estimates of the fracture stress for most types of industrial fibre composite, and even where more complex theories are developed, the TCD is often retained as an essential element.

Ongoing issues in which more research can usefully be conducted on composite materials include the continuing debate about possible changes in *L* with notch size, and the current lack of test data on anything but simple holes and notches in tensile plates. More work could also be carried out on natural materials. Many natural materials have structures similar to those of fibre composites: we have seen above that the effects of stress concentrations in bone can usefully be studied using the same approach. Another natural material with fibre-composite structure is wood; though it can be analysed using LEFM in some cases, the fracture of wood displays phenomena such as size effects and stable crack growth under monotonic loading (Smith and Vasic, 2003) which would lead us to think that a TCD-type analysis may be useful.

This chapter concludes a series of four chapters concerned with brittle fracture and other forms of failure under monotonic loading. In the next chapter, we will consider the application of the TCD to the prediction of fatigue failure under cyclic loading, in metals and other materials.

References

Afaghi-Khatibi, A., Ye, L., and Mai, Y.W. (1996) Evaluations of the effective crack growth and residual strength of fibre-reinforced metal laminates with a sharp notch. *Composites Science and Technology* **56**, 1079–1088.

Ahlstrom, L.M. and Backlund, J. (1992) Shape optimisation of openings in composite pressure vessels. *Composite Structures* **20**, 53–62.

Aktas, A. and Dirikolu, M.H. (2004) An experimental and numerical investigation of strength characteristics of carbon-epoxy pinned-joint plates. *Composites Science and Technology* **64**, 1605–1611.

Antti, M.L., Lara-Curzio, E., and Warren, R. (2004) Thermal degradation of an oxide fibre (Nextel 720) aluminosilicate composite. *Journal of the European Ceramic Society* **24**, 565–578.

Awerbuch, J. and Madhukar, M.S. (1985) Notched strength of composite laminates: Predictions and experiments – A review. *Journal of Reinforced Plastics and Composites* **4**, 3–159.

Belmonte, H.M.S., Manger, C.I.C., Ogin, S.L., Smith, P.A., and Lewin, R. (2001) Characterisation and modelling of the notched tensile fracture of woven quasi-isotropic GFRP laminates. *Composites Science and Technology* **61**, 585–597.

Bonfield, W. and Datta, P.K. (1976) Fracture toughness of compact bone. *Journal of Biomechanics* **9**, 131–134.

Chang, F.K., Scott, R., and Springer, G.S. (1982) Strength of mechanically fastened composite joints. *Journal of Composite Materials* **16**, 470–494.

Clark, C.R., Morgan, C., Sonstegard, D.A., and Matthews, L.S. (1977) The effect of biopsy-hole shape and size on bone strength. *Journal of Bone and Joint Surgery* **59A**, 213–217.

Cowley, K.D. and Beaumont, P.W.R. (1997a) Damage accumulation at notches and the fracture strss of carbon-fibre/polymer composites: combined effects of stress and temperature. *Composites Science and Technology* **57**, 1211–1219.

Cowley, K.D. and Beaumont, P.W.R. (1997b) Modelling problems of damage at notches and the fracture stress of carbon-fibre/polymer composites: matrix, temperature and residual stress effects. *Composites Science and Technology* **57**, 1309–1329.

Currey, J.D. (2002) *Bones*. Princeton University Press, USA.

Engelstad, S.P. and Actis, R.L. (2003) Development of p-version handbook solutions for analysis of composite bonded joints. *Computers and mathematics with applications* **46**, 81–94.

Govindan Potti, P.K., Nageswara Rao, B., and Srivastava, V.K. (2000) Notched tensile strength for long- and short-fiber reinforced polyamide. *Theoretical and Applied Fracture Mechanics* **33**, 145–152.

Hollmann, K. (1991) Failure analysis of a shear loaded graphite/epoxy beam containing an irregular cutout. *Engineering Fracture Mechanics* **39**, 159–175.

Huh, J.S. and Hwang, W. (1999) Fatigue life prediction of circular notched CRFP laminates. *Composite Structures* **44**, 163–168.

Karlak, R.F. (1977) Hole effects in a related series of symmetrical laminates. In *Proceedings of failure modes in composites IV* pp. 105–117. The metallurgical society of AIME, Chicago.

Kennedy, T.C., Cho, M.H., and Kassner, M.E. (2002) Predicting failure of composite structures containing cracks. *Composites Part A* **33**, 583–588.

Khondker, O.A., Herszberg, I., and Hamada, H. (2004) Measurements and prediction of the compression-after-impact strength of glass knitted textile composites. *Composites Part A* **35**, 145–157.

Lakes, R.S., Nakamura, S., Behiri, J.C., and Bonfield, W. (1990) Fracture mechanics of bone with short cracks. *Journal of Biomechanics* **23**, 967–975.

Leguillon, D., Marion, G., Harry, R., and Lecuyer, F. (2001) The onset of delamination at stress-free edges in angle-ply laminates - analysis of two criteria. *Composites Science and Technology* **61**, 377–382.

Lindhagen, J.E. and Berglund, L.A. (2000) Application of bridging-law concepts to short-fibre composites Part 2: Notch sensitivity. *Composites Science and Technology* **60**, 885–893.

Maa, R.H. and Cheng, J.H. (2002) A CDM-based failure model for predicting strength of notched composite laminates. *Composites Part B* **33**, 479–489.

Mar, J.W. and Lin, K.Y. (1977) Fracture mechanics correlation for tensile failure of filamentary composites with holes. *Journal of Aircraft* **14**, 703–704.

McNulty, J.C., He, M.Y., and Zok, F.W. (2001) Notch sensitivity of fatigue life in a Sylramic/SiC composite at elevated temperature. *Composites Science and Technology* **61**, 1331–1338.

Moholkar, K., Taylor, D., O'Reagan, M., and Fenelon, G. (2002) A biomechanical analysis of four different methods of harvesting bone-patellar tendon-bone graft in porcine knees. *Journal of Bone and Joint Surgery* **84A**, 1782–1787.

Moyle, D.D. and Gavens, A.J. (1986) Fracture properties of bovine tibial bone. *Journal of Biomechanics* **9**, 919–927.

Naik, N.K. and Ganesh, V.K. (1993) Optimum design studies on FRP beams with holes. *Composite Structures* **24**, 59–66.

Pereira, A.B., deMorais, A.B., Marques, A.T., and deCastro, P.T. (2004) Mode II interlaminar fracture of carbon/epoxy multidirectional laminates. *Composites Science and Technology* **64**, 1653–1659.

Pipes, R.B., Wetherhold, R.C., and Gillespie, J.W. (1979) Notched strength of composite materials. *Journal of Composite Materials* **12**, 148–160.

Poe, C.C. and Sova, J.A. (1980) Fracture toughness of boron/aluminum laminates with various proportions of 0 and 45 plies. Langley, NASA. NASA Technical Paper 1707. Ref Type: Report

Reifsnider, K., Case, S., and Duthoit, J. (2000) The mechanics of composite strength evolution. *Composites Science and Technology* **60**, 2539–2546.

Smith, I. and Vasic, S. (2003) Fracture behaviour of softwood. *Mechanics of Materials* **35**, 803–815.

Soriano, E. and Almeida, S. (1999) Notch sensitivity of carbon/epoxy fabric laminates. *Composites Science and Technology* **59**, 1143–1151.

Tan, S.C. (1988) Mixed-mode fracture of notched composite laminates under uniaxial and multi-axial loading. *Engineering Fracture Mechanics* **31**, 733–746.

Vellaichamy, S., Prakash, B.G., and Brun, S. (1990) Optimum design of cutouts in laminated composite structures. *Computers and Structures* **37**, 241–246.

Waddoups, M.E., Eisenmann, J.R., and Kaminski, B.E. (1971) Macroscopic fracture mechanics of advanced composite materials. *Journal of Composite Materials* **5**, 446–454.

Wang, J., Banbury, A., and Kelly, D.W. (1998) Evaluation of approaches for determining design allowables for bolted joints in laminated composites. *Composite Structures* **41**, 167–176.

Wang, J., Callus, P.J., and Bannister, M.K. (2004) Experimental and numerical investigation of the tension and compression strength of un-notched and notched quasi-isotropic laminates. *Composite Structures* **64**, 297–306.

Wetherhold, R.C. and Mahmoud, M.A. (1986) Tensile strength of notched composite materials. *Materials Science and Engineering* **79**, 55–65.

Whitney, J.M., Daniel, I.M., and Pipes, R.B. (1982) *Experimental Mechanics of Fiber Reinforced Composite Materials*. Society for Experimental Stress Analysis, Connecticut.

Whitney, J.M. and Nuismer, R.J. (1974) Stress fracture criteria for laminated composites containing stress concentrations. *Journal of Composite Materials* **8**, 253–265.
Wu, T.J. and Hahn, H.T. (1998) The bearing strength of E-glass/vinyl-ester composites fabricated by VARTM. *Composites Science and Technology* **58**, 1519–1529.
Xu, X.W., Man, H.C., and Yue, T.M. (2000) Strength prediction of composite laminates with multiple elliptical holes. *International Journal of Solids and Structures* **37**, 2887–2900.
Yao, X.F., Kolstein, M.H., Bijlaard, F.S.K., Xu, W., and Xu, M.Q. (2003) Tensile strength and fracture of glass fiber-reinforced plastic (GFRP) plate with an eccentrically located circular hole. *Polymer Testing* **22**, 955–963.
Zetterberg, T., Astrom, B.T., Backlund, J., and Burman, M. (2001) On design of joints between composite profiles for bridge deck applications. *Composite Structures* **51**, 83–91.
Zhang, S.Y. (1999) Micro- and macroscopic characterisations of the viscoelastic fracture of resin-based fibre composites. *Composites Science and Technology* **59**, 317–323.

CHAPTER 9

Fatigue

Predicting Fatigue Limit and Fatigue Life

9.1 Introduction

Fatigue is by far the most common cause of mechanical failure in engineering components; the prevention of fatigue failure is a major preoccupation of designers in many industries, such as power generation and transport. When fatigue occurs in critical components (Colour Plate 4), the potential exists not only for economic loss but for loss of life. Fatigue cracks usually initiate at stress concentration features, and this is only to be expected because such features are almost inevitable in the design of engineering components. Even in those rare cases where the stress is constant throughout the part (e.g. cables and tie-bars loaded in pure tension) fatigue cracks will probably initiate from stress-concentrating defects such as inclusions or porosity. It follows then that the ability to predict the effect of stress concentrations on fatigue life and fatigue strength is crucial in engineering design; any improvements in prediction methods will inevitably pay major dividends in terms of more efficient design and reduced incidence of failure.

However, failure under cyclic loading conditions is a more complex phenomenon than failure under static or monotonic loading. From the start, one can see that there are many more parameters to take into account. Fatigue is dependant on the entire history of cyclic loading experienced by the component. In general, this will be a complex, variable-amplitude loading involving cycles of both high and low magnitude; even in the simple case of constant-amplitude cycles, one must consider not only the stress amplitude but also the mean stress of the cycle and, in some circumstances, the frequency. The problem is further complicated by the fact that fatigue is a multi-stage process. We can divide the total fatigue life into a period of crack initiation followed by one of crack propagation; the initiation period is normally assumed to include a stage of short-crack growth. These different stages involve distinctly different mechanisms, controlled and affected in different ways by the loading history. Different stages may dominate in different circumstances.

Early work on fatigue, which goes back to the middle of the nineteenth century, defined the total life (i.e. the number of cycles to failure for an initially uncracked specimen) in terms of the stress range or strain range. With the advent of fracture mechanics the crack propagation stage began to be studied in greater detail. The 1960s and 1970s were decades of great advances in our understanding of the growth of long cracks, the realisation that short cracks behaved differently, and the discovery of the important mechanism of crack closure. Since that time, improvements have been made at a slower pace; work has concentrated on refining our knowledge of existing mechanisms and on developing numerical simulations which take advantage of modern computer technology.

There are many excellent textbooks on fatigue, such as Suresh's comprehensive work *Fatigue of Materials* and the very practical contribution of Stephens and Fuchs: *Metal Fatigue in Engineering* (Stephens and Fuchs, 2001; Suresh, 1998). Practical advice on state-of-the-art methods for designers can be found in works such as the ASM handbook *Failure Analysis and Prevention* (ASM, 2002) and the Society of Automotive Engineers' *Fatigue Design Handbook* (SAE, 1997). The basic approaches to characterising a material's fatigue behaviour – stress–life curves, strain–life curves and crack propagation curves – have been described earlier, in Chapter 1, and so will not be repeated here. In what follows, I will briefly summarise the well-known methods which are currently being used to assess stress-concentration features and then proceed to demonstrate the application of the TCD in this area.

9.1.1 Current methods for the fatigue design of components

Structures which already contain long cracks can be assessed using standard crack-propagation data of the type described in Chapter 1. The exact distinction between 'long cracks' and 'short cracks' will be explained below: suffice it to say that if a crack is long enough, its propagation rate, da/dN, will be a unique function of the applied stress intensity range ΔK and the stress ratio R; propagation will effectively cease altogether below a threshold, ΔK_{th}, whose value is also dependent on R. Such situations are relatively unusual, but include some important cases, such as aircraft fuselages and offshore oil rigs, in which frequent inspection allows relatively long cracks to be monitored without the risk of sudden failure. Even if inspection cannot occur during life, it may be used as part of quality-assurance procedures during manufacture of the component. If cracks are introduced during manufacture, then these can also be assessed using crack-propagation data, provided they are large enough to be classified as long cracks. Initial ΔK values can be calculated and kept below ΔK_{th}, or alternatively the total life of the crack up to failure can be estimated by integrating the propagation curve.

In practice, such situations are rare. Fatigue in most components starts from initially uncracked material, usually in the vicinity of a stress concentration feature. The growth of long cracks cannot be monitored in most components, either for practical and economic reasons (e.g. car components) or because critical crack lengths are too small (e.g. jet engines), so most designers concentrate on keeping the total life of the component above an acceptable level. Though there are some important exceptions, most loading histories involve more than one million cycles, and this implies that we must keep below the fatigue limit, essentially designing for infinite life. Important exceptions include occasional, high-amplitude cycles (e.g. start/stop cycles in engines) which may dominate over more frequent, low amplitude cycles.

Given a component which does not already contain a long crack, there are three commonly used methods for predicting fatigue life: the stress-life method; the strain-life method and fracture mechanics. The stress-life method simply involves estimating the local stress amplitude and R ratio in the region of the feature where stresses are highest (e.g. the stress at the root of a notch) and estimating the total number of cycles to failure, N_f, using data in the form of an S/N curve generated from standard test specimens. The strain-life method uses the same approach, except it is the strain at the notch which is used: this method is often called the 'local strain approach'. If applied stresses are fairly low, within the elastic deformation regime of the material, then these two methods will be identical; in practice the stress-life method is then normally used. The strain-life method is used at higher local stress levels, where plastic strain is occurring on every cycle and, inevitably, N_f is relatively low (usually less than 10^4 cycles). Both of these methods are based on the assumption that material in the region close to the stress concentration feature will behave identically to material in a standard test specimen; this is true for relatively large, blunt features, but if the notch is sharp, or small, then these methods encounter major errors as we shall see below. Many companies, using these methods in conjunction with FEA, have discovered that they greatly underestimate the fatigue lives of real components.

The alternative method for assessing notches and other such features is a fracture mechanics approach, taking account of crack propagation in both the short- and the long-crack stages. An initial crack length must be assumed in order to calculate a stress intensity; some methods also estimate the number of cycles needed for this initiation phase using the local strain approach (Dowling, 1979), whilst others assume that this is negligible if the size of the initial crack is chosen to be small enough. Correction factors are incorporated to allow for the different rates of propagation that occur for short cracks, of which more will be said later in this chapter. This approach, though it is computationally more complex, can give good estimates in some circumstances, especially those very sharp notches for which the stress-life and strain-life methods have such difficulty. These methods are still rarely used in engineering practice, because they require extensive numerical modelling, and because there is still much disagreement on how to simulate the initiation and short-crack propagation phases. An alternative fracture mechanics approach is to assume that the notch is itself a crack; this method, which is appropriate for sharp notches, will be discussed in more detail below.

9.1.2 Crack closure

The discovery of the phenomenon of crack closure marked a turning point in our understanding of the mechanisms of fatigue crack growth (Elber, 1970, 1971): no discussion of this subject would be complete without mention of crack closure, which is now incorporated explicitly into some methods of life prediction. However, a complete description of the phenomenon is beyond the scope of this chapter, and the reader is referred to any of the excellent textbooks on fatigue which are available; what follows is a brief explanation.

A crack, subjected to cyclic loading, will open and close: the crack opening displacement δ, defined as the distance between the two crack faces, will vary cyclically in phase with the applied stress. Obviously, if the loading cycle includes a period of negative (i.e. compressive) applied stress, then during this period δ will be equal to zero as the crack faces are pressed together. It turns out, however, that δ can also be zero at other times, even when the applied stress is positive. This can occur for various different reasons, the most obvious of which is if some foreign material (such as corrosion debris or a viscous fluid) enters the crack whilst it is open. Oxide layers forming on the crack faces can also cause closure. Less obviously, closure can be induced by the residual stresses set up in the plastic zone of the crack. Closure can be measured directly by microscopic techniques, but this is difficult; usually it is inferred by measuring remote strains (e.g. deflections at the crack mouth or strains on the specimen face opposite the crack) and watching for non-linearities which mark the start and finish of the closure phase.

Crack closure plays a major role in several aspects of fatigue crack growth, including the effect of R ratio (there is less closure at higher R) and short cracks (which display less closure than long cracks for the same cyclic stress intensity). The effect of closure is to reduce crack propagation rates and increase threshold values. A simple view of the situation is that the crack does not really exist during the closure phase, because stresses can be transmitted across the crack faces. This view leads to the definition of an 'effective' (or 'closure-free') value of the stress-intensity range, which is that part of the cycle during which the crack is open. This concept is summarised in Fig. 9.1. This idea is clearly simplistic, because we know that fatigue behaviour is affected by the negative part of the cycle; otherwise, for example, two stress cycles at $R = 0$ and at $R = -1$ with the same maximum stress would have the same effect, when in reality the $R = -1$ cycle is invariably worse. However, it has been shown time and again that replotting data using effective quantities of stress and stress-intensity can account for many phenomena, including the effect of R, the short-crack effect, the role of overloads in variable-amplitude loading and some effects of heat treatment and of corrosive environments.

The following section, which constitutes the bulk of this chapter, considers the use of the TCD, and some other methods, for predicting the fatigue limits of bodies containing notches or cracks. The approach will be similar to that used in the preceding chapters: experimental evidence will be presented to show the success of the TCD when applied to a variety of different materials and features. Practical aspects will be emphasised; any explanation or justification for the success of the theory will be postponed for a later chapter (Chapter 13) when it will be discussed along with experience gained by examining other types of failure. Subsequent sections in this chapter will consider the prediction of fatigue life in the high-cycle and low-cycle regimes, under conditions of variable-amplitude loading and in non-metallic materials. A final section will consider some other more recently developed theories, which approach the problem of fatigue prediction in quite different ways.

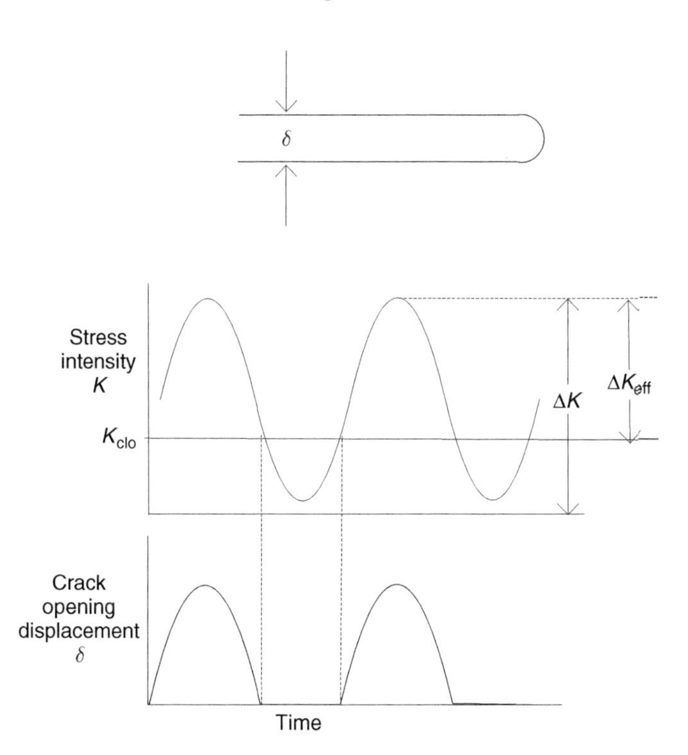

Fig. 9.1. Typical variation of crack opening displacement, δ, with time during a sinusoidal cycle of K. Even though K is always positive, crack closure occurs when $K < K_{clo}$, allowing us to define an effective stress intensity range, ΔK_{eff}, less than the normal applied value ΔK.

9.2 Fatigue Limit Predictions

The aim of this section is to predict the loading conditions under which fatigue failure will not occur, for a given stress-concentration feature. The basic terminology used has already been described in the section on fatigue in Chapter 1. A precise definition of the term 'fatigue limit' is difficult, because some materials do not show clear asymptotic behaviour on the S/N curve. In what follows, we will define the fatigue limit as the value of cyclic stress range corresponding to a specific number of cycles to failure in the range 10^6–10^7. Experience shows that the exact choice of N_f, within this range, is not important, but that it is important to be consistent, for example to use the same value when comparing data from two different notches in the same material. Likewise the fatigue crack propagation threshold ΔK_{th} is, in principle, the range of cyclic stress intensity at which an existing crack will not propagate, but in practice it is usually defined at values of the propagation rate (da/dN) between 10^{-8} and 10^{-7} mm/cycle. Both of these properties are affected by the stress ratio R. The fatigue limit for a specimen containing a notch will be denoted by $\Delta\sigma_{on}$ and will, unless otherwise specified, refer to the nominal stress applied to the gross section of the specimen.

9.2.1 Notches

Figure 9.2 shows some typical data on the effect of notch root radius on fatigue limit, in this case for a 0.15% carbon steel tested in tension-compression loading at $R = -1$ (Frost et al., 1974). Circular bars were used with a circumferential notch of depth $a = 5$ mm. The general appearance of the data is very similar to the results presented in previous chapters (Chapters 5–7) which showed the change in measured K_c as a function of root radius for notches in various materials. The fatigue limit is constant below a critical value of ρ, in this case about 1 mm, and rises steeply thereafter. The figure shows predictions made using the PM and LM. These, and most of the succeeding predictions of notched-specimen behaviour, have been made using finite element models of the specimens, to ensure precision in the stress analysis. In some cases it would be acceptable to use the equations of Creager and Paris to estimate stresses near the notch, as was done in previous chapters (see, for example, Section 5.2.2); one can simply replace the monotonic properties σ_o and K_c with the corresponding cyclic ones $\Delta\sigma_o$ and ΔK_{th}. However, this will not be sufficiently accurate for most fatigue specimens, especially if the notches are relatively blunt, or relatively large compared to specimen width.

The PM and LM predictions shown on the figure have been made using known values of $\Delta\sigma_o$ and ΔK_{th} for this material, so no best-fit procedure was needed. It is evident that the PM prediction describes the data extremely well. Note that the value of critical stress range used here was the plain specimen fatigue limit, $\Delta\sigma_o$, and therefore L is calculated as follows:

$$L = \frac{1}{\pi}\left(\frac{\Delta K_{th}}{\Delta\sigma_o}\right)^2 \tag{9.1}$$

Therefore we may draw a parallel between the fatigue-limit behaviour of this metal and the brittle fracture behaviour of ceramic materials (Chapter 5) and of fibre-composites (Chapter 8): in both cases the characteristic strength parameter was equal to the measured

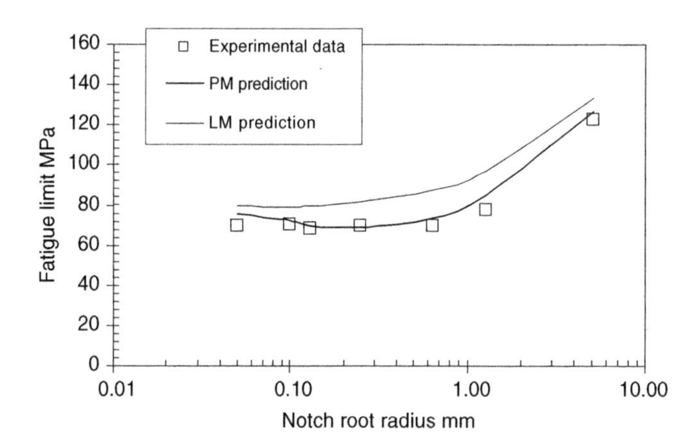

Fig. 9.2. Data on the effect of notch radius on fatigue limit (Frost et al. 1974) for carbon steel tested at $R = -1$; predictions using the PM and LM.

plain specimen strength. By contrast, we saw that to obtain predictions of brittle fracture in polymers and metals (Chapters 6 and 7), a higher strength value was needed.

The LM slightly over-predicts the experimental data, though the difference is only about 20%, which is within the limits of error in the experimental measurement of fatigue limit and the construction of the FE model, so this prediction is also acceptable. Figure 9.3 shows the same data plotted in a different way, using the stress concentration factor of the notch, K_t. In this plot it is possible to add the plain specimen result as well. Two other prediction lines are shown, which together constitute the approach used by Smith and Miller for the prediction of notched fatigue limit (Smith and Miller, 1978). These workers, building on earlier work by Frost (Frost, 1960; Frost and Dugdale, 1957), noticed that the fatigue limit of the notched specimens ($\Delta\sigma_{on}$) could be predicted at low K_t using the following equation:

$$\Delta\sigma_{on} = \frac{\Delta\sigma_o}{K_t} \tag{9.2}$$

This amounts to saying that fatigue failure occurs if the stress range at the root of the notch is equal to the plain specimen fatigue limit. At high K_t, where the value of $\Delta\sigma_{on}$ becomes constant, Smith and Miller noted that a prediction could be made simply by assuming that the notch is a crack of the same length, thus:

$$\Delta\sigma_{on} = \frac{\Delta K_{th}}{F\sqrt{\pi a}} \tag{9.3}$$

Here F is the geometry correction factor for the particular crack; for the data of Figs 9.2 and 9.3, this is a circumferential crack of the same length, a, as the actual notch. This equation gives a horizontal line on the figure since a is constant; being an LEFM

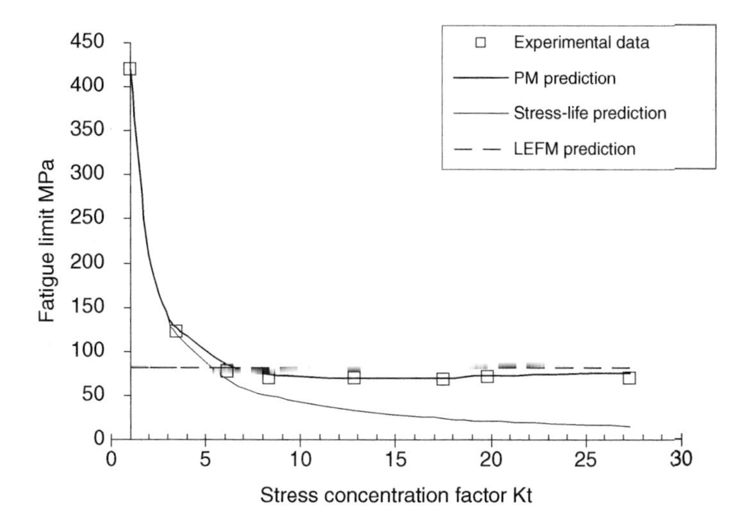

Fig. 9.3. The same data and predictions as in Fig. 9.2 (plus the plain-specimen data) replotted using the notch K_t factor. Predictions using the PM, LEFM and stress-life methods.

prediction, it will necessarily be correct as K_t approaches infinity, when the notch will become a sharp crack. As Fig. 9.3 shows, the overall prediction using these two equations is very good, though it is interesting to note that the data points fall slightly below the LEFM line implying that notches of finite size are slightly worse than cracks; this small effect is predicted by the PM.

The approach of Smith and Miller accords very well with the known fact that there are two different mechanisms of fatigue failure. Blunt notches tend to fail in much the same way as plain specimens, but in sharp notches a different phenomenon occurs, that of the so-called 'non-propagating cracks' (Frost, 1960). Cracks are found to initiate very easily from sharp notches, but may stop after a small amount of growth, usually less than 1 mm. For these notches the fatigue limit is defined by the ability of this small notch-root crack to propagate. The use of Eq. (9.3) approximates this condition provided the original notch length is considerably larger than the length of the non-propagating crack. For this reason the Smith and Miller method does not work well for small notches, as we shall see below. Another practical problem with the method is its application to stress concentration features which are not notches, for example component features such as corners, bends, keyways and so on. In that case, Eq. (9.3) cannot be formulated because a and F are not defined. I found a way to overcome this problem a few years ago, using the elastic stress field ahead of the feature to define an equivalent K by employing a modelling approach. This method, which I called the Crack Modelling Method, allowed predictions to be made of the fatigue limits of stress concentrations of any shape, provided they came within the category of 'sharp notches' (Lawless and Taylor, 1996). Several engineering components have been successfully analysed using this approach (Taylor, 1996; Taylor et al., 1997).

Returning to the phenomenon of non-propagating cracks, some workers have attempted to predict their behaviour more precisely, using the resistance curve approach. This method recognises the fact that non-propagating cracks are short cracks, whose propagation behaviour cannot be defined by normal LEFM methods. More will be said about short cracks below, in Section 9.2.3; one feature of these cracks is that ΔK_{th} is not constant but rather increases with crack length (see Fig. 9.4). The applied ΔK for the crack is also increasing as the crack grows as can also be represented by a line on this figure. If the two lines cross, then a non-propagating crack will occur. The fatigue limit condition is that in which the two lines are tangential at some point. Various workers have used this approach to predict fatigue limits of notched specimens and the lengths of non-propagating cracks; the main difficulty with the method is that it is very sensitive to the exact shape of the ΔK_{th} line, which is usually not known with any great accuracy.

Figure 9.5 shows another example of experimental data from Frost et al. (1974), in this case for notched samples of steel loaded in rotating bending. This was the same material as shown in Fig. 9.2 but heat-treated to give a higher strength, which reduced the L value from 0.3 to 0.05 mm. Again the PM and LM give acceptable predictions. Figures 9.2 and 9.5 are two examples of data collected as part of a large validation exercise which we conducted, using results from our own laboratories and from published literature (Taylor and Wang, 2000). The full details of this work have been published elsewhere: Table 9.1 summarises our findings, from the examination of a total of 47 fatigue-limit results, spanning a wide range of different materials, different notches, loading types,

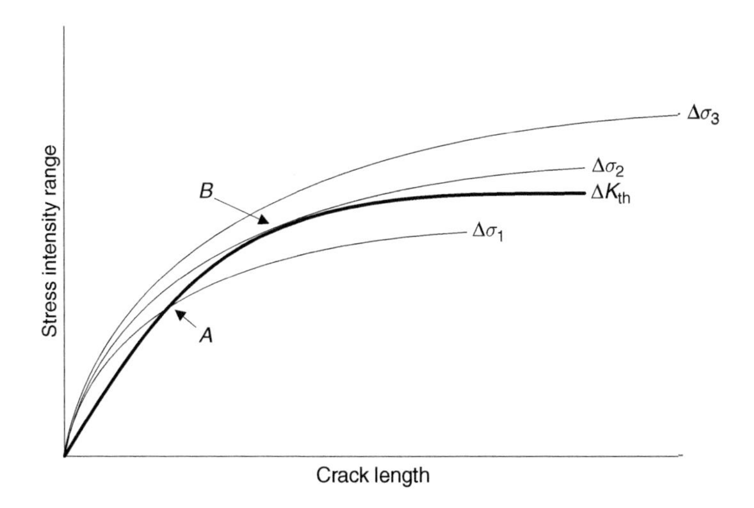

Fig. 9.4. The resistance curve approach. Threshold stress intensity ΔK_{th} is a function of crack length as shown by the thick line. The thin lines show how ΔK increases with crack growth from the notch for three different stress ranges: $\Delta\sigma_1$, $\Delta\sigma_2$ and $\Delta\sigma_3$. At $\Delta\sigma_1$ the crack initially grows but stops when the lines cross at point *A*, creating a non-propagating crack. At $\Delta\sigma_2$ the stress intensity remains just above the threshold, so this corresponds to the fatigue limit situation: point *B* gives the maximum possible length for a non-propagating crack.

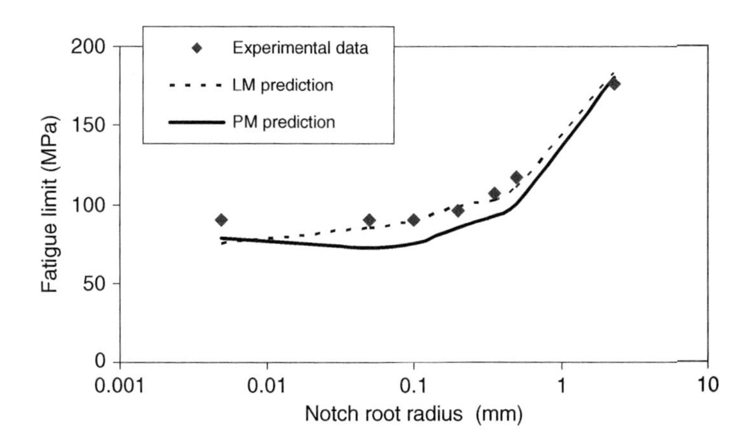

Fig. 9.5. Further experimental data from Frost et al. (1974), with PM and LM predictions.

and *R* ratios. It was felt that any prediction method could be defined as successful if the predicted fatigue limit was within 20% of the experimental value, since errors of at least 10% arise in both the experimental measurement and the stress analysis. On this criterion, the PM was found to be the best method, with a 94% success rate (which rose to 100% if we widened our accuracy criterion to 30% error); the LM was also found to be very good but with a slight tendency to over-estimation which reduced its accuracy somewhat. The Smith and Miller method was very successful for large notches

Table 9.1. Summary of the verification exercise described by Taylor and Wang (2000), showing the percentage of all results which fell within a specified error

Method of prediction	Percentage falling within 20% error	Percentage falling within 30% error
Point method	94	100
Line method	81	100
Area method	92	100
Stress-life method	18	25
LEFM	44	53
Smith and Miller	42	56

(i.e. for notches having $a >> L$) but gave very poor predictions for short notches (see Section 9.2.2). Another similar study (Susmel and Taylor, 2003) also found that the TCD was very successful. Most materials had values of L in the range 0.1–1 mm, but some smaller values occurred in high-strength alloys, whilst cast irons displayed values up to 3 mm.

9.2.2 Size effects in notches

There is ample evidence in the published literature to show that the effect of a notch on the fatigue limit is influenced by the absolute size of the notch. For a constant notch shape, decreasing the size increases the fatigue limit. Figure 9.6 shows an example of this effect, in data on steel specimens containing circular holes (DuQuesnay et al., 1986). We already saw a very similar phenomenon in the monotonic strength of composite materials: see, for example, Fig. 8.2. Both the PM and the LM are able to predict this effect. The value of L for this material was 0.13 mm. Knowing how the TCD works, we can anticipate the general dependence here: if the hole radius is much smaller

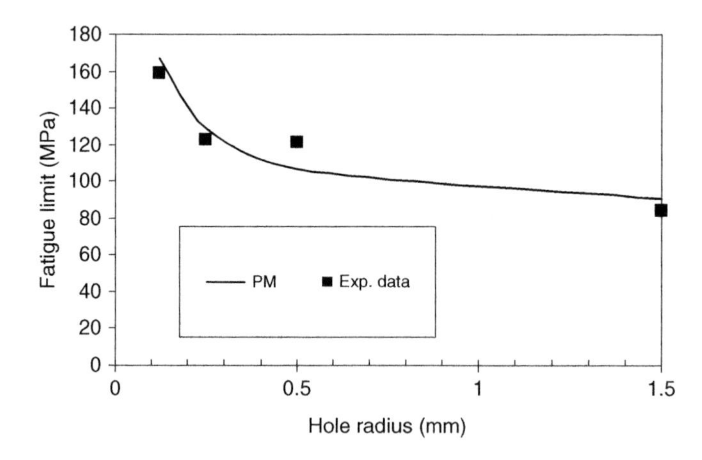

Fig. 9.6. Data from DuQuesnay et al. (1986) showing the size effect for circular holes in aluminium alloy 2024-T351. Predictions using the PM.

than L, we can expect the hole to have almost no effect since we will be sampling stresses which are, relatively speaking, a long way from the hole; in that case the fatigue limit will be the same as that of a plain specimen. This can be demonstrated experimentally: see, for example, the data from Murakami (2002) shown in Fig. 9.9. At the other extreme, a hole with a radius much larger than L will be expected to exert the full effect of K_t, reducing the fatigue limit by about a factor of 3.

Figure 9.7 shows a second example of this size effect, in this case for semi-circular notches of varying radius; data were taken from a study on a Cr–Mo steel containing circumferential notches and loaded in tension at $R = -1$ (Lukas et al., 1986); the value of L in this case was 0.24 mm. Also shown are the two prediction lines that make up the Smith and Miller method which was discussed in Section 9.2.1. It is clear that neither the stress-life prediction (Eq. 9.2) nor the LEFM prediction (Eq. 9.3) is suitable here. The prediction from Eq. (9.2) will approach the data only at very large radii, when $\rho \gg L$; Eq. (9.3) predicts much larger values of $\Delta\sigma_{on}$ because it does not take account of the short crack effect, which will be discussed in the next section. As an extreme example of the size effect, McCullough et al, testing an aluminium foam material, found no effect on fatigue limit for holes up to 4mm radius (McCullough et al., 2000). The value of L in that material is likely to be very large because the relevant microstructural dimension is the cell size, which was of the order of 1 mm. This size effect illustrates in a very simple and obvious way the need for a theory which contains a material length constant: it is clear that any approach which does not use such a parameter, for example the stress-life and strain-life methods, will inevitably predict fatigue limits which are independent of the absolute size of the notch. In most materials this effect becomes significant only for notches which are quite small, typically less than 1 mm in size, so its industrial importance will be mainly in the assessment of defects such as inclusions and pores, or in design features in very small components such as electronic circuits,

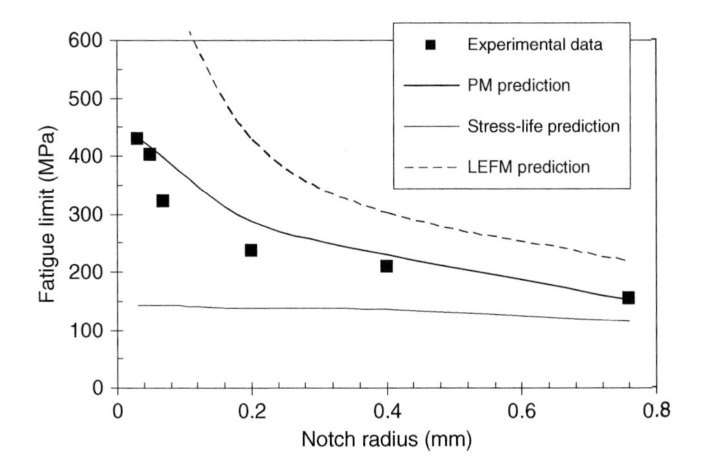

Fig. 9.7. The size effect demonstrated for specimens containing semi-circular notches. Data from Lukas et al. (1986): PM predictions compared with the two prediction lines used by the Smith and Miller method.

biomedical devices and microscopic MEMS devices. However, the larger values of L in cast irons will mean that features up to 10 mm in size will fall into this category.

Murakami has developed a method specifically for the analysis of small defects in steels and other metals: this method is now quite widely used (for an example of its industrial application, see Beretta et al., 1997), so it is useful to make some comparison between it and the TCD. In Murakami's method, the effective size of a defect is defined as the square root of the area which it projects on a plane normal to the stress axis (Fig. 9.8). The fatigue limit is described by the following equation:

$$\Delta\sigma_{on} = C/\left(\sqrt{area}\right)^{1/6} \tag{9.4}$$

This relationship is an empirical one, determined by consulting a large amount of experimental data. It is expected to be valid only within a certain range of sizes, since if the defect becomes very large it must behave like a macroscopic notch or crack in which case Eq. (9.4) will clearly not apply. At the other extreme, a point must be reached at which the defect becomes so small that it is no longer responsible for initiating the fatigue failure, other features in the material being larger. Murakami's method does not consider these extreme values: it is a practical tool designed to help engineers assess the kinds of defects which commonly occur in steels and other metallic materials. Figure 9.9 shows some typical experimental data from tests conducted on samples containing small blind holes drilled on the surface, for a 0.46% carbon steel, tested in two different heat treatments: annealed and quenched. Murakami's method predicts the data for the annealed material very well, and also the data for the smaller holes in the quenched material. Both the PM and the LM can predict this data fairly well, though the LM prediction (shown in the figure) is slightly better. The great advantage of these TCD predictions over Murakami's method is that they can be used for any hole size, including very small holes, and large holes, for which the fatigue limit will become asymptotic at values of the plain-specimen limit, or its value divided by 3, respectively.

Another feature which can be considered in the category of short notches is surface roughness. Any real surface is not perfectly flat but contains a series of undulations and marks as a legacy of the manufacturing process used to create the surface. The scale of these features may be less than 1 μm on highly polished surfaces, but more

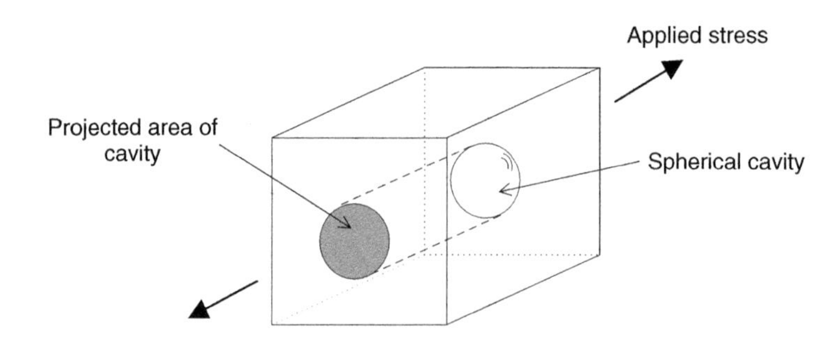

Fig. 9.8. Murakami's method uses the projected area of the defect in the plane normal to the stress axis (illustrated here for a spherical cavity).

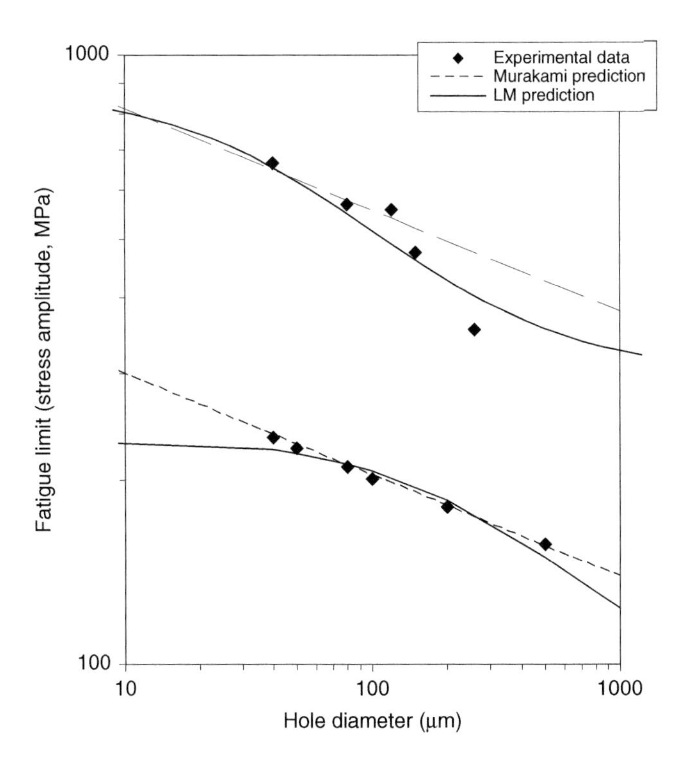

Fig. 9.9. Data and predictions for small surface holes in 0.46%C steel, in the annealed condition (lower data points) and as-quenched (upper points).

common machined surfaces contain grooves and indentations in the 10–100 μm range. The precise measurement of surface roughness is difficult but it can be thought of as a series of notches of varying size and shape. Typically the depth-to-width ratio of these notches will be large, giving quite low K_t factors of the order of 1.5 for typical machined surfaces. Previous work has shown that surface roughness features can be modelled as short cracks or short notches (Suhr, 1986; Taylor and Clancy, 1991) so it is likely that the TCD will be useful here.

9.2.3 Short cracks

In the 1970s, following the successful application of LEFM to the behaviour of long fatigue cracks, it was noticed that shorter cracks displayed anomalous behaviour; their growth rates were much higher than predicted from the Paris equation and growth could occur at stress intensity ranges less than the threshold (Pearson, 1975). In some cases, cracks were seen to slow down as they grew, passing through a minimum in growth rate or sometimes arresting. Figure 9.10 shows some typical early data (Lankford, 1982; Taylor and Knott, 1981). This was a subject of great concern because it meant that LEFM could not be used to assess the early growth of fatigue cracks – a very critical stage which often takes up the majority of the total life – nor could it be used to predict the effect of small crack-like defects such as manufacturing flaws. Since this early work, a huge amount of research has been carried out, but the problem has by no means been solved.

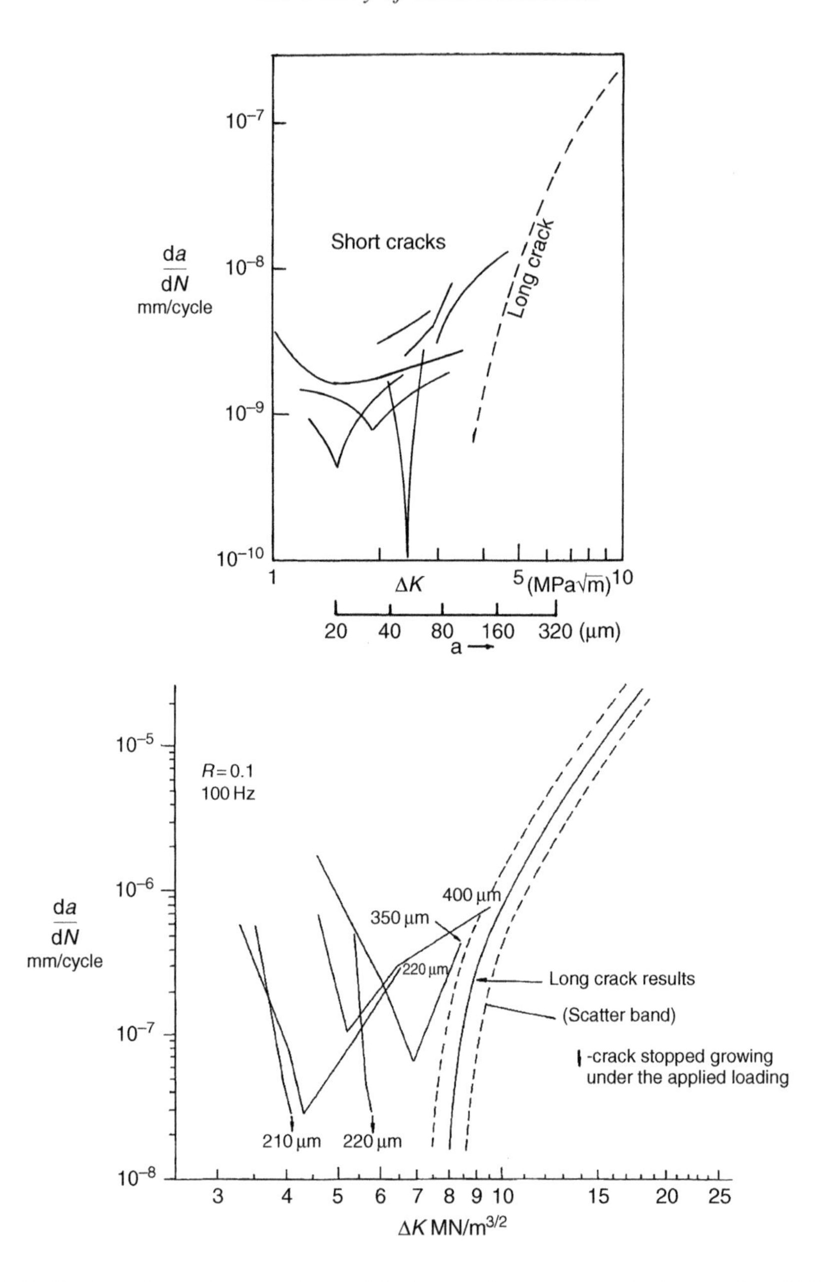

Fig. 9.10. Early results on short-crack growth behaviour. The upper graph, from Lankford (1982), shows data on an aluminium alloy; the lower graph shows data on aluminium bronze (Taylor and Knott, 1981).

A first step in making progress was to realise that this anomalous behaviour was being caused by several different mechanisms, working at different size scales. Very small cracks are strongly influenced by the local microstructure: for instance, in low-strength steels cracks tend to initiate at lengths less than the grain size (from inclusions, for example). These cracks grow relatively quickly at first, slowing down when they meet a grain boundary. In this case the grain boundary is clearly the principal barrier to growth, effectively determining the fatigue limit of the material. Miller has done a lot of work in this area (e.g. Miller and Akid, 1996): he has proposed that microstructural barriers of this kind operate in all materials. However, it is more difficult to identify the barriers in high-strength materials where the process happens at a much smaller scale.

Slightly larger cracks, in the size range 1–10 grain diameters, display lower thresholds than long cracks and faster growth rates for the same applied ΔK. There are several mechanisms at work at this size scale: microstructural barriers continue to exert an influence; crack closure is known to be less than that experienced by long cracks and the absolute stress levels tend to be high – often exceeding $0.7\sigma_y$, which implies that LEFM theory may be invalid. Many workers have tried to develop models which incorporate these various mechanisms, with limited predictive success. Practical solutions tend to be empirical, the most widely used one being that proposed by El Haddad et al., in which the normal LEFM equation (Eq. 9.3) is modified by the addition of a constant a_o (El Haddad et al., 1979a). This gives the following prediction for the fatigue limit of a specimen containing a crack, $\Delta\sigma_{oc}$:

$$\Delta\sigma_{oc} = \frac{\Delta K_{th}}{F\sqrt{\pi(a+a_o)}} \tag{9.5}$$

This can be recognised as a model of the 'imaginary crack' type, as described in Chapter 3; the approach can be stated in words as 'the crack behaves as if it were longer by an amount a_o and the laws of LEFM were applicable'. This method was advocated as a purely empirical approach a_o having no physical meaning. It was a found to be very successful for describing short-crack thresholds and growth rates and is still used extensively for that purpose. In fact the value of a_o is almost the same as that of our critical distance L, the difference being the geometry factor F; comparing Eqs (9.1) and (9.5) and letting $a = 0$, we find

$$a_o = L/F^2 \tag{9.6}$$

In the case where $F = 1$, it can be shown that predictions from this type of model become exactly the same as those from the LM (see Chapter 3): for other F values the two models will still agree at extreme values but will differ somewhat when a is of the same magnitude as L. Other workers have developed different equations to describe short crack behaviour; most of these, such as the model of Tanaka and co-workers, attempt to take account of the reduced amount of crack closure experienced by the short crack (Akiniwa et al., 1996).

Figure 9.11 shows a plot of $\Delta\sigma_{oc}$ as a function of crack length, including two straight lines corresponding to the long-crack LEFM prediction and the plain-specimen fatigue limit respectively. This kind of plot, which was pioneered by Kitagawa and Takahashi, is a useful way to represent short-crack effects, emphasising as it does that $\Delta\sigma_{oc}$ must inevitably deviate from the LEFM line, because logically it cannot rise above $\Delta\sigma_o$, and in fact must approach $\Delta\sigma_o$ as the crack length approaches zero (Kitagawa and Takahashi,

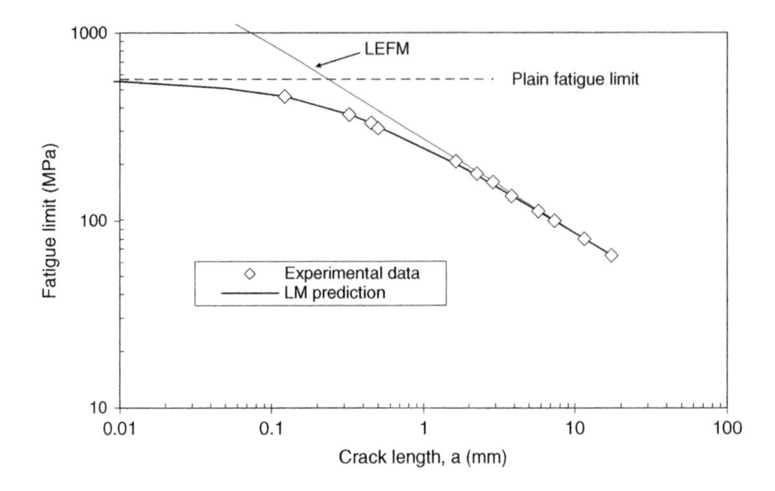

Fig. 9.11. Typical experimental data for the effect of crack length on fatigue limit, in this case for a G40.11 steel (El Haddad et al., 1979b). At low crack lengths the data deviate from the LEFM prediction towards the plain fatigue limit; the LM gives very accurate predictions.

1976). The behaviour of very short cracks (less than one grain in size) is difficult to represent on this type of plot (though some attempts have been made) because values of $\Delta\sigma_{oc}$ become hard to define at that size scale. Nevertheless the plot is a useful way to compare data and predictions for many cracks of practical lengths. The reader will recall that a very similar plot was used in studying the effect of crack length on brittle-fracture strength in ceramic materials (Section 5.2.1).

Figure 9.11 shows some typical experimental data (El Haddad et al., 1979b) along with predictions using the LM, which are very accurate; in this type of plot the PM predictions are almost identical except for lengths close to L at which they are about 5% higher than the LM predictions. In some materials (e.g. Fig. 9.12) the predictions tend to underestimate the results, in which case the data points fall in the region between the prediction lines and the two straight lines corresponding to $\Delta\sigma_o$ and the LEFM prediction (i.e. a prediction assuming constant ΔK_{th}). This tends to happen in higher-strength materials.

This problem, along with a possible solution, will be discussed in Section 9.2.5; however, it is worth pointing out that the prediction errors will never be very large in any case, since the upper bound of the data will always be the two straight lines. The largest difference between these lines and the LM line, for example, is only a factor of 1.4, occurring when $a = L$.

Figure 9.13 shows results from a validation exercise which we conducted, using short crack data from various different sources (Taylor and O'Donnell, 1994). The prediction error using the LM (defined as positive if the prediction was conservative) is plotted as a function of a/L. Errors found using the PM were of similar magnitude. Most errors are within the acceptable level of 20%, though there is an overall tendency for the predictions to be conservative and for the errors to be higher in high-strength steels,

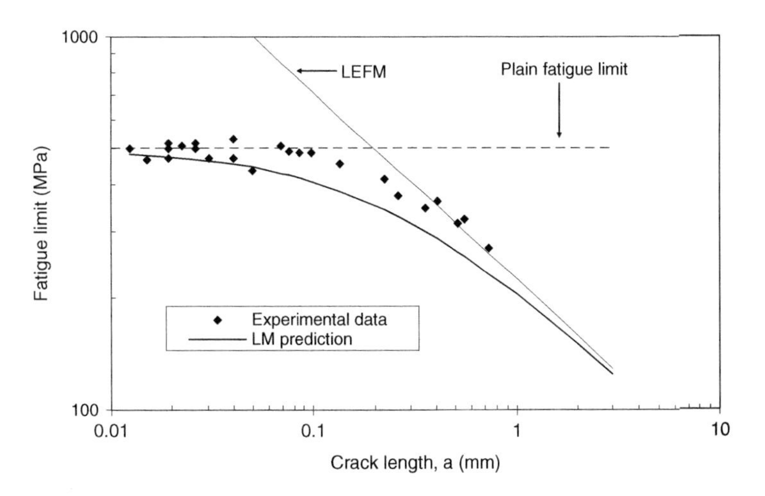

Fig. 9.12. In some materials, such as this Cr–Mo steel (Lukas et al., 1986), the data points lie above the LM prediction, closer to the two straight lines (LEFM and plain fatigue limit).

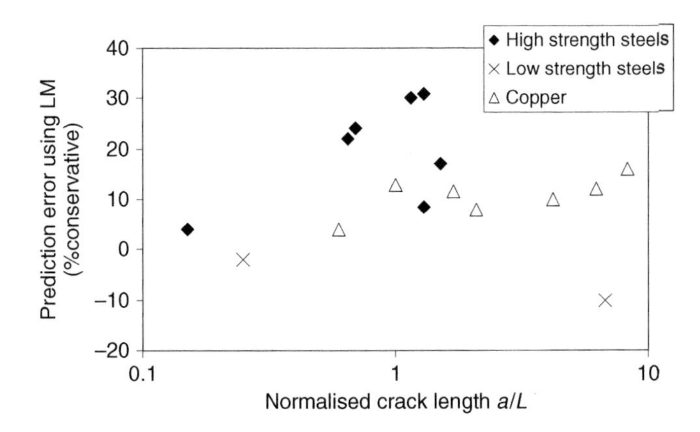

Fig. 9.13. Summary of predictions for the fatigue limits of specimens containing short cracks in various materials (Taylor and O'Donnell 1994).

especially near $a/L = 1$, reflecting the points made above. It should also be pointed out that this kind of data is strongly affected by the experimental technique used. We have concentrated on data which records the fatigue limit (i.e. minimum stress range to cause total specimen failure) for specimens containing small cracks or crack-like defects, introduced using methods which leave relatively little residual stress or other history effects. Cracks which have been grown at higher stress intensity levels and which have not been stress relieved will tend to show higher fatigue limits due to residual stresses causing larger amounts of closure. On the other hand, methods to determine short crack thresholds by recording the cessation of growth in a small crack subjected to constant $\Delta\sigma$ or ΔK (e.g. Tabernig and Pippan, 1998) tend to give smaller threshold values.

9.2.4 The effect of R ratio

The TCD is valid over a wide range of R ratios as shown, for example, in Fig. 9.14 which analyses data on notched specimens of cast iron (Taylor et al., 1996) for R ratios from -1 to 0.7. It might be expected that if L is a material property, then it should remain constant at different R ratios. However, we know that both $\Delta\sigma_o$ and ΔK_{th} change with R, and that the dependencies are different for these two parameters. The reduction in fatigue limit often approximates to the Goodman law, whereby $\Delta\sigma_o$ decreases linearly as the mean stress increases towards the UTS: Fig. 9.15 shows this dependence using some values which are typical for steel. The threshold, on the other hand, is often found to decrease linearly with increasing R up to some limiting value R^*, usually in the range 0.5–0.7, beyond which ΔK_{th} becomes constant; Fig. 9.15 shows some typical values, using $R^* = 0.5$. This behaviour can be argued on the basis of crack closure, assuming that the effective threshold is constant and that closure occurs below a fixed value of K in the cycle, irrespective of R. Using these values to calculate L, we find that it is indeed not constant but tends to decrease slightly with R up to R^* and to increase significantly thereafter (Fig. 9.15). It should be pointed out, however, that the trends in $\Delta\sigma_o$ and ΔK_{th} represented here are rather simplistic and are certainly not followed by all materials. Small departures from these trends can lead to large changes in the value of L.

Examining the experimental data, Atzori et al. have concluded that L tends to decrease with increasing $\Delta\sigma_o$ whether that change is caused by increasing R or by other means, such as changing the material or heat treatment (Atzori et al., 2005). Within this general trend they found quite a lot of scatter, concluding that it is not possible to estimate L accurately from $\Delta\sigma_o$ alone. It is instructive to look at data from two specific examples: a grey cast iron and a typical ferritic steel (Table 9.2). The cast iron shows no significant change in L even at quite high R ratios. The steel shows a large variation, with L decreasing by about a factor of 2 between $R = -1$ and 0.4. However, as Fig. 9.16 shows, we find that we can make accurate predictions of the fatigue limit (in this case for a notch of radius 0.16 mm, depth 3 mm) even if we assume a constant value of L. In this case the value used was that measured at $R = 0$. At first sight this seems surprising: the reason is that the notch-root stress is not a very strong function of distance, even

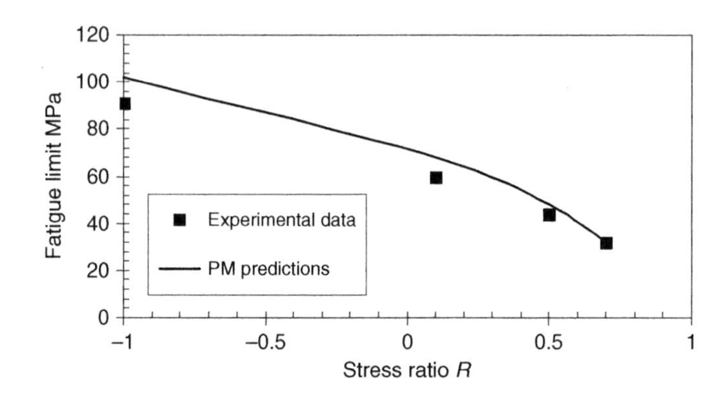

Fig. 9.14. Data from notched specimens of cast iron (Taylor et al., 1996); predictions using the PM, calculating L from $\Delta\sigma_o$ and ΔK_{th} at each R ratio.

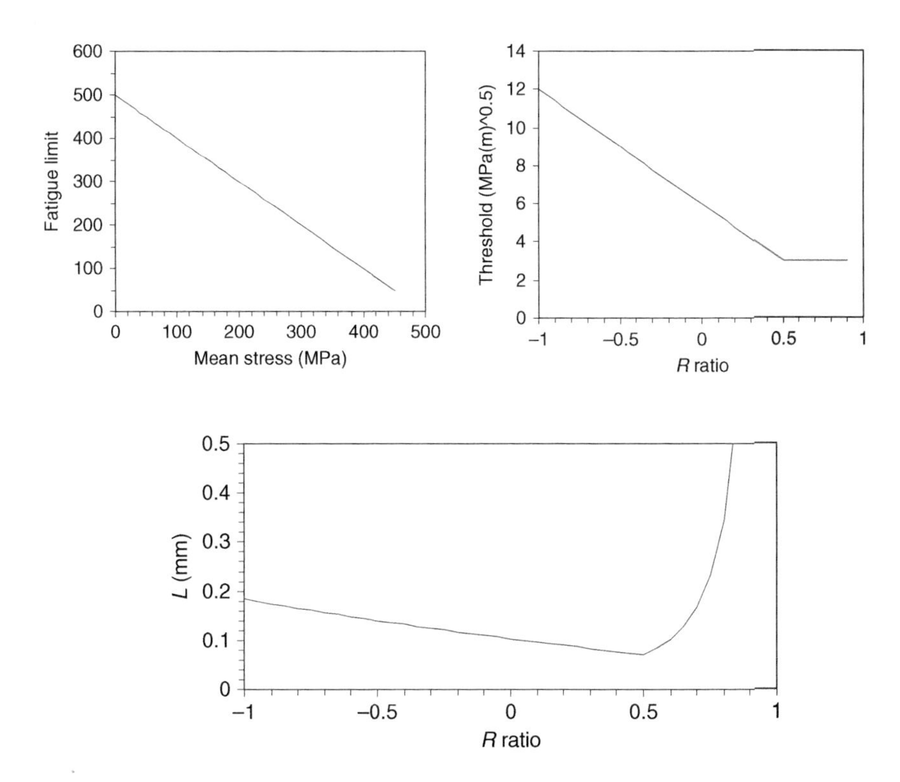

Fig. 9.15. Typical variation of fatigue limit with mean stress (top right) and threshold with R ratio (top left) and the resulting variation of the calculated value of L with R (bottom).

for sharp notches; so the value of the critical distance does not need to be known with great precision. Lanning et al applied an elastic–plastic stress analysis to their data on Ti-6Al-4V (Lanning et al., 2005). They showed that the critical distance lies outside the plastic zone of the notch at low R (therefore justifying a purely elastic analysis) but at high R the large monotonic plastic zone changes both stress range and R; when they

Table 9.2. Values of fatigue limit and threshold as a function of R for two materials (from Tanaka and Nakai, 1983 and Taylor et al., 1996): Calculated values of L

Material	R ratio	$\Delta\sigma_o$ (MPa)	ΔK_{th} (MPa(m))$^{1/2}$	L (mm)
Steel SN41B	−1	326	12.4	0.46
	0	247	8.4	0.30
	0.4	244	6.4	0.22
Cast iron	−1	160	15.9	3.2
	0.1	99	11.2	4.1
	0.5	68	8.0	4.4
	0.7	48	5.2	3.7

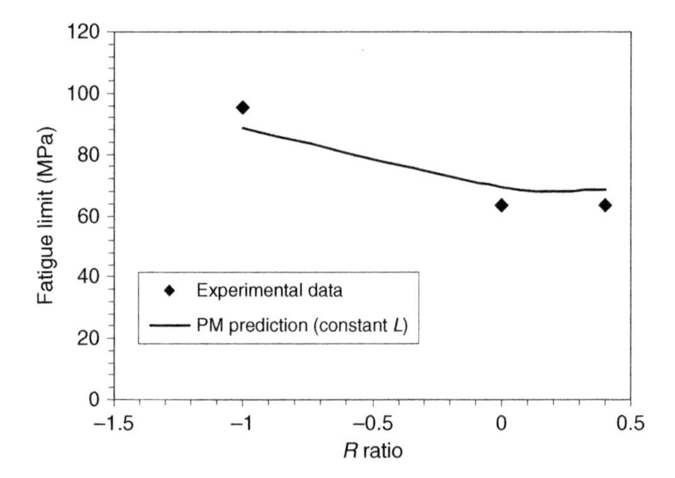

Fig. 9.16. Data on the fatigue limit of notched specimens of SN41B steel (Tanaka and Nakai, 1983) with predictions using the PM, with a constant value of L, equal to 0.3 mm.

took these effects into account they found that a constant value of L could be used with both PM and LM approaches.

9.2.5 Discussion on fatigue limit prediction

The above section has demonstrated that the TCD is able to give accurate predictions of the fatigue limits of specimens containing notches and cracks of all kinds in a variety of metallic materials. The relevant stress parameter to be used with the model is $\Delta\sigma_o$, the plain fatigue limit, and this allows predictions to be made for the whole range of stress concentration factors. Size effects, by which the fatigue limit increases towards $\Delta\sigma_o$ as feature size decreases, are also well predicted for both notches and cracks. All this suggests that the TCD may be useful in a very important aspect of industrial design – the prediction of long-term fatigue behaviour in engineering components – which will be addressed specifically in Chapter 12.

The accuracy of the TCD in this field should come as no surprise, because methods such as the PM and LM have been advocated for many years. As pointed out in Chapter 3, the first researchers to discover the TCD were Neuber and Peterson, whose work was well publicised during the 1950s and 1960s (Neuber, 1958; Peterson, 1959). However, at that time the methods could not be applied explicitly to components, because the stress fields near features could not be predicted accurately, FEA and suchlike computer methods being in their infancy. Therefore Neuber and Peterson used simplified stress analysis to develop formulae which could be used by the designers of the day. They relied on the fact that, for typical notches, the root radius is the most important factor controlling the stress field. So, knowing K_t and ρ for a given notch, one can make a reasonable approximation of the stress–distance curve. They assumed that other, more complex, stress-concentration features on components, such as corners and keyways, would perform in a similar way. Using this reasoning, Neuber converted the LM into the following equation, which defines an effective stress concentration factor for fatigue, K_f, as a function of K_t, ρ and a critical distance constant ρ^*:

$$K_f = 1 + \frac{K_t - 1}{1 + \left(\frac{\rho^*}{\rho}\right)^{1/2}} \tag{9.7}$$

Peterson, starting from the PM and using similar reasoning, developed the following equation using a critical distance constant ρ'.

$$K_f = 1 + \frac{K_t - 1}{1 + \frac{\rho'}{\rho}} \tag{9.8}$$

Equations with a similar form, though slightly different in detail, were developed by Siebel and Stieler, who used the local stress gradient as their controlling parameter (Siebel and Stieler, 1955), and later by Klesnil and Lukas, who used a model of the imaginary-crack type (Klesnil and Lukas, 1980). All of these equations work reasonably well if one knows the appropriate value of the length constant (ρ^* or ρ'), which is found by fitting predictions to the experimental data. They are still used to this day by many engineers, and can be found quoted in textbooks and datasheets, but most of the people who use them probably are not aware of the theoretical models on which they are based. Also the equations do have some limitations which the original theory does not: for example, Eqs (9.7) and (9.8) are clearly not applicable to sharp cracks; they would predict $K_f = 1$ when $\rho = 0$ which is clearly incorrect. Given that Neuber's and Peterson's equations are really approximations to the TCD anyway, it clearly makes more sense these days to use the PM and LM explicitly, taking stress data from FEA.

The link between the Neuber/Peterson methods and fracture mechanics, allowing L to be calculated explicitly as a function of the fatigue limit and threshold values (Eq. 9.1), was first made by Tanaka (1983), though Whitney and Nuismer had already applied the same theory to brittle fracture in composites (Whitney and Nuismer, 1974). The approach then seems to have lain dormant, at least as regards its use in fatigue studies, until its reintroduction by the present author (Taylor, 1999) and other workers (Atzori et al., 2001; Fujimoto et al., 2001; Kfouri, 1997; Lazzarin et al., 1997; Livieri and Tovo, 2004) in several groups working independently of each other.

The above work invariably assumes that the critical stress range is identical with the plain-specimen fatigue limit $\Delta\sigma_o$. It is worthwhile questioning this assumption, especially since we saw that for some cases of brittle fracture, in polymers and metals, a higher critical stress (and consequently lower L value) was appropriate (see Sections 6.2.1 and 7.2.1). In the case of fatigue it is clear that two different mechanisms operate – initiation and propagation – and that these dominate under different circumstances. For example, the high-cycle life of a plain specimen or blunt notch is dominated by the mechanism of initiation (including short-crack growth to the first grain boundary), whilst the behaviour of a sharply notched specimen is dominated by the propagation of an easily initiated crack. Our use of the TCD ignores this distinction since we make predictions for both blunt and sharp notches using the same theory. This suggests that there may

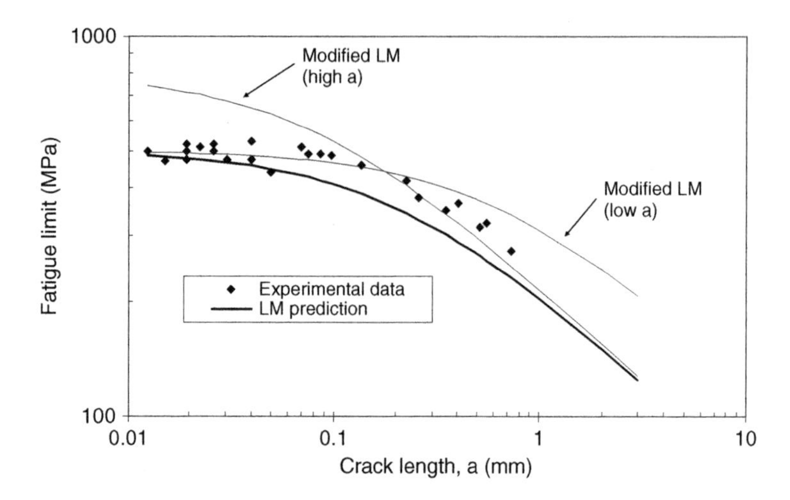

Fig. 9.17. Short-crack data as shown in Fig. 9.12; predictions using the normal LM (with a constant value of L) and also a modified approach which leads to two prediction lines, one valid at low a and one at high a.

be an argument for using two different sets of constants (i.e. different values of L and $\Delta\sigma_o$) to represent these two different mechanisms. The most obvious case where this might improve predictions is the short crack data, mentioned above, which in some cases lies above our TCD predictions. Figure 9.17 shows again the data of Fig. 9.12, which illustrates a case in point.

The normal LM prediction (which in this case uses values of $\Delta\sigma_o = 500\,\text{MPa}$, $\Delta K_{th} = 10\,\text{MPa(m)}^{1/2}$ and $L = 0.127\,\text{mm}$) distinctly underestimates the data around $a = L$. An accurate prediction can be obtained using two different prediction lines, one at relatively low crack lengths (keeping $\Delta\sigma_o$ equal to 500 MPa but increasing L to 0.4 mm) and another at higher crack lengths (keeping $\Delta K_{th} = 10\,\text{MPa(m)}^{1/2}$ but decreasing L to 0.05 mm which changes $\Delta\sigma_o$ to 798 MPa). It would be interesting to investigate the physical significance of these two different sets of constants in relation to the operative mechanisms. However, it is worth pointing out that the errors involved in using the normal TCD approach (with a single L value) are small in any case. Re-examining the data used for our validation exercise (Taylor and Wang, 2000), it was found that the optimum value of L (i.e. the one which minimised the total error in all the data) was very close to the normal value as calculated using Eq. (9.1): the difference was only a factor of 1.02. My colleague Luca Susmel considered whether it was better to use L or El Haddad's a_o (which is related to L through the geometry factor F, see Eq. 9.6): he found that L gave better predictions, the use of a_o leading to large errors in the range 40–80% when F was large.

A final point which merits some discussion is the relationship between the TCD and the phenomenon of crack closure. When we make predictions using the TCD we do not include the effects of crack closure, even though it is clear that this phenomenon does exist, that it exerts a strong effect, and that the degree of crack closure varies with crack length and R ratio. If closure occurs during the fatigue cycle, this will obviously affect the stress field near the crack tip; furthermore, closure is essentially a process which occurs in the region behind the crack tip, whereas when using the TCD we consider only the region in front of the crack or notch. Nevertheless, we can make very accurate predictions using the TCD as we do. It would seem that the two approaches are incompatible: either one 'believes' in the TCD or one 'believes' in crack closure. In this respect, it is useful to recall the data and analyses presented in earlier chapters, which showed that short cracks and notches exert similar effects on the static fracture strength of materials, despite the fact that the failure mechanisms which operate are very different. For example, the short-crack plot shown in Fig. 9.11 is very like that presented in Fig. 5.4 and elsewhere. It seems then that these effects are rather independent of the operative mechanisms: certainly there can be no crack closure at work in the case of the data of Fig. 5.4. And if the data appear similar, then obviously a method of prediction which works for one set of data will also work for the other, that is the TCD is an approach which is independent of the underlying failure mechanism.

That having been said, it is clear that if a mechanism affects the outcome, for example if it affects the fatigue limit, then this effect must be reflected in the values of the constants used in the prediction. To take the case of crack closure, this is known to proceed by several different mechanisms, which further complicates the picture, but one commonly occurring mechanism, known as roughness-induced closure, occurs due to roughness of the crack faces which is usually of the same order of magnitude as the grain size, due to deflections of the crack as it grows across grain boundaries. This, then, might explain why L values for fatigue in metallic materials are typically of the same order of magnitude as the grain size. Furthermore the presence of closure will certainly have an effect on the plain fatigue limit and long-crack threshold, thus influencing the other parameters used in the TCD.

9.3 Finite Life Predictions

So far we have considered only the fatigue limit, defined as the stress range giving a fatigue life of 10^6–10^7 cycles. Until recently, 10^7 cycles was a practical upper limit in fatigue test programmes and was generally assumed to be equivalent to infinite life. Recent work, however, has shown that fatigue failures can occur after much larger numbers of cycles, in excess of 10^9: in some materials the stress range required for these so-called 'ultra-long lives' is considerably lower than the conventionally defined fatigue limit. The mechanism of failure may be different: initiations often occur inside the specimen, at inclusions. At the time of writing, this is still an emerging field; the reader is directed to a recent book on this subject (Bathias and Paris, 2005). To my knowledge, the use of the TCD at these very long fatigue lives has not been investigated, though there seems to be no reason in principle why the theory should not be applicable.

As pointed out in Chapter 1, the conventional S/N curve, below 10^7 cycles, is divided up into two regimes: HCF and LCF. The dividing line occurs typically around 10^4 cycles, though the important difference is that HCF occurs under conditions of nominally elastic loading, whereas general yield occurs on every cycle during LCF. The situation is complicated in many materials by the existence of cyclic hardening and softening phenomena, which mean that, for example, though the first cycle may cause general yielding, the material's yield strength may increase on subsequent cycles, returning the specimen to the elastic regime. In notched specimens, as described earlier in this chapter, a crack may initiate easily in the essentially LCF conditions near the notch, but it may then slow down or even stop as it grows away from the notch. In the case of a specimen containing a sharp crack, the total fatigue life becomes the number of cycles needed to propagate the crack from some initial length a_i to a final length a_f. Thus, whilst LEFM cannot be applied to the case of LCF in a plain specimen, it may be applicable to short fatigue lives if a pre-crack is present.

Given that the TCD can be used to predict monotonic fracture (i.e. $N_f = 1$) in metals and other materials, can it also be used to predict finite fatigue lives, bridging the gap between monotonic failure and the fatigue limit? One can identify at least one potential difficulty here. Consider the case of a specimen containing a sharp crack, initial length a_i. The number of cycles to failure can be calculated assuming that the crack growth rate conforms to the Paris equation (see Chapter 1):

$$\frac{da}{dN} = A(\Delta K)^n \tag{9.9}$$

Integrating this equation between limits of a_i and a_f, and assuming for simplicity that $a_i >> a_f$, we obtain N_f as a function of the initial stress intensity range ΔK_i, as follows:

$$N_f = \frac{a_i}{A(\Delta K_i)^n \left(\frac{n}{2} - 1\right)} \tag{9.10}$$

Now in this case of a long, sharp crack, the two theories of LEFM and the TCD are exactly equivalent. So for two cracks of different lengths but having the same value of ΔK_i, the TCD would predict the same value of N_f. But Eq. (9.10) clearly shows that this is incorrect: the longer of the two cracks will in fact have a larger N_f, due to the term a_i in the equation. The physical meaning of this is that, for a relatively long crack, a small amount of growth will not change ΔK very much, whilst for a shorter crack the same amount of growth will cause a larger increase in ΔK; therefore the smaller crack will accelerate away more quickly and fail sooner.

The behaviour of the different cracks will, however, coincide at the two extremes of the S/N curve: at the fatigue limit where behaviour is uniquely characterised by ΔK_{th}, and at monotonic fracture where K_c applies. For intermediate fatigue lives, errors will arise, proportional to a_i, in the estimation of N_f, and proportional to $a_i^{1/n}$ in the estimation of fatigue strength. This problem certainly bears more scrutiny, but it should be emphasised that a worst-case scenario has been chosen here. The errors may be significant when analysing sharp cracks with N_f values in the mid-range (e.g. 10^3–10^4), but will be less important for notches and for larger or smaller N_f values. An interesting theoretical aspect of the problem is that it might be solved if more features of the stress–distance curve were considered, from which an estimate of crack length might be obtained. In practical situations the problem can be largely avoided by using data from test specimens in which the dimensions of the notch are similar in magnitude to those of component features to be analysed. The use of smaller notches will ensure conservatism in the predictions.

As a practical test of the use of the TCD for finite N_f, my colleague Luca Susmel has applied it to data from our laboratories on a plain carbon steel En3B and two sets of data taken from the literature, on SAE1045 steel and 2024 T351 aluminium alloy (DuQuesnay et al., 1986). He considered the medium and high-cycle ranges, from 10^4 to 10^7 cycles to failure. We can expect that the value of L in a given material will be different in monotonic fracture from its value at the fatigue limit, since there will be different mechanisms operating. It was assumed that L would change gradually with N_f according to a function which was chosen arbitrarily to have the following form:

$$L(N_f) = AN_f^{\ B} \tag{9.11}$$

Here A and B are two constants which were found by obtaining L values from the data, using stress–distance curves for a sharply notched specimen and assuming that the relevant stress parameter was the fatigue strength of plain specimens at the appropriate life. For the EN3B material the constants turned out to be $A = 67.4$, $B = -0.342$, giving a result for $L(N_f)$ in units of mm. This gave a value of L which changed from 0.27 mm at 10^7 cycles to 2.9 mm at 10^4 cycles. It is unlikely that this same relationship would apply at smaller numbers of cycles, since if extended to $N_f = 1$ we obtain an L value of 67.4 mm, which is very large. However, the relationship appears to work very well within the range of N_f values considered, as shown in Fig. 9.18 which compares the experimental fatigue lives in the three materials with predicted values using the LM.

Thus it seems that, despite the theoretical problem raised above, the TCD is capable of predicting finite-life fatigue behaviour in features typical of those found on engineering components. We have also used the TCD to predict the behaviour of an automotive component in this life range, as will be described in detail in Section 12.3. Clearly more work is needed in this area, particularly to extend this investigation to the LCF regime.

9.4 Multiaxial and Variable Amplitude Loading

So far we have considered only the simplest type of cyclic loading, in which the stress range and R ratio remain constant in time and the load regime is essentially tensile in character, whether applied through axial loading or bending. Real components experience much more complex loading patterns. In general, any element of the stress tensor can be present, and can vary with time in any fashion. Under these circumstances it becomes much more difficult to estimate the number of cycles to failure, indeed in some cases even the definition of the individual cycles of stress becomes problematic.

For convenience we can divide the general problem into two areas, though in practice there is much interaction between them. The first area is multiaxial loading, in which other elements of the stress tensor (in addition to the maximum principal stress) become important. For example, a material element may be subjected to torsion, creating local shear stress, or to mixtures of tension and torsion. This problem is considered in Chapter 11, in respect of both fatigue and static failure modes, where it is shown that

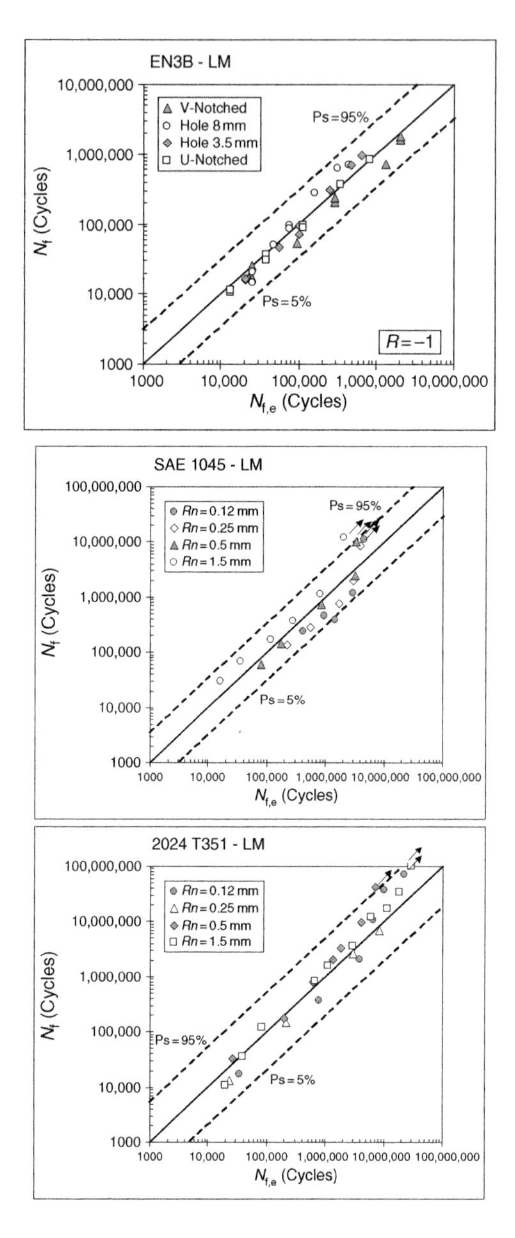

Fig. 9.18. Measured fatigue life N_f versus estimated value N_{fe}, using the LM, for three different materials (Susmel and Taylor, 2005). Predictions lie within a scatter band of a factor of 3 on life, which is comparable to the scatter in the original data.

the TCD can be successfully applied, using some existing methods by which these extra stress terms can be taken into account.

The second area is known as variable amplitude loading, by which we mean the imposition of stress cycles of different amplitudes, as opposed to the simpler, constant-amplitude loading. We will not consider variable amplitude loading in this book, because, to the best of my knowledge, no one has yet attempted to apply the TCD, or any similar approaches, in this area. In principle, I can see no reason why the TCD should not be used in situations of variable amplitude loading; a number of strategies exist for identifying stress cycles and taking account of the existence of cycles of different amplitudes, and I see no reason why these methods cannot be applied to consider stresses at a critical point, or averaged over a critical line. This would be a very interesting area for future study.

9.5 Fatigue in Non-Metallic Materials

Many non-metallic materials show fatigue behaviour which has similar characteristics to that of metals. In very brittle materials such as ceramics, the phenomenon of fatigue does occur, but the difference between the fatigue limit and the static strength can be small, giving only a small range of stress levels over which fatigue operates; however, the fatigue regime is considerably greater in some of the tougher ceramics which have been developed. Fatigue is common in many polymers and composites, where further complications arise such as effects of frequency and temperature which will not be discussed here.

Several studies exist to show that fatigue in composite materials can be predicted using the TCD, employing exactly the same approach as described above for metals. For example, McNulty et al. applied the PM to fatigue data from a ceramic composite material at elevated temperature, and Huh and Wang applied Whitney and Nuismer's model (which is identical to the TCD) to a carbon fibre–reinforced polymer composite containing circular holes (Huh and Hwang, 1999; McNulty et al., 2001). Good predictions were achieved in both cases.

On the other hand, the TCD has not previously been applied to fatigue problems in polymeric materials; indeed the whole question of the effect of notches has been only lightly treated; for example, Hertzberg and Manson's book *Fatigue of Engineering Plastics*, whilst being very comprehensive in other respects, contains only one example of the effect of notches on fatigue strength (Hertzberg and Manson, 1980). Tests in our own laboratories, on PMMA in the form of orthopaedic bone cement, revealed an interesting effect (Taylor and Hoey, 2006). We found that, whilst the TCD can be used successfully, the relevant stress range value is not the plain-specimen fatigue limit $\Delta\sigma_o$ but rather a higher value of approximately $2\Delta\sigma_o$. This is exactly the same factor which we found for the ratio σ_o/σ_u when predicting monotonic brittle fracture in this material (see Section 6.2.2). It is not clear if this will be a general feature of fatigue in polymers, though it is also evident in some data on PMMA reported in Hertzberg and Manson. The phenomenon certainly merits further investigation.

One interesting consequence is the prediction, as in the case of brittle fracture, of non-damaging notches. We predicted that notches with K_t factors less than or equal to 2 would have no effect on fatigue life, and this was borne out by tests using small hemispherical surface notches and large, blunt notches with $K_t = 1.5$ (see Fig. 9.19). Such behaviour never occurs in the HCF of metals. We also noticed that defects in the form of millimetre-sized spherical pores did not preferentially act as failure sites.

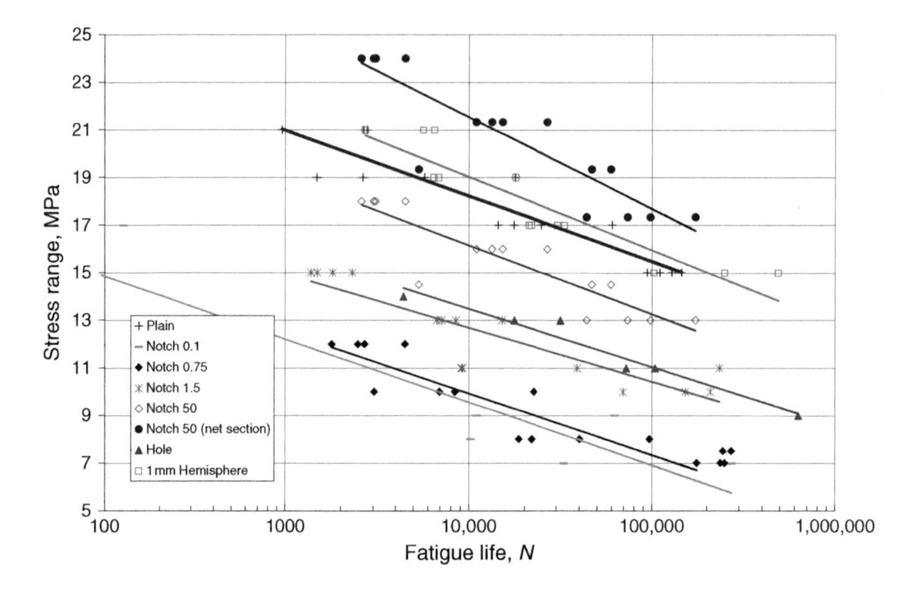

Fig. 9.19. S/N curves for PMMA bone cement containing various notches: the number in the legend indicates the root radius. Gross section stress is used, with the data for the blunt notch ($K_t = 1.5$) replotted using net section stress. Note that the blunt notch and the hemispherical notch do not reduce fatigue life, compared to that of the plain specimen, when the net section stress is accounted for.

This finding is potentially very important because bone cement fails in service due to fatigue, and much effort is being expended in devising techniques to reduce the number and size of pores and other defects in the material (James et al., 1992).

Ceramic materials also display fatigue failures; in the more brittle ceramics the fatigue regime is small, that is the HCF strength is only slightly lower than the static strength, so fatigue is not a practical problem. However, some of the tougher engineering ceramics show appreciable fatigue regimes. Another phenomenon present in many ceramic materials is the so-called 'static fatigue' – environmentally induced slow crack growth – which can be a significant problem. Unfortunately, very little work has been done on the effect of notch geometry on fatigue in ceramic materials, so there is no significant database from which we might attempt to make predictions. Sonsino, in a recent review article on fatigue design in structural ceramic parts, suggested that notches always exert their full theoretical K_t factors and that there were no effects of stress gradient or stressed volume to consider (Sonsino, 2003). However, the data presented only covered K_t factors between 1 and 2.2 and did not consider physically small defects. Given our findings with regard to the static strength of ceramics (Chapter 5) and the fatigue behaviour of metallic materials, it is highly likely that notch gradient effects will play a role in ceramics, for notches and defects whose sizes and/or root radii are sufficiently small. This would be a very interesting area for future research, which could have significant industrial value.

9.6 Other Recent Theories

At this point it is appropriate to mention some other recent theoretical developments in the field of fatigue life prediction for notched and cracked bodies. Models of the 'introduced-crack' type, in which fatigue behaviour is predicted by assuming that a crack is always present, have been advocated by several workers: examples are Chapetti et al., who related crack size to multiples of grain size (Chapetti et al., 1998) and Ostash and co-workers, for whom the physical meaning of the introduced crack is a surface layer of material having a lower yield strength (Ostash and Panasyuk, 2001; Ostash et al., 1999). These models are slightly different from the 'imaginary crack' models mentioned above, such as El Haddad (for short cracks) and Klesnil and Lukas (for notches), because an actual physical crack is assumed to be present; see Chapter 3 for more discussion of the theoretical issues involved.

Several approaches focus on the process zones ahead of the crack; in fatigue, two plastic zones occur: a monotonic zone related to the spread of plasticity at the maximum stress in the cycle, and a smaller cyclic plastic zone which remains even at the minimum stress. The near-threshold region of the growth-rate curve is associated with a different type of fracture mechanism, known as 'structure-sensitive growth' in which ductile facets form as the crack grows across each separate grain. Irving and Beevers showed that this behaviour occurred in long cracks when the cyclic plastic zone size was similar to the grain size (Irving and Beevers, 1974); Usami and Shida used equivalence of cyclic plastic zone size, combined with an elastic–plastic analysis, to predict short-crack thresholds (Usami and Shida, 1979). Another theoretical model of this general type is the approach of Navarro and De Los Rios, in which the plastic zone ahead of the crack is treated as a pile-up of dislocations reaching from the crack tip to the next grain boundary (Vallellano et al., 2000a; Vallellano et al., 2000b). This model can predict short-crack behaviour very elegantly; it can also predict the effect of notches, though some difficulties are encountered in taking account of the notch stress field.

Pluvinage has made extensive use of two different methods. The first is the NSIF approach in which notches of zero root radius can be analysed using a modified form of LEFM (see Chapter 4). The second method of Pluvinage is a critical distance method rather like the LM but with two differences. First, an elastic–plastic analysis is conducted and this is used to define the critical distance, which is determined not as a material parameter but from the shape of the stress–distance curve. Secondly, the relevant stress parameter is calculated as a weighted average of distance over this critical region, using a special weighting function. Good predictions have been obtained using this model; a major disadvantage is that it will not be able to predict size effects, since it contains no fixed length parameter. Descriptions of the two methods and of the relationship between the second method and the TCD can be found in the following references (Adib and Pluvinage, 2003; Pluvinage, 1998).

Finally, a recent model by McEvily analyses the growth of short cracks by combining several concepts mentioned at various other places in this book (McEvily and Ishihara, 2001; McEvily et al., 2003). A fictitious radius is employed (after Neuber) to avoid the singularity at the crack tip (it was shown in Chapter 3 that this is equivalent to the use of the TCD); crack closure concepts are used to define an effective stress intensity range from which da/dN is estimated, allowing for the existence of a threshold, and elastic–plastic behaviour is taken into account (following Irwin) by adding a fictitious crack length equal to the size of the plastic zone radius.

9.7 Concluding Remarks

This chapter has demonstrated that there is a wealth of evidence to support the use of the TCD to predict the fatigue limit and HCF strength of notched specimens of metallic materials and composites. In Chapter 12, we will see that this can be successfully extended to stress concentration features on components. There is also extensive practical support for the use of the TCD at lower numbers of cycles to failure, at least down to 10^4. The TCD has not been attempted in the LCF regime, and one can envisage some theoretical problems arising, but the fact that the TCD can be applied to predict static failure in metals (Chapter 7) suggests that it may well be possible to bridge this gap. Currently our experience with polymeric materials is limited to a couple of studies on PMMA, which suggest that the TCD can be used, though with some modification to the critical stress range.

What of other mechanisms of long-term failure? One might expect that the TCD will also be valid for other failure modes which involve crack growth processes, especially under conditions of small-scale yielding. These would include stress-corrosion cracking as experienced by many metallic and polymeric materials, and gradual crack extension at constant load such as the 'stick-slip' process in brittle polymers (see Section 13.3). I have not been able to find reference to the use of the TCD for these failure modes: this is an interesting area for further study. Creep failure also involves cracking but the central role of plastic deformation here implies the need for a different approach. Critical distance concepts have been used in conjunction with strain-based models of creep crack initiation and growth (Yatomi et al., 2006; Zhao et al., 2006).

This chapter concludes with Table 9.3, which lists mechanical property values for a range of the materials to be found above and in the accompanying references.

Table 9.3. Values of L, $\Delta\sigma_o$ and ΔK_{th} for various materials at various R ratios

Material	L (mm)	$\Delta\sigma_o$ (MPa)	ΔK_{th} (MPa.m$^{1/2}$)	R	Reference
0.46% C steel quenched	0.023	1780	15.0	−1	(Murakami, 2002)
Stainless steel AISI 304	0.088	720	12.0	−1	(Susmel and Taylor, 2003)
0.46% C steel annealed	0.15	480	10.4	−1	(Murakami, 2002)
Steel SAE1045	0.166	608	13.9	−1	(DuQuesnay et al., 1986)

(Continued)

Table 9.3. Continued

Material	L (mm)	$\Delta\sigma_o$ (MPa)	ΔK_{th} ($MPa.m^{1/2}$)	R	Reference
Steel 15313 2.25Cr 1Mo	0.237	440	12.0	−1	(Lukas et al., 1986)
Mild steel 0.15%C	0.3	420	12.8	−1	(Frost et al., 1974)
Steel SM41B	0.458	326	12.4	−1	(Tanaka and Nakai, 1983)
	0.296	274	8.4	0	
	0.218	244	6.4	0.4	
S.G. cast iron	0.51	590	23.5	−1	(Taylor et al., 1997)
Grey cast iron	3.15	155	15.9	−1	(Taylor et al., 1996)
	4.07	99	11.2	0.1	
	4.41	68	8.0	0.5	
	3.74	48	5.2	0.7	
Aluminium alloy 7075 T6	0.028	428	4.0	−1	(Vallellano et al., 2003)
Al alloy L65	0.062	300	4.2	−1	(Susmel and Taylor, 2003)
Al alloy 2024 T351	0.129	248	5.0	−1	(DuQuesnay et al., 1986)
Al alloy AA356 T6	0.16	231	4.0	−1	(Susmel and Taylor, 2003)
Al alloy 6060 T6	0.4	109.6	6.1	0.1	(Susmel and Taylor, 2003)
Al alloy LM25	1.87	77.5	5.9	−1	(Taylor and Wang, 1999)
PMMA bone cement	0.1	28*	0.5	0.1	(Taylor and Hoey, 2006)

Note: All data refer to the Fatigue limit at 10^6–10^7 cycles to failure, except for the PMMA bone cement which was at 10^5 cycles.

* In this material, unlike all the metallic materials examined, the value of $\Delta\sigma_o$ which gave successful predictions, was different from the plain-specimen value, which was 15 MPa.

References

Adib, H. and Pluvinage, G. (2003) Theoretical and numerical aspects of the volumetric approach for fatigue life prediction in notched components. *International Journal of Fatigue* **25**, 67–76.

Akiniwa, Y., Tanaka, K., and Zhang, L. (1996) Prediction of fatigue thresholds of notched components based on Resistance-curve method. In *Fatigue 96* (Edited by Lutjering, G. and Nowack, H.) pp. 449–454. Pergamon, Oxford, UK.

ASM (2002) *Failure analysis and prevention.* ASM International, Ohio USA.

Atzori, B., Lazzarin, P., and Filippi, S. (2001) Cracks and notches: Analogies and differences of the relevant stress distributions and practical consequences in fatigue limit predictions. *International Journal of Fatigue* **23**, 355–362.

Atzori, B., Meneghetti, G., and Susmel, L. (2005) Material fatigue properties for assessing mechanical components weakened by notches and defects. *Fatigue and Fracture of Engineering Materials and Structures* **28**, 83–97.

Bathias, C. and Paris, P.C. (2005) *Gigacycle fatigue in mechanical practice.* Dekker, New York USA.

Beretta, S., Blarasin, A., Endo, M., Giunti, T., and Murakami, Y. (1997) Defect tolerant design of automotive components. *International Journal of Fatigue* **19**, 319–333.

Chapetti, M.D., Kitano, T., Tagawa, T., and Miyata, T. (1998) Fatigue limit of blunt-notched components. *Fatigue and Fracture of Engineering Materials and Structures* **21**, 1525–1536.

Dowling, N.E. (1979) Notched member fatigue life predictions combining crack initiation and propagation. *Fatigue of Engineering Materials and Structures* **2**, 129–138.

DuQuesnay, D.L., Topper, T.H., and Yu, M.T. (1986) The effect of notch radius on the fatigue notch factor and the propagation of short cracks. In *The Behaviour of Short Fatigue Cracks (EGF1)* (Edited by Miller, K.J. and delosRios, E.R.) pp. 323–335. MEP, London.

El Haddad, M.H., Smith, K.N., and Topper, T.H. (1979a) Fatigue crack propagation of short cracks. *Journal of Engineering Materials and Technology (Trans. ASME)* **101**, 42–46.

El Haddad, M.H., Topper, T.H., and Smith, K.N. (1979b) Prediction of non propagating cracks. *Engineering Fracture Mechanics* **11**, 573–584.

Elber, W. (1970) Fatigue crack closure under cyclic tension. *Engineering Fracture Mechanics* **2**, 37–45.

Elber, W. (1971) The significance of fatigue crack closure. In *Damage Tolerance in Aircraft Structures (ASTM STP 486)* pp. 230–242. ASTM.

Frost, N.E. (1960) Notch effects and the critical alternating stress required to propagate a crack in an aluminium alloy subject to fatigue loading. *Journal of Mechanical Engineering Science* **2**, 109–119.

Frost, N.E. and Dugdale, D.S. (1957) *Journal of the Mechanics and Physics of Solids* **5**, 182–192.

Frost, N.E., Marsh, K.J., and Pook, L.P. (1974) *Metal fatigue.* Oxford University Press, London.

Fujimoto, Y., Hamada, K., Shintaku, E., and Pirker, G. (2001) Inherant damage zone model for strength evaluation of small fatigue cracks. *Engineering Fracture Mechanics* **68**, 455–473.

Hertzberg, R.W. and Manson, J.A. (1980) *Fatigue of engineering plastics.* Academic Press, London.

Huh, J.S. and Hwang, W. (1999) Fatigue life prediction of circular notched CRFP laminates. *Composite Structures* **44**, 163–168.

Irving, P.E. and Beevers, C.J. (1974) Microstructural influences on fatigue crack growth in Ti-6Al-4V. *Materials Science and Engineering* **14**, 229–238.

James, S.P., Jasty, M., Davies, J., Piehler, H., and Harris, W.H. (1992) A fractographic investigation of PMMA bone cement focusing on the relationship between porosity reduction and increased fatigue life. *Journal of Biomedical Materials Research* **26**, 651–662.

Kfouri, A.P. (1997) Limitations on the use of the stress intensity factor, K, as a fracture parameter in the fatigue propagation of short cracks. *Fatigue and Fracture of Engineering Materials and Structures* **20**, 1687–1698.

Kitagawa, H. and Takahashi, S. (1976) Application of fracture mechanics to very small cracks or the cracks in the early stage. In *Second International Conference on Mechanical Behaviour of Materials* pp. 627–630. ASM.

Klesnil, M. and Lukas, P. (1980) *Fatigue of metallic materials.* Elsevier, Amsterdam.

Lankford, J. (1982) The growth of small fatigue cracks in 7075-T6. *Fatigue of Engineering Materials and Structures* **5**, 233–248.

Lanning, D., Nicholas, T., and Haritos, G.K. (2005) On the use of critical distance theories for the prediction of the high cycle fatigue limit stress in notched Ti-6Al-4V. *International Journal of Fatigue* **27**, 45–57.

Lawless, S. and Taylor, D. (1996) Prediction of fatigue failure in stress concentrators of arbitrary geometry. *Engineering Fracture Mechanics* **53**, 929–939.

Lazzarin, P., Tovo, R., and Meneghetti, G. (1997) Fatigue crack initiation and propagation phases near notches in metals with low notch sensitivity. *International Journal of Fatigue* **19**, 647–657.

Livieri, P. and Tovo, R. (2004) Fatigue limit evaluation of notches, small cracks and defects: An engineering approach. *Fatigue and Fracture of Engineering Materials and Structures* **27**, 1037–1049.

Lukas, P., Kunz, L., Weiss, B., and Stickler, R. (1986) Non-damaging notches in fatigue. *Fatigue and Fracture of Engineering Materials and Structures* **9**, 195–204.

McCullough, K.Y.G., Fleck, N.A., and Ashby, M.F. (2000) The stress-life fatigue behaviour of aluminium alloy foams. *Fatigue and Fracture of Engineering Materials and Structures* **23**, 199–208.

McEvily, A.J., Endo, M., and Murakami, Y. (2003) On the root-area relationship and the short fatigue crack threshold. *Fatigue and Fracture of Engineering Materials and Structures* **26**, 269–278.

McEvily, A.J. and Ishihara, S. (2001) On the dependence of the rate of fatigue crack growth on the Sna(2a) parameter. *International Journal of Fatigue* **23**, 115–120.

McNulty, J.C., He, M.Y., and Zok, F.W. (2001) Notch sensitivity of fatigue life in a Sylramic/SiC composite at elevated temperature. *Composites Science and Technology* **61**, 1331–1338.

Miller, K.J. and Akid, R. (1996) The application of microstructural fracture mechanics to various metal surface states. *Proceedings of the Royal Society of London* **452**, 1411–1432.

Murakami, Y. (2002) *Metal fatigue: Effects of small defects and nonmetallic inclusions.* Elsevier, Oxford.

Neuber, H. (1958) *Theory of notch stresses: principles for exact calculation of strength with reference to structural form and material.* Springer Verlag, Berlin.

Ostash, O.P. and Panasyuk, V.V. (2001) Fatigue process zone at notches. *International Journal of Fatigue* **23**, 627–636.

Ostash, O.P., Panasyuk, V.V., and Kostyk, E.M. (1999) A phenomenological model of fatigue macrocrack initiation near stress concentrators. *Fatigue and Fracture of Engineering Materials and Structures* **22**, 161–172.

Pearson, S. (1975) Initiation of fatigue cracks in commercial aluminium alloys and the subsequent propagation of very short cracks. *Engineering Fracture Mechanics* **7**, 235–247.

Peterson, R.E. (1959) Notch-sensitivity. In *Metal Fatigue* (Edited by Sines, G. and Waisman, J.L.) pp. 293–306. McGraw Hill, New York.

Pluvinage, G. (1998) Fatigue and fracture emanating from notch; the use of the notch stress intensity factor. *Nuclear Engineering and Design* **185**, 173–184.

SAE (1997) *Fatigue design handbook.* Society of Automotive Engineers, Warrendale USA.

Siebel, E. and Stieler, M. (1955) Dissimilar stress distributions and cyclic loading. *Z.Ver.Deutsch.Ing* **97**, 121–131.

Smith, R.A. and Miller, K.J. (1978) Prediction of fatigue regimes in notched components. *International Journal of Mechanical Science* **20**, 201–206.

Sonsino, C.M. (2003) Fatigue design of structural ceramic parts by the example of automotive intake and exhaust valves. *International Journal of Fatigue* **25**, 107–116.

Stephens, R.I. and Fuchs, H.O. (2001) *Metal fatigue in engineering.* Wiley, New York.

Suhr, R.W. (1986) The effect of surface finish on high cycle fatigue of a low alloy steel. In *The Behaviour of Short Fatigue Cracks (EGF1)* (Edited by Miller, K.J. and delosRios, E.R.) pp. 69 86. MEP, London.

Suresh, S. (1998) *Fatigue of materials.* Cambridge University Press, Cambridge UK.

Susmel, L. and Taylor, D. (2003) Fatigue design in the presence of stress concentrations. *International Journal of Strain Analysis* **38**, 443–452.

Susmel, L. and Taylor, D. (2005) The theory of critical distances to predict fatigue lifetime of notched components. In *Advances in Fracture and Damage Mechanics IV* (Edited by Aliabadi, M.H.) pp. 411–416. EC, Eastleigh UK.

Tabernig, B. and Pippan, R. (1998) Resistance curves for the threshold of fatigue crack propagation. In *Fatigue Design 98* pp. 127–134.

Tanaka, K. (1983) Engineering formulae for fatigue strength reduction due to crack-like notches. *International Journal of Fracture* **22**, R39–R45.

Tanaka, K. and Nakai, Y. (1983) Propagation and non-propagation of short fatigue cracks. *Fatigue and Fracture of Engineering Materials and Structures* **6**, 315–327.

Taylor, D. (1996) Crack modelling: a technique for the fatigue design of components. *Engineering Failure Analysis* **3**, 129–136.

Taylor, D. (1999) Geometrical effects in fatigue: a unifying theoretical model. *International Journal of Fatigue* **21**, 413–420.

Taylor, D., Ciepalowicz, A.J., Rogers, P., and Devlukia, J. (1997) Prediction of fatigue failure in a crankshaft using the technique of crack modelling. *Fatigue and Fracture of Engineering Materials and Structures* **20**, 13–21.

Taylor, D. and Clancy, O.M. (1991) The fatigue performance of machined surfaces. *Fatigue and Fracture of Engineering Materials and Structures* **14**, 329–336.

Taylor, D. and Hoey, D. (2006) The role of defects in the fatigue strength of bone cement. In *Proc. Fatigue 2006* Atlanta, USA.

Taylor, D., Hughes, M., and Allen, D. (1996) Notch fatigue behaviour in cast irons explained using a fracture mechanics approach. *International Journal of Fatigue* **18**, 439–445.

Taylor, D. and Knott, J.F. (1981) Fatigue crack propagation behaviour of short cracks: The effect of microstructure. *Fatigue of Engineering Materials and Structures* **4**, 147–155.

Taylor, D. and O'Donnell, M. (1994) Notch geometry effects in fatigue: A conservative design approach. *Engineering Failure Analysis* **1**, 275–287.

Taylor, D. and Wang, G. (1999) A critical distance approach which unifies the prediction of fatigue limits for large and small cracks and notches. In *Proc. Fatigue 99* pp. 579–584. Higher Education Press (China) and EMAS (Warley, UK).

Taylor, D. and Wang, G. (2000) The validation of some methods of notch fatigue analysis. *Fatigue and Fracture of Engineering Materials and Structures* **23**, 387–394.

Usami, S. and Shida, S. (1979) Elastic-plastic analysis of the fatigue limit for a material with small flaws. *Fatigue of Engineering Materials and Structures* **1**, 471–481.

Vallellano, C., Dominguez, J., and Navarro, A. (2003) On the estimation of fatigue failure under fretting conditions using notch methodologies. *Fatigue and Fracture of Engineering Materials and Structures* **26**, 469–478.

Vallellano, C., Navarro, A., and Dominguez, J. (2000a) Fatigue crack growth threshold conditions at notches. Part 1:Theory. *Fatigue and Fracture of Engineering Materials and Structures* **23**, 113–121.

Vallellano, C., Navarro, A., and Dominguez, J. (2000b) Fatigue crack growth threshold conditions at notches. Part II: Generalisation and application to experimental results. *Fatigue and Fracture of Engineering Materials and Structures* **23**, 123–128.

Whitney, J.M. and Nuismer, R.J. (1974) Stress fracture criteria for laminated composites containing stress concentrations. *Journal of Composite Materials* **8**, 253–265.

Yatomi, M., O'Dowd, N.P., Nikbin, K.M., and Webster, G.A. (2006) Theoretical and numerical modelling of creep crack growth in a carbon-manganese steel. *Engineering Fracture Mechanics* **73**, 1158–1175.

Zhao, L.G., O'Dowd, N.P., and Busso, E.P. (2006) A coupled kinetic-constitutive approach to the study of high temperature crack initiation in single crystal nickel-base superalloys. *Journal of the Mechanics and Physics of Solids* **54**, 288–309.

CHAPTER 10

Contact Problems

Failure Processes at Points of Contact Between Bodies

10.1 Introduction

So far in this book we have considered stress concentration effects caused by notches and cracks, but this is not the only way to create a local stress and stress gradient. Other geometric features can act as stress concentrators on engineering components: features such as corners, bends and joints; these will be considered in Chapter 12. In this chapter, we will turn our attention to the stresses that arise due to local contact between bodies. Such contact can give rise to stresses which are very high, and highly localised, creating significant stress gradients. Many industrial components rely on such contacts, obvious examples being bearings and joints, not only in machine components but also in the joints of the human body, and their replacement parts (e.g. Colour Plate 5).

Many mechanical devices would be impossible without contact between parts, and yet it is particularly difficult to predict the failures which occur at these locations. The basic stress analysis of contact situations is problematic in itself, being strongly affected by factors which are difficult to estimate, such as the degree of friction and adhesion between moving bodies. Cracking is a common feature of contact-related failures, suggesting the use of fracture mechanics in their solution. Short cracks and non-propagating cracks are frequently involved as a result of crack growth through a rapidly decaying stress field. This suggests that methods such as the TCD may be applicable.

After a brief introduction to the general field of contact mechanics, this chapter reviews the existing methods used to predict failure under monotonic and cyclic contact, with particular emphasis on the problem of fretting fatigue. Several workers have used short-crack concepts and NSIF approaches, but to date there have been only a very few studies using the PM, LM or similar stress-based TCD methods. These studies have been successful in predicting fretting fatigue limits, suggesting a role for the TCD in contact-related failure. The chapter concludes with suggestions for other contact problems to which the TCD might be applied in the future.

Two useful textbooks in this area are *Engineering Tribology* (Williams, 1994) and *Fracture of Brittle Solids* (Lawn, 1993). Williams provides a clear introduction to the mechanics of contact and sliding, whilst Lawn gives an exceptional treatment of the problem of contact failure in brittle materials and its analysis using fracture mechanics.

10.2 Contact Situations

We can classify the various types of contact which can lead to failure, as follows:

Static Contact: Two bodies pressed together with a constant, or monotonically increasing, force; for example, a standard hardness test using a diamond indenter. The area of mutual contact is known as the contact patch. In ductile materials the result is plastic deformation, leaving a permanent indentation. In brittle materials cracking may occur, either beneath the indenter or around the edges of the contact patch. Cracks may form during loading or unloading.

Cyclic Contact: If the contact force is cyclic in nature, then fatigue cracks may form in or near the contact patch. The cracking patterns are quite similar to those found in brittle materials in static contact, though there are some differences.

Rolling Contact: This is a variant of cyclic contact in which the cyclic nature arises due to one surface rolling over the other, as in gears and bearings. This produces different cracking patterns as the stress field moves across the surface of the body.

Sliding Contact: Forces tangential to the surface can arise if one body is sliding over another, due to friction or adhesion between the two surfaces. Relative sliding under friction is termed 'slip'; adhesion is termed 'stick'. Sliding contact is an essential ingredient in the failure modes of wear and fretting fatigue and also in machining processes. However, there is also an element of sliding contact to be found in other types of monotonic and cyclic contact: for example, as an indenter is pressed into a material, sliding will occur at the interface. This turns out to be very important because it leads to an elastic singularity in the stress field at the edge of the contact patch.

10.3 Contact Stress Fields

Analytical solutions exist for many simple cases of contact, beginning with the work of Hertz, who developed the theory to describe the pressure distribution on a curved interface between two bodies (Hertz, 1895). The entire stress field in the body can be estimated from knowledge of the forces in the contact patch, which can be divided into normal forces P and tangential forces Q (see Fig. 10.1). Figure 10.2 shows an example of such an analysis, in this case for two identical cylinders in elastic contact: note that the stresses σ_x and σ_z are compressive. The shear stress τ rises to a maximum value at a sub-surface point, as does the Von Mises effective stress (not shown). Therefore yielding, if it occurs, will begin below the surface. The compressive force P causes small tensile stresses on radial lines (Fig. 10.1). Though these stresses are small compared to the compressive and shear stresses, they can be responsible for failure in brittle materials and for fatigue cracking.

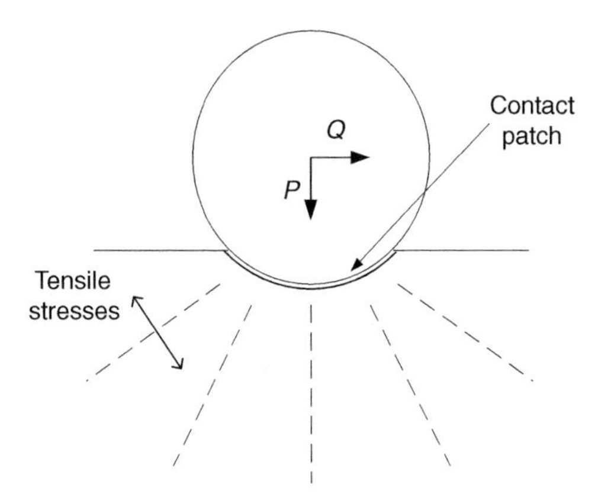

Fig. 10.1. A typical contact situation with normal forces P and tangential forces Q. Stresses due to P are mainly compressive (see Fig. 10.2) but tension also arises on radial lines.

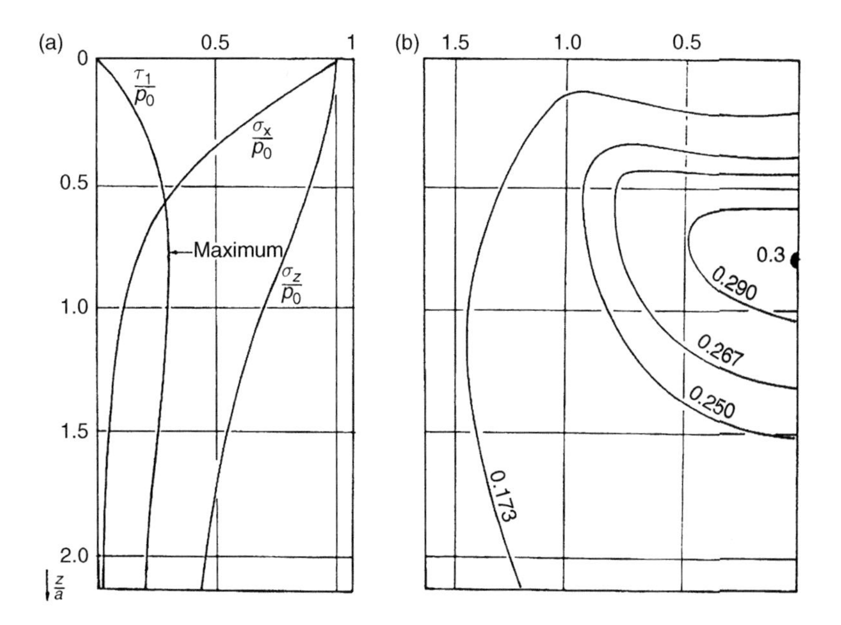

Fig. 10.2. Elastic contact of cylinders (reproduced from Williams 1994). (a) subsurface compressive and shear stresses along the line of symmetry; (b) contours of constant principal shear stress.

An important point to note is that whilst these elastic stresses can be large in magnitude they are never singular, that is the stress is never predicted to rise to infinity. Therefore the magnitude of the stress gradient will be directly proportional to the size of the contacting bodies: changing the scale by a certain factor for a given geometry will decrease the stress gradient by the same factor. This implies that significant stress-gradient effects may occur when the scale of the problem is small, such as a microscopic hardness indenter or the very local stress fields produced by surface asperities.

More severe stress gradients arise as a result of slip, that is sliding contact between bodies which experience friction at their interface. This creates significant tangential (Q) forces, which change abruptly at the edges of the slip zone. This leads to elastic singularities, as shown in Fig. 10.3 for the simple case of a uniform tangential stress over a contact patch. These stresses are very often the most damaging, because, in practice, contact situations almost always include some element of friction and sliding. In principle, the stress field can be determined as above, from the surface forces P and Q, but in practice these forces are very difficult to estimate, depending as they do on the amount of friction between the bodies. If there is sufficient pressure, and only small amounts of movement, between the bodies, as occurs during fretting, then parts of the interface may develop adhesion (stick) which again changes the distribution of forces. The amounts of stick and slip which occur depend not only on the mechanical forces applied but also on the nature of the two materials and their environment. Many stress analyses, both analytical solutions and FE models, have been carried out, revealing considerable disagreement. One crucial element which is virtually impossible to determine accurately is the variation in the friction factor within the contact patch.

It will be clear from the preceding discussion that the stress fields of interest here are highly multiaxial, containing complex mixtures of tension and shear in critical regions. A more detailed analysis of multiaxial problems and the special failure criteria used to solve them will be found in the next chapter (Chapter 11). The other source of stress,

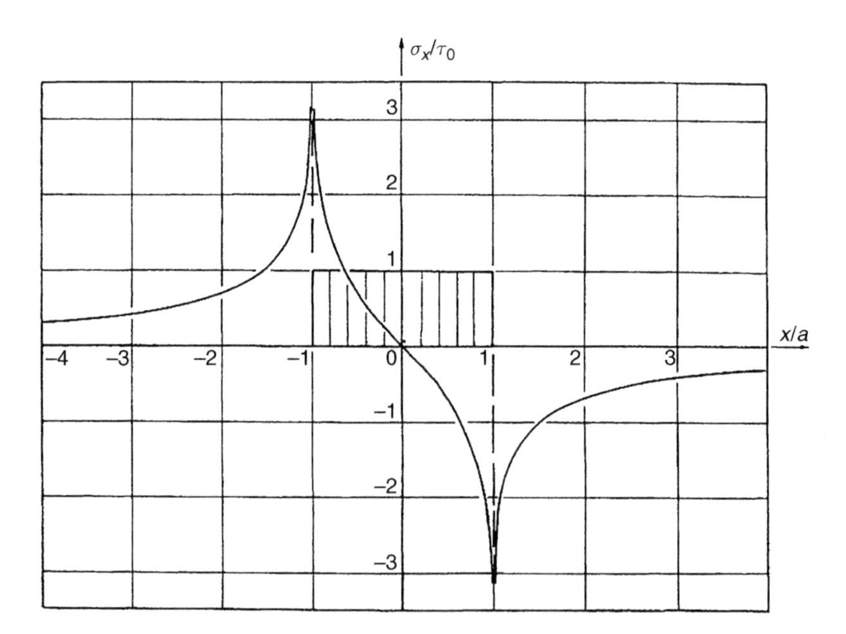

Fig. 10.3. Surface tensile stress σ_x resulting from a uniform shear loading of τ_0 applied over a patch of size $2a$, centred at $x = 0$. Note that the stresses are singular at $x = +a$ and $x = -a$.

which completes the picture of the contact stress field, is the stress which arises due to body forces. A component such as an automotive crankshaft will be loaded not only by contact forces at the bearings but also by body forces creating bending and torsion in the entire shaft. In these cases a crack which has initiated under the action of contact forces may be induced to continue propagating by the body forces. This leads to the phenomenon of fretting fatigue which can be thought of as a normal fatigue failure process in which the initiation and early growth stages have been facilitated by the action of the contact forces. In the absence of body forces the cracks which form due to contact will normally stop growing at quite short lengths, having propagated out of the stress field. However, these cracks may be long enough to cause loss of material by delamination, an important wear mechanism in gears and bearings. Sub-surface cracks may also grow to macroscopic lengths if rolling contact is present, leading to massive wear by a spalling mechanism.

We will now consider some specific types of failure and the approaches used to predict them. In what follows it is worth bearing in mind the points made above about the inherent uncertainties of the stress analysis, which naturally place a limit on the accuracy of any predictive theory. We will begin with the case of fretting fatigue, since this problem has been researched extensively, including a few studies which have used the TCD. After discussing this phenomenon in some detail, the chapter will conclude by mentioning some other cases of contact-induced failure, cases in which the TCD has not yet been used but to which it might usefully be applied in the future.

10.4 Fretting Fatigue

Fretting fatigue occurs as a result of local contact forces, which cause crack initiation and early crack growth into the body, combined with body forces which cause this crack growth to continue to failure. Lindley has written an excellent introduction to this subject (Lindley, 1997). Tests for fretting fatigue are normally conducted as shown in Fig. 10.4: a conventional fatigue test is carried out, with the addition of a pad (in fact usually a pair of pads) clamped to the sides of the specimen. This set up has the advantage of simplicity; a conventional fatigue-testing machine can be used, combined with some means of maintaining a constant clamping force. Cyclic strain in the specimen causes

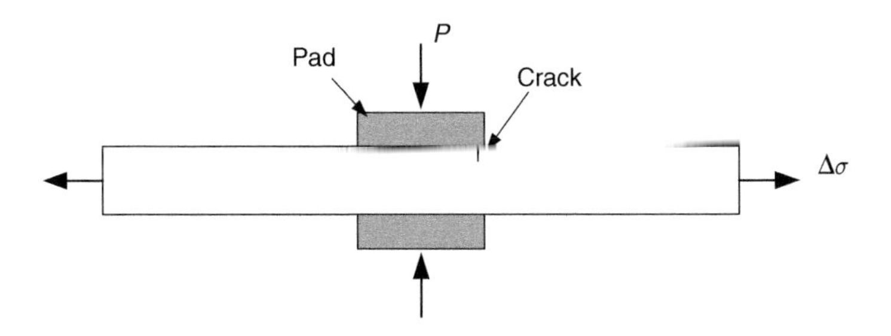

Fig. 10.4. Elements of a fretting fatigue test: a cyclic stress $\Delta\sigma$ is applied to a specimen which has pads pressed onto its surface with pressure P. Relative slip occurs between specimen and pads: fatigue cracks are initiated near the ends of the pads.

relative movement between it and the pads, creating the slip necessary for high local stresses. Usually the pad is flat, with sharp corners at the edges of the contact surface, but other shapes have been used, including spherical pads.

Enormous reductions in fatigue strength are possible under these conditions: the fatigue limit can be reduced by a factor of three or, in extreme cases, as much as five, compared to the normal plain-specimen value. There is an optimum amount of slip, of the order of 20 μm, at which fatigue life is lowest. For larger amounts of slip the mechanism changes from fretting fatigue to wear. Opinions differ on the question of whether the process is dominated by crack initiation and early growth, or by crack propagation. For example, Lindley (1997) remarks that the number of cycles to initiate a crack is usually small, whereas others assume that the initiation stage takes up the majority of life (Lykins et al., 2001). This difference of opinion probably arises due to the interplay between the two forces present: contact forces and body forces. In cases where the body forces are relatively high, a crack, once initiated, will be quickly accelerated to failure, so initiation may take up most of the life.

Conversely, if the body forces are low there will be a longer propagation stage, perhaps resulting in a non-propagating crack. Faanes and Harkegard carried out a very interesting analysis of crack growth, showing that, unlike the case of notch fatigue, non-propagating cracks can occur at two different lengths. This idea is illustrated in Fig. 10.5 taken from their paper (Faanes and Harkegard, 1994); this shows an *R*-curve analysis in which the estimated stress intensity range for the crack is compared to the threshold value, both being functions of crack length. The first type of non-propagating crack (labelled α_1) is the same type which occurs in notch fatigue, arising due to the increasing value of the threshold ΔK_{th} with increasing crack length in the short-crack regime. Typical crack lengths here would be of the order of 100 μm. But in fretting there is a second opportunity

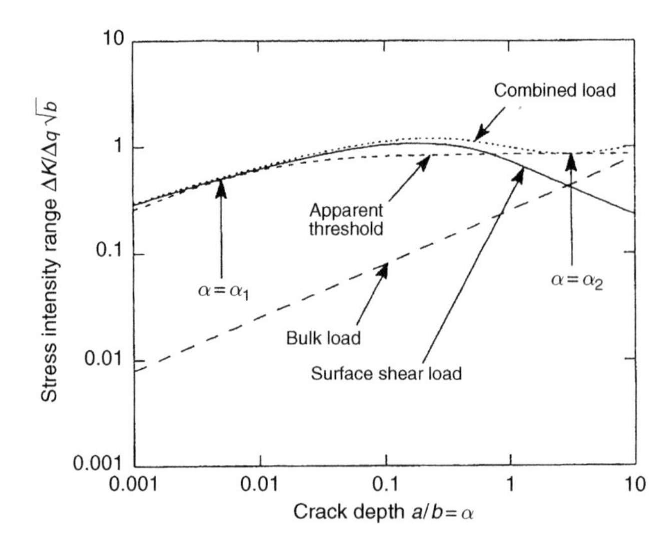

Fig. 10.5. The resistance curve of Faanes and Harkegard (1994) showing two possible lengths for a non-propagating crack: α_1 and α_2.

for crack arrest, α_2, which occurs as the stress intensity of the crack decreases with increasing length due to the rapid fall-off in stress when the crack moves out of the contact zone. These second non-propagating cracks will occur if the body forces are relatively low.

Another important feature of the experimental data is the existence of strong size effects. Other things being equal, large pads creating large contact areas will tend to result in shorter fatigue lives (Araujo and Nowell, 1999; Ciavarella, 2003; Fouvry et al., 2000). For example, tests were carried out on an aluminium alloy, using a constant value of the nominal applied pressure, varying the contact area by changing the pad size (Araujo and Nowell, 1999). In one case, the fatigue life was of the order of 1 million cycles for contact lengths in the range 0.38–1.14 mm, but greater than 10 million (no failures occurring) for contact lengths in the range 0.1–0.28 mm. It is interesting that this effect occurred at contact lengths of the same order of magnitude as the critical distance L in this material, suggesting a possible role for the TCD.

A variety of methods of prediction have been developed for fretting fatigue: given the above remarks, we can expect that these different methods may be applicable in different circumstances, reflecting the relative importance of the two sources of stress. The approaches taken are broadly similar to those described in the previous chapter, on fatigue, and suffer from the same limitations and simplifications. Some workers have concentrated on the crack initiation phase, normally using a multiaxial fatigue criterion such as a critical plane approach (Namjoshi et al., 2002): these multiaxial criteria will be discussed in more detail in the next chapter. One group (Kondo and Bodai, 2001) has made successful use of the 'hot spot stress' concept originally developed for welded joints (see Chapter 12).

Many researchers have developed crack growth models using LEFM or modified LEFM approaches. Most of these ignore the multiaxial nature of the stress field, assuming that the mode I stress intensity dominates. However, some authors have developed methods for predicting the shear contributions (Kimura and Sato, 2003). Nix and Lindley describe a carefully constructed model which allows for the change in R ratio during crack growth, but assumes long-crack behaviour throughout (Nix and Lindley, 1985). Other authors take account of the accelerated growth and lower thresholds that occur in short cracks, usually by incorporating the El Haddad imaginary-crack idea (as described in Section 9.2.3): various methods are used to determine the number of cycles to initiate the crack (Navarro et al., 2003). In some cases this is ignored and an initial crack of a specified small length is assumed (Tanaka et al., 1985). Vallellano et al. used some realistic stress intensity factor approximations to predict fatigue limits and non-propagating crack lengths for fretting fatigue under a spherical indenter (Vallellano et al., 2003).

A different, and very interesting approach has been taken by some workers who have modelled the fretting fatigue situation as a crack or notch (Ciavarella, 2003; Giannakopoulos et al., 1998; Giannakopoulos et al., 1999). Giannakopoulos et al. suggested that the stress field created near the contact pad is similar to that created ahead of a sharp crack. If this analogy can be made, then it might be possible to predict the behaviour in the contact problem by finding an equivalent crack. This approach is similar to the CMM for notch fatigue described in the previous chapter (Section 9.2.1). The problem is then to find the size and geometry of the equivalent crack. This approach, which is represented schematically in Fig. 10.6, is an elegant one, but suffers from some problems. First, it is by no means certain that the contact stress field is indeed similar to that of a crack. Giannakopoulos et al. claimed that the $1/r^{1/2}$ stress singularity which occurs ahead of a crack also occurs at the edge of the contact pad, but other workers disagree. For example, Hattori, whose work will be discussed further below, conducted FEAs of contact pads with different angles of inclination to the specimen surface (Hattori and Nakamura, 1994); he found that the exponent of the singularity varied with angle (as one would expect from knowledge of sharp V-shaped notches – see Section 1.4); even at the conventional angle of 90° the exponent was 0.39 rather than 0.5 as assumed by Giannakopoulos et al.

A second difficulty with this approach is that in order for the crack to grow in the appropriate direction it must turn during growth, becoming a doubly kinked crack (see Fig. 10.6). This, combined with the fact that the applied stress in the specimen takes the role of a T-stress in the analogue, makes the analysis quite complex. Ciavarella has developed a relatively simple approach using crack and notch analogies, which successfully predicted the size-effect data of Nowell (Ciavarella, 2003).

Hattori developed a model which included crack initiation and propagation stages, applying it to data on the effect of the contact angle between the side of the pad and the specimen surface. As noted above, this angle would be expected to affect the order

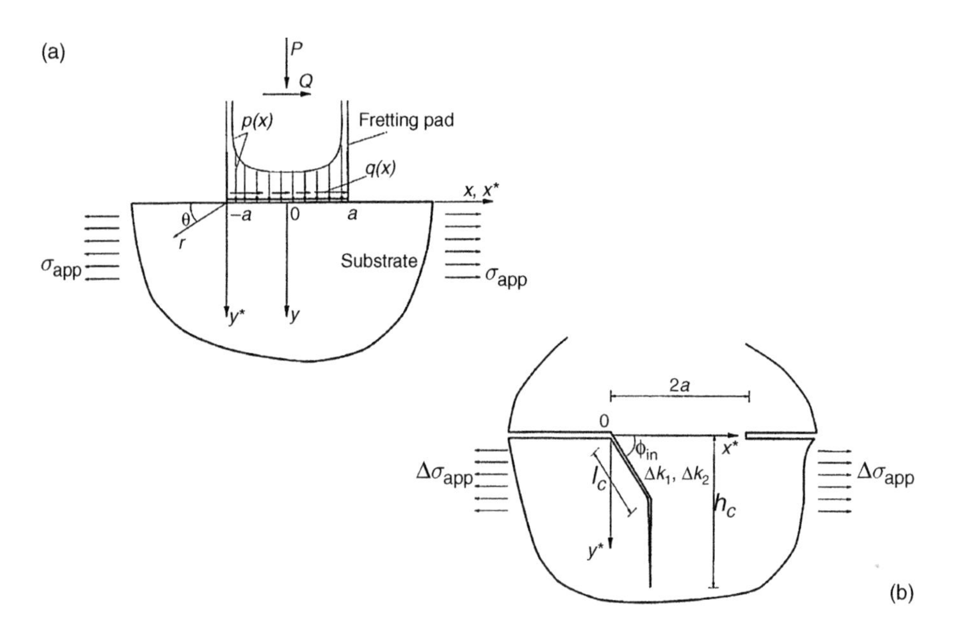

Fig. 10.6. The crack analogue approach of Giannakopoulos et al.; the original fretting problem (a) is reposed as a cracked body; (b) further cracking has a double-kinked form.

of the singularity: drawing an analogy with a sharp V-shaped notch (see sections 1.4 and 6.2.3) one might expect the stress–distance curve to take the form:

$$\sigma(r) = \frac{\Psi}{r^{\lambda}} \tag{10.1}$$

Here λ is a constant which varies with angle and Ψ is an equivalent stress intensity factor which becomes identical to the conventional K value of a crack when $\lambda = 0.5$. Hattori showed that this equation did indeed apply to stress fields generated by FEA. He proposed a criterion for crack initiation using a critical value Ψ_c, which had the following form:

$$\Psi_c = \Delta\sigma_o(d)^{\lambda} \tag{10.2}$$

Here $\Delta\sigma_o$ is the conventional plain-specimen fatigue limit and d is a material constant. This approach is essentially the same as the NSIF method (see Section 3.5); Hattori offers no experimental proof of this criterion, but includes it in more complex models of the entire crack initiation and growth process.

10.4.1 The use of the TCD in fretting fatigue

A few investigations have used the TCD or similar approaches to predict fretting fatigue behaviour. To date the total number of studies has been small, but sufficient to suggest that the TCD may be a useful tool in this field.

Fouvry et al. used a multiaxial fatigue criterion, averaging stresses over an elemental volume (Fouvry et al., 2000); this approach successfully predicted size effects. Vallellano et al. used the PM and LM to predict fretting fatigue limits for specimens of an aluminium alloy (Al 7075-T6) tested using a spherical indenter (Vallellano et al., 2003). They estimated the local stress fields using an analytical solution; the necessary material constants ($\Delta\sigma_o = 428\,\text{MPa}$, $\Delta K_{th} = 4.0\,\text{MPa(m)}^{1/2}$) were obtained from the published literature on this material. The critical distance was calculated in the normal way (Eq. 2.6) giving a value of $L = 27.8\,\mu\text{m}$. Table 10.1 shows their experimental results and predictions. Five different tests were conducted, in which the pressure applied to the spherical pad was kept constant whilst the axial cyclic load was varied. The table shows the outcome as either failure (F) or non-failure (NF). The predictions are shown, along with a factor which gives the calculated stress parameter (i.e. the stress at $L/2$ for the PM or the average stress over $2L$ for the LM) divided by the material's fatigue limit. Thus, if this factor is greater than unity, one would predict failure. Both methods gave good predictions, with errors of the order of 10%: the PM was slightly conservative and the LM slightly non-conservative – this difference between the PM and the LM was also found in our analysis of notch fatigue (see Chapter 9). Non-propagating cracks were observed in those specimens which did not fail after a large number of cycles: the lengths of these cracks were 110–150 μm, which corresponds to a few times the value of L, and also a few times the grain size, which was 35 μm.

This is a very impressive result considering the inherent difficulties of the stress analysis and determination of material properties. Nowell reported an attempt to apply the TCD to his experimental data: he found that the critical distance which successfully predicted

Table 10.1. Summary of data and predictions of fretting fatigue from Vallellano et al. (2003)

Axial stress amplitude (MPa)	Cycles and result (F = failure, NF = no failure)	LM prediction (ratio of average stress over fatigue limit)	PM prediction (ratio of point stress over fatigue limit)
83	549,000 [F]	F [1.00]	F [1.20]
70	516,000 [F]	F [0.93]	F [1.13]
56	1,540,000 [NF]	NF [0.85]	F [1.05]
63	2,940,000 [NF]	NF [0.89]	F [1.09]
59	1,777,000 [NF]	NF [0.87]	F [1.07]

the fretting fatigue data was different from that calculated using conventional fatigue test specimens, by a factor of two. This may have been because, like Vallellano et al. above, he used data from the literature for $\Delta\sigma_o$ and ΔK_{th}, which might have been slightly different for his own material. In fact, to date no one has carried out a TCD analysis using test data on both fretting fatigue and conventional fatigue from the same batch of material in the same laboratory, which would be necessary in order to apply the method with confidence.

Araujo et al. also applied the TCD, using previously published experimental data on fretting fatigue under cylindrical contacts (Araujo et al., 2006). They combined the PM with a multiaxial failure criterion due to Susmel and Lazzarin: this criterion, which is of the critical plane type, will be discussed in detail in the next chapter (Section 11.3.3). It uses two parameters: the shear stress amplitude τ_a and the ratio between the normal and the shear stresses ρ. These authors incorporated the PM by calculating the stresses at a point $L/2$ from the point of maximum stress. Figure 10.7 summarises their results. In these diagrams each data point represents a single test, characterized by particular values of τ_a and ρ; the symbols indicate either failures or run-outs. The prediction line (labelled SU = 0%) should lie below the failure points and above the run-outs. The SU is a measure of the prediction error. It can be seen that all the data are correctly predicted, with errors of 10–25%.

To my knowledge, the publications described above are the only ones in which the TCD has been used to predict fretting fatigue, or indeed any kind of contact failure problem. The results are certainly promising, but clearly more work is needed in this area.

10.5 Other Contact-Related Failure Modes: Opportunities for the TCD

In this section, I will briefly consider a number of other mechanisms of failure related to contact stresses. It seems to me that the TCD might usefully be applied to these problems, but, as far as I know, this has not yet been attempted.

10.5.1 Static indentation fracture

Lawn (1993) has provided an excellent treatment of the cracking of brittle materials caused by the application of monotonic loads through indenters. Relatively blunt

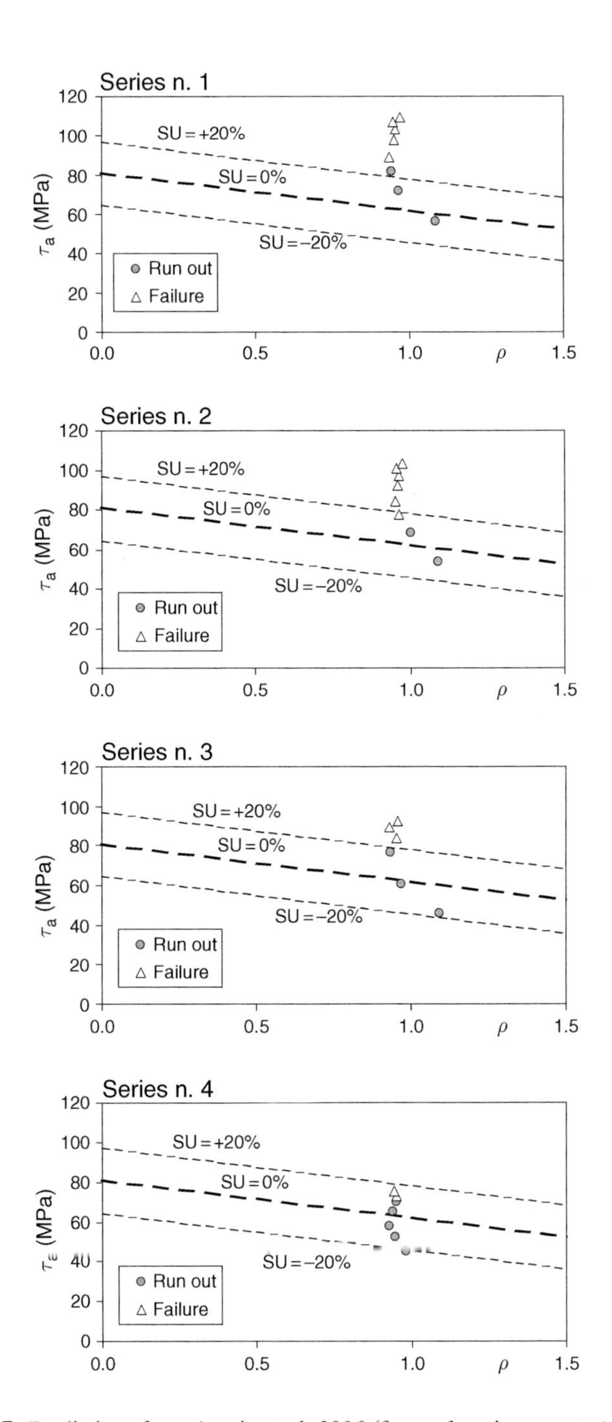

Fig. 10.7. Predictions from Araujo et al. 2006 (for explanation, see text above).

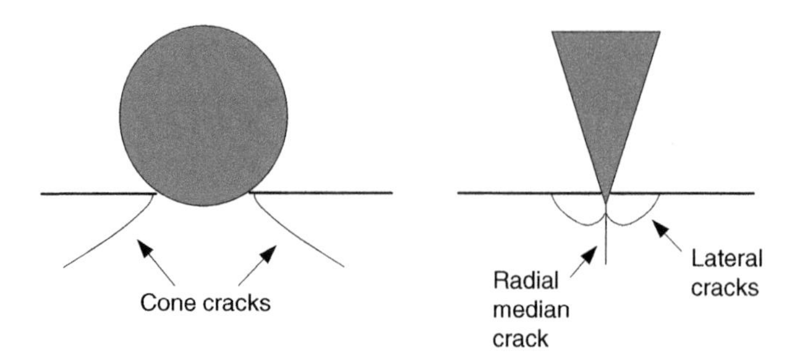

Fig. 10.8. Indentation cracking of brittle materials. A blunt indenter (on the left) gives rise to a cone crack: a sharp indenter (on the right) causes a radial median crack on loading and lateral cracks on unloading.

indenters (such as a sphere) generate only elastic stresses before failure. Tensile stresses at the surface near the edge of the contact patch give rise to circumferential cracks, known as cone cracks (see Fig. 10.8) which spread downwards and outwards. Sharp indenters, such as the diamond pyramid used in the Vickers hardness test, generate plastic deformation. Normally it can be assumed that brittle ceramic materials do not undergo plastic deformation; however, it can occur under the intense compressive stresses at the tip of the indenter. Tensile stresses just below the indenter cause the growth of a radial-median crack, and the plastic deformation gives rise to residual stresses on unloading, which can cause lateral cracks (Fig. 10.8).

Normally these various cracks will stop growing as they propagate away from the contact stress field. The final crack length will depend on the applied load and the material's toughness, so this is a convenient way to measure toughness in brittle materials. In principle, these cracking phenomena can be predicted using fracture mechanics, though the details of the analysis can be complex.

One essential aspect of such an analysis is the recognition that toughness is not a material constant but varies with crack length when the crack is small. Lawn uses a resistance-curve approach (the theory of which was discussed previously in respect of fatigue, in Chapter 9.2.1). The fact that the crack lengths are small enough to lie within this short-crack regime suggests that the TCD might be applicable here. One particular parameter which fracture mechanics is not able to predict but for which the TCD should be suitable is the limiting load needed to initiate a crack using a blunt indenter.

10.5.2 Contact fatigue

The simplest way to generate contact fatigue is to apply a cyclic load to an indenter, creating the so-called 'standing contact fatigue' (Alfredsson and Olsson, 1999; Alfredsson and Olsson, 2003). The cracking patterns so generated are broadly similar to those described in the previous section – not surprising since both are generated by the same field of elastic stresses – though there are some differences in the details. Given the success of the TCD in predicting fretting fatigue, I have no doubt that it could be successfully applied here also.

In practice, however, contact fatigue normally involves a moving point of contact, as in rolling contact between gear teeth and bearing components. Whilst the initiation and early growth of the crack may be similar to that in standing fatigue, the moving force has a considerable effect on the subsequent crack growth and the tendency for the crack to turn back to the surface and cause spalling. Thus, whilst the TCD may be useful in modelling the early stages of this process (perhaps predicting the limit below which only non-propagating microscopic cracks will occur), a crack-propagation analysis would be needed to describe the entire process. Examples of such analyses can be found in the literature (Frolish et al., 2002; Glodez et al., 1999).

10.5.3 Mechanical joints

Many types of mechanical joints involve local contact forces, both static and cyclic in nature, often with superimposed residual stresses. The TCD might certainly be used to investigate these situations, and in fact is already being applied to pin-loaded joints in composite materials (see Section 8.5).

10.5.4 Wear

Wear can occur by a variety of mechanisms (see Williams 1994 for a thorough treatment of this subject); some of these mechanisms may be suitable for analysis using the TCD, as follows:

Fatigue Wear: This is essentially another name for the rolling-contact fatigue that was mentioned above. It tends to occur in high-quality materials such as hardened steels, in situations where the loading is entirely elastic. Wear takes the form of loss of large particles – typically a large fraction of a millimetre – by the spalling mechanism mentioned above. As already noted, the TCD might play a partial role in describing this mechanism, especially its lower limit.

Delamination Wear: This is similar to fatigue wear, except at a smaller scale. Fatigue cracks initiate typically about 10 μm below the surface, often at inclusions and voids, as a result of local plastic deformation in the surface layer. Cracks subsequently grow until they meet the surface, causing sheets of material to fall off. There have been many models of delamination wear, some of which have viewed it as a LCF problem, controlled by the amount of cyclic plastic strain. One model, due to Suh, was developed in the 1980s but has since fallen from popularity. This model uses terms for the volume fraction of voids and inclusions which can be reinterpreted as length scales, perhaps leading to a TCD-type model.

Abrasive Wear: This kind of wear is essentially a microscopic machining process and so will be discussed in the following section.

10.5.5 Machining

Machining is essentially a process of controlled failure of material. Theoretically, machining processes are difficult to predict because of the variety of failure modes involved and the highly non-linear deformation and damage that the material suffers.

A detailed treatment of this topic is certainly beyond the scope of this book. It is worth noting, however, that there are essentially two types of machining: plastic deformation (ploughing) and fracture; the latter process involves the controlled propagation of a crack. Fleck et al. studied chip formation in surface machining (Fleck et al., 1996): in some cases a continuous chip of material is formed, leading to long pieces of swarf, whilst in other cases the process of chip formation is discontinuous, leading to many small chips. Fleck noted that the condition for transition from one mechanism to the other depended on the size of the plastic zone ahead of the cutting tool, which itself is proportional to the ratio $(K_c/\sigma_y)^2$. This parameter is similar to L, except for the use of σ_y rather than σ_o. This suggests that the TCD might be useful in analysing certain types of machining process.

References

Alfredsson, B. and Olsson, M. (1999) Standing contact fatigue. *Fatigue and Fracture of Engineering Materials and Structures* **22**, 225–237.

Alfredsson, B. and Olsson, M. (2003) Inclined standing contact fatigue. *Fatigue and Fracture of Engineering Materials and Structures* **26**, 589–602.

Araujo, J.A. and Nowell, D. (1999) Analysis of pad size effects in fretting fatigue using short crack arrest methodologies. *International Journal of Fatigue* **21**, 947–956.

Araujo, J.A., Susmel, L., Taylor, D., Ferro, J.T.C., and Mamiya, E.N. (2006) On the use of the theory of critical distances and the modified Wohler curve method to estimate fretting fatigue strength of cylindrical contacts. *International Journal of Fatigue* **29**, 95–107.

Ciavarella, M. (2003) A 'crack-like' notch analogue for a safe-life fretting fatigue design methodology. *Fatigue and Fracture of Engineering Materials and Structures* **26**, 1159–170.

Faanes, S. and Harkegard, G. (1994) Simplified stress intensity factors in fretting fatigue. In *Fretting Fatigue ESIS 18* (Edited by Waterhouse, R.B. and Lindley, T.C.) pp. 73–81. Mechanical Engineering Publications, London.

Fleck, N.A., Kang, K.J., and Williams, J.A. (1996) The machining of sintered bronze. *International Journal of Mechanical Science* **38**, 141–155.

Fouvry, S., Kapsa, P., and Vincent, L. (2000) A multiaxial fatigue analysis of fretting contact taking into account the size effect. In *Fretting fatigue: Current Technology and Practices ASTM STP 1367* (Edited by Hoeppner, D., Chandrasekaran, V., and Elliot, C.B.) pp. 167–182. ASTM, West Conshohoken, PA, USA.

Frolish, M.F., Fletcher, D.I., and Beynon, J.H. (2002) A quantitative model for predicting the morphology of surface initiated rolling contact fatigue cracks in back-up roll steels. *Fatigue and Fracture of Engineering Materials and Structures* **25**, 1073–1086.

Giannakopoulos, A.E., Lindley, T.C., and Suresh, S. (1998) Aspects of equivalence between contact mechanics and fracture mechanics: theoretical connections and a life-prediction methodology for fretting-fatigue. *Acta Materialia* **46**, 2955–2968.

Giannakopoulos, A.E., Venkatesh, T.A., Lindley, T.C., and Suresh, S. (1999) The role of adhesion in contact fatigue. *Acta Materialia* **47**, 4653–4664.

Glodez, S., Ren, Z., and Flasker, J. (1999) Surface fatigue of gear teeth flanks. *Computers and Structures* **73**, 475–483.

Hattori, T. and Nakamura, M. (1994) Fretting fatigue evaluation using stress singularity parameters at contact edges. In *Fretting Fatigue, ESIS 18* (Edited by Waterhouse, R.B. and Lindley, T.C.) pp. 453–460. Mechanical Engineering Publications, London.

Hertz, H. (1895) *Gesammelte Werke Vol.1*. Leipzig, Germany.

Kimura, T. and Sato, K. (2003) Simplified method to determine contact stress distribution and stress intensity factors in fretting fatigue. *International Journal of Fatigue* **25**, 633–640.

Kondo, Y. and Bodai, M. (2001) The fretting fatigue limit based on local stress at the contact edge. *Fatigue and Fracture of Engineering Materials and Structures* **24**, 791–801.
Lawn, B. (1993) *Fracture of brittle solids*. Cambridge University Press, Cambridge.
Lindley, T.C. (1997) Fretting fatigue in engineering alloys. *International Journal of Fatigue* **19**, S39–S49.
Lykins, C.D., Mall, S., and Jain, V. (2001) A shear stress-based parameter for fretting fatigue crack initiation. *Fatigue and Fracture of Engineering Materials and Structures* **24**, 461–473.
Namjoshi, S.A., Mall, S., Jain, V.K., and Jin, O. (2002) Fretting fatigue crack initiation mechanism in Ti-6Al-4V. *Fatigue and Fracture of Engineering Materials and Structures* **25**, 955–965.
Navarro, C., Garcia, M., and Dominguez, J. (2003) A procedure for estimating the total life in fretting fatigue. *Fatigue and Fracture of Engineering Materials and Structures* **26**, 459–468.
Nix, K.J. and Lindley, T.C. (1985) The application of fracture mechanics to fretting fatigue. *Fatigue and Fracture of Engineering Materials and Structures* **8**, 143–160.
Tanaka, K., Mutoh, Y., Sakoda, S., and Leadbeater, G. (1985) Fretting fatigue in 0.55C spring steel and 0.45C carbon steel. *Fatigue and Fracture of Engineering Materials and Structures* **8**, 129–142.
Vallellano, C., Dominguez, J., and Navarro, A. (2003) On the estimation of fatigue failure under fretting conditions using notch methodologies. *Fatigue and Fracture of Engineering Materials and Structures* **26**, 469–478.
Williams (1994) *Engineering tribology*. Oxford Science Publishers, Oxford.

CHAPTER 11

Multiaxial Loading

Fracture and Fatigue Under Complex Stress States

11.1 Introduction

Most laboratory experiments are conducted using the simple loading states of axial tension or in-plane bending, but we all know that engineering components are subjected to much more complex systems of loading. Torsion and pressure occur frequently, and many components experience more than one source of loading, these inputs occurring at different points in time; this is especially true of components in engines and vehicle suspension systems, for which even measuring the load/time characteristics can be a challenging task.

Strictly speaking, we should define as 'multiaxial' any stress state in which the stress tensor is not dominated by a single, tensile, principal stress. Using this definition, we realise that the presence of a notch or crack creates a multiaxial stress field, even when the applied loading is simple tension. The fact that fatigue and brittle fracture of cracked components in tension can be successfully analysed by considering only the tensile stress intensity parameter, K_I, should encourage us to think that multiaxial problems are, after all, not entirely intractable.

Having said that, there is no doubt that the prediction of failure under conditions of generalised multiaxial loading is a major problem. It has received considerable attention in the research literature, but despite this there are still no universally agreed procedures for failure prediction. The problem contains many variables – variables both in the mechanical state of stress and in the material responses – and this is especially true in fatigue when considering multiple, out-of-phase, load inputs.

The nature of the stress field affects the orientation of the crack plane, as illustrated in the shaft failure shown in Colour Plate 6. In general, cracks tend to adjust themselves so as to maximise the tensile (Mode I) loading across the crack faces. Complex cracking patterns may develop, whose interaction with the applied loading system may be difficult to predict. For example, under torsion loading of a circular shaft, cracks tend to grow in the two planes at 45° to the loading axis, leading to a so-called 'factory roof' fracture surface; interaction of this surface with the shear loading causes crack-face contact which tends to reduce the effective stress intensity.

Material response can be broadly thought of as varying between two extremes: 'brittle', in which the controlling parameter is the maximum principal stress, and 'ductile', in which shear stress determines behaviour. However, hydrostatic stress also plays a role in some failure mechanisms such as yielding and crazing in polymers. In anisotropic materials such as fibre composites and bone, cracking may follow weak material directions.

In this chapter, it is not intended to provide a comprehensive coverage of multiaxial fracture and fatigue – indeed such an endeavour would require a whole book to itself. Rather, the aim is to ask the question 'Can the TCD be used in situations of multiaxial loading?' In considering this question we will confine ourselves to two cases: the monotonic fracture of brittle materials and the HCF limits of metals. Current methods used for the assessment of notches and cracks will be outlined, and it will be seen that the basic theoretical approaches are the same in these two cases, though with some extra subtleties in the case of fatigue. We will then develop a strategy for applying TCD methods such as the PM and LM, and use these methods to predict experimental data.

In attempting to use the TCD in this way, we are entering into almost virgin territory. Even though, as we have seen in previous chapters, the TCD has been known for the last 50 years, and used extensively in some fields for the last 30 years, a literature search yielded only a handful of papers which address its use in multiaxial situations, all of these being in the field of monotonic brittle fracture. Consequently, my colleague Luca Susmel and I embarked on a project to investigate the use of the TCD in multiaxial fatigue: our findings are presented below.

11.2 A Simplified View

At this stage it will be useful to take a simplified view of the problem: a view which encompasses the main issues associated with the presence of notches in multiaxial stress fields, without getting tangled up in the complications that arise when one takes a more general, more comprehensive view. Consider Fig. 11.1, which shows the effect of notch stress concentration factor, K_t, on the failure stress. The length of the notch is assumed to be constant, so K_t is changed by changing only the root radius ρ. We have seen plots of this kind in previous chapters: here we can consider it as representing either brittle fracture or fatigue. Under simple tensile loading, the failure stress decreases from the plain specimen value on the left to the value appropriate for a cracked body (and therefore depending also on the length and geometry of the crack) on the right. If we now take specimens of the same geometry but apply a loading state of pure shear, then we will create a second line on this figure, which will be almost always lower than the first. The shape of this notch, and its orientation with respect to the applied loading, is a consideration which we will return to later on. The stress used on the vertical axis is the maximum principal stress: in what follows, we will normally refer to this as the

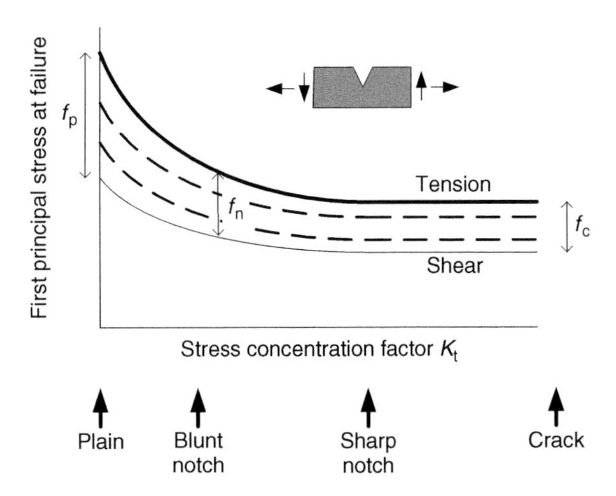

Fig. 11.1. Failure stress as a function of K_t for tension and shear loading; the dashed lines indicate intermediate cases of mixed tension and shear.

'first principal stress' to avoid confusion with the maximum stress in a cyclic load. For the shear case the first principal stress is numerically equal to the applied shear stress.

We can also imagine intermediate states in which a mixture of tension and shear is applied; results from such tests will lie on a series of lines which fall in between the two extreme cases. In this way we can investigate the whole range of multiaxial loading states, from pure tension to pure shear. This approximates to many loading states which exist in engineering components, though not all. It omits situations where two or more principal stresses are positive, such as in pressure vessels; more importantly it does not include out-of-plane shear, which will be considered separately below.

This view can be completed if we add two more diagrams: a plot showing the effect of notch size (Fig. 11.2) and one showing the effect of notch angle (Fig. 11.3). If we can predict these three effects, we will certainly be a long way towards understanding multiaxial failure problems. The differences between the tension and the shear lines can be expressed as a series of factors, as shown on the figures. Thus f_p can be defined as the ratio between the strength of plain specimens in shear and in tension: this appears on all three figures. The factors f_c and f_n describe the behaviour of a specimen containing a crack or notch respectively: unlike f_p, these factors will not be constants but will depend on the geometry of the feature. So, in Fig. 11.1, for example, f_n is the spacing between the tension and the shear lines at all points except the two extreme ends.

11.3 Material Response: The Factor f_p

In practice, it is very unlikely that we would be able to predict the behaviour in shear knowing only the behaviour in tension. The reason for this is that different materials respond differently to shear stresses. Therefore criteria which have been developed to predict mixed tension/shear conditions generally start by assuming that the experimental

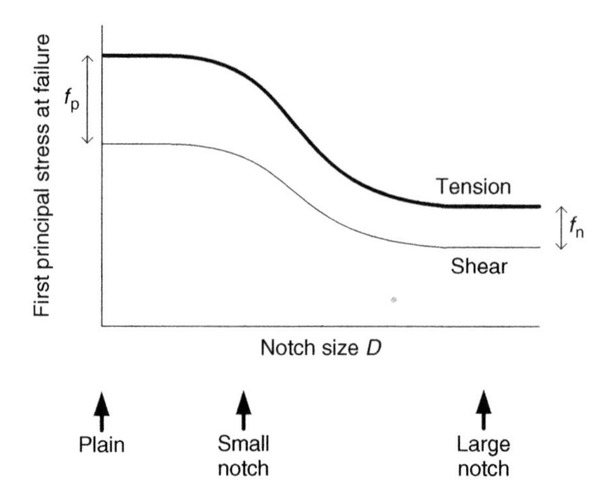

Fig. 11.2. Failure stress as a function of notch size for tension and shear loading.

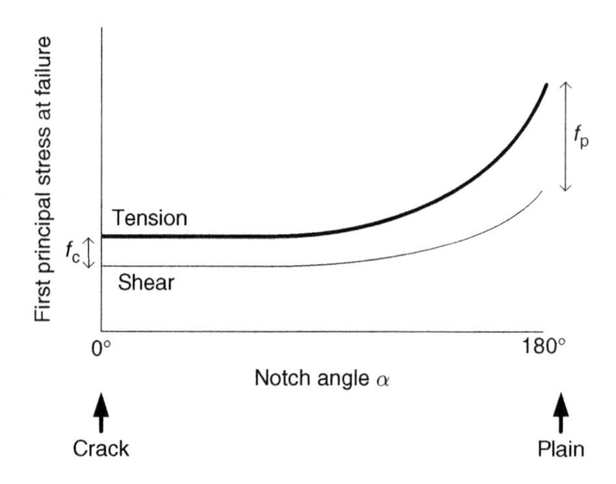

Fig. 11.3. Failure stress as a function of notch angle for sharp V-shaped notches in tension and shear.

strengths in pure tension and pure shear are both known. We can consider two extremes of material behaviour which can, in a rather loose fashion, be described by the terms 'brittle' and 'ductile'. Here a brittle material is one whose failure is controlled by the first principal stress only: for this material $f_p = 1$. The monotonic fracture behaviour of ceramics, and some polymers, conforms to this behaviour. At the other extreme, a ductile material which is entirely controlled by the maximum shear stress will have $f_p = 0.5$: this behaviour would be equivalent to the Tresca yield criterion. In practice, real materials display behaviour which lies between these two extremes. The Von Mises criterion would give $f_p = 0.58$, and in fact the behaviour of many ductile metallic materials lies close to this value.

11.3.1 Multiaxial fatigue criteria

Many different criteria have been developed for predicting the fatigue behaviour of plain specimens under multiaxial loading. This work includes a large effort in the prediction of LCF, which is complicated by the need to predict the material's responses to cyclic plastic deformation. This will not be considered in the present analysis: here we will concentrate exclusively on HCF and in particular on the fatigue limit. A number of different approaches have been taken to the problem of defining parameters from the stress tensor which characterise the fatigue limit. However, the great majority of this work has been conducted and validated on plain specimens only; relatively few workers have considered the effects of notches.

11.3.2 Scalar invariants

Multiaxial fatigue criteria can be divided into two types. The first type uses parameters that can be expressed as scalar quantities. We can write the general stress tensor as:

$$[\sigma] = \begin{bmatrix} \sigma_x & \tau_{xy} & \tau_{xz} \\ \tau_{xy} & \sigma_y & \tau_{yz} \\ \tau_{xz} & \tau_{yz} & \sigma_z \end{bmatrix} \tag{11.1}$$

Most criteria use two parameters, the first of which is the hydrostatic stress, defined as:

$$\sigma_H = \frac{1}{3}[\sigma_x + \sigma_y + \sigma_z] \tag{11.2}$$

The second parameter is more difficult to define. Clearly this parameter should reflect the applied shear stress, but it is not so easy to find a scalar quantity in this case. The deviatoric stress tensor is defined as:

$$[\sigma_D] = \begin{bmatrix} \sigma_x - \sigma_H & \tau_{xy} & \tau_{xz} \\ \tau_{xy} & \sigma_y - \sigma_H & \tau_{yz} \\ \tau_{xz} & \tau_{yz} & \sigma_z - \sigma_H \end{bmatrix} \tag{11.3}$$

One scalar quantity which can be obtained from this is its second invariant, which is

$$J_2 = \frac{1}{2}[\sigma_D] \cdot [\sigma_D] \tag{11.4}$$

Because σ_H and J_2 are scalar quantities, their variation in time can be simply described using amplitude, mean or maximum values. Thus, for example, the criterion of Crossland is:

$$\sqrt{J_{2,a}} + A\sigma_{H,\max} = B \tag{11.5}$$

Here $J_{2,a}$ is the amplitude (i.e. half the range) of J_2 and $\sigma_{H\max}$ is the maximum value of the hydrostatic stress. The maximum value is used here instead of the range in order to attempt to account for effects of R ratio. A and B are material constants which can, in principle, be determined from experimental data on the material tested using any two different types of loading. In practice, these constants are usually expressed in terms of the fatigue limits in pure tension and pure torsion; writing these also as amplitude values, σ_{oa} and τ_{oa}, it can be shown that

$$A = \frac{3\tau_{oa}}{\sigma_{oa}} - \sqrt{3};\ B = \tau_{oa} \tag{11.6}$$

Other workers have used different scalar quantities to define the deviatoric component (for example, Dang Van used the maximum shear stress) and different approaches to characterise the effect of R ratio; for example, Sines used the mean hydrostatic stress rather than the range, which leads to a dependence on the material's tensile strength.

11.3.3 Critical plane theories

The second type of approach which is used for the prediction of multiaxial fatigue is the so-called 'critical plane approach'. This method uses a different solution to the problem of reducing the stress tensor to scalar quantities. The approach is to refer all stresses to a single plane, the so-called 'critical plane', which is assumed to be the plane on which the initial fatigue crack will form and grow. Papadopoulos discusses critical plane theories and their application in a particularly clear and thorough manner (Papadopoulos, 1998). At this point it is worthwhile recalling the observation that, for plain specimens and blunt notches, fatigue crack initiation is usually controlled by shear stress, because it arises from local dislocation motion. The initial growth of the crack, known as Stage 1 growth, usually occurs on a plane of high shear stress (e.g. a 45° plane under uniaxial tension). After some amount of growth, which is usually similar in magnitude to the material's grain size, the crack turns to grow on a plane perpendicular to the tensile stress: this is known as Stage 2 growth. Critical distance theories define the plane of Stage 1 growth, and therefore the criterion often used is that it is the plane of maximum shear stress amplitude.

This method of referring stresses to a single plane greatly simplifies the mathematics. We can define the shear stress amplitude on this plane, τ, and the tensile stress normal to the plane, σ_n, unambiguously. Many different critical plane theories exist; for example, Matake's criterion uses the amplitude of shear stress and the maximum value of normal stress, giving:

$$\tau_a + \left(\frac{2\tau_{oa}}{\sigma_{oa}} - 1\right)\sigma_{n,\max} = \tau_{oa} \tag{11.7}$$

A new critical plane theory was proposed recently by Susmel and Lazzarin, which is

$$\tau_a + \left(\tau_{oa} - \frac{\sigma_{oa}}{2}\right)\frac{\sigma_{n\max}}{\tau_a} = \tau_{oa} \tag{11.8}$$

The Susmel–Lazzarin (S–L) criterion has been shown to have very good predictive accuracy for a wide range of materials, loading types and R ratios (Susmel and Lazzarin, 2002). In attempting to apply the TCD to multiaxial problems, we have chosen to use this criterion, as will be described in detail below. The philosophy (as with many critical plane criteria) is that fatigue is controlled by shear stress, but that this control is mitigated by the presence of tensile stress on the shear plane, to a degree which depends on the type of material we are dealing with. The term $(\tau_{oa} - \sigma_{oa}/2)$ expresses material behaviour, taking a value that lies between the two extremes of 0 (for a Tresca-controlled or 'ductile' material) and $\sigma_{oa}/2$ (for a completely brittle material). The importance of this term in the equation is dictated by the factor σ_{nmax}/τ_a, which expresses the relative magnitudes of tensile and shear stress on the plane.

11.4 Cracked Bodies: The Factor f_c

Much work has been done to extend LEFM into the field of multiaxial loading. Once a crack is introduced into the problem, some of the complexities of the stress field multiaxiality disappear, because (as noted in Section 1.5) there are only three types of loading which can cause crack propagation: Mode I (tension across the crack faces); Mode II (in-plane shear) and Mode III (out-of-plane shear). We will return to the case of Mode III later: for now we consider only mixed mode I/II situations. Given a mixed-mode loading state $K_I + K_{II}$, the problem is to predict whether the crack will propagate and, if so, in what direction. The solution relies on developing a criterion for crack propagation: many different criteria have been proposed, of which the three most popular are as follows:

(a) *Maximum Hoop Stress*. According to this criterion, the crack will grow in such a way as to maximise the tensile stress across its faces, that is to maximise the Mode I loading. The direction of crack growth can be found by examining the stress field a small distance from the crack tip; it is necessary to move away from the crack tip itself where the stress field is asymptotic. Drawing a circle of radius r, we can examine the tangential stress on this circle, finding the point where it is maximum. We assume that crack propagation occurs when this hoop stress reaches a certain constant value, which is independent of the type of loading.

(b) *Maximum Strain-Energy Release Rate*. This criterion considers a small amount of crack extension, calculating the release of strain energy which occurs. It is assumed that the direction of crack growth will be that which achieves the maximum strain-energy release, since this will provide the greatest driving force for crack growth.

(c) *Minimum Strain-Energy Density*. According to this criterion the crack will grow in the direction in which the strain-energy density is the lowest.

For the case of pure Mode II loading the maximum hoop stress theory predicts that the fracture toughness in pure shear, K_{IIC}, will be smaller than its tensile value K_{IC}, the ratio K_{IIC}/K_{IC} (which is equivalent to our f_p) being 0.866. This theory predicts that, under pure shear loading, the crack will propagate at an angle of 70.5° to the direction of the original crack. The corresponding values for the maximum strain-energy release rate theory are quite similar, being 0.816 and 77.4°, whilst for the minimum strain-energy density theory they are 1.054 and 79.2° (assuming a Poisson's ratio of 0.22). Most experimental data follow the predictions of the hoop stress and strain-energy release rate theories quite accurately (Maccagno and Knott, 1989), and these two theories are generally so close as to be indistinguishable within the scatter in data. These methods have been used also for predicting fatigue crack growth, though here they display some inaccuracy: for example, Suresh shows experimental data in which $f_c = 0.65$, rather lower than the values predicted above (Suresh, 1998).

11.5 Applying the TCD to Multiaxial Failure

We now consider how to predict the effect of a notch subjected to multiaxial loading and in particular the possible use of the TCD in conjunction with the approaches reviewed in the previous two sections for predicting f_p and f_c.

The reader may have noticed the similarity between several of these approaches and the TCD. For example, two of the criteria used for predicting f_c – the maximum hoop stress criterion and the maximum strain-energy release rate criterion – bear great similarity to the PM and FFM, the only difference being in the choice of the distance (or crack extension increment) involved. Likewise one can imagine adapting the critical plane theory, simply by carrying out the calculations at a distance $L/2$ (or averaged over $2L$) rather than at the surface point as would be the normal procedure. On the other hand, it is possible to argue on theoretical grounds (Susmel and Taylor, 2006) that approaches using the scalar invariants cannot be used in conjunction with the TCD because, when applied to bodies containing cracks or notches, the values of the constants A and B change, becoming functions of notch geometry. In the following sections, we consider the use of the TCD for predicting brittle fracture and fatigue under mixed mode I/II situations, first for notches of macroscopic size and then for smaller notches. Finally, we consider the issue of out-of-plane shear, examining the data on tension/torsion loading of circumferential notches.

11.6 Multiaxial Brittle Fracture

The literature on the use of the TCD for multiaxial brittle fracture is quite sparse. As with other uses of the TCD, the greatest progress has been made in the field of fibre composites, where a number of workers have combined the TCD (in the form proposed by Whitney and Nuismer) with multiaxial criteria commonly applied to these materials, such as the Yamada-Sun criterion. These activities were already mentioned in Section 8.5.

Regarding brittle materials, the principal contribution has been by Seweryn and co-workers, who proposed several critical distance theories which were applied to mixed-mode brittle fracture in PMMA (Seweryn, 1998; Seweryn and Lukaszewicz, 2002). PMMA is often used in fracture mechanics studies because it displays classic brittle behaviour, which can be well predicted by theories of mixed-mode crack propagation such as those discussed above (e.g. Maccagno and Knott, 1989; Smith et al., 2001). Seweryn and colleagues considered only long, sharp cracks and sharp V-shaped notches of zero root radius, so their findings cannot be considered to be an exhaustive test of the TCD. However, their results were very encouraging. Figure 11.4 shows an example

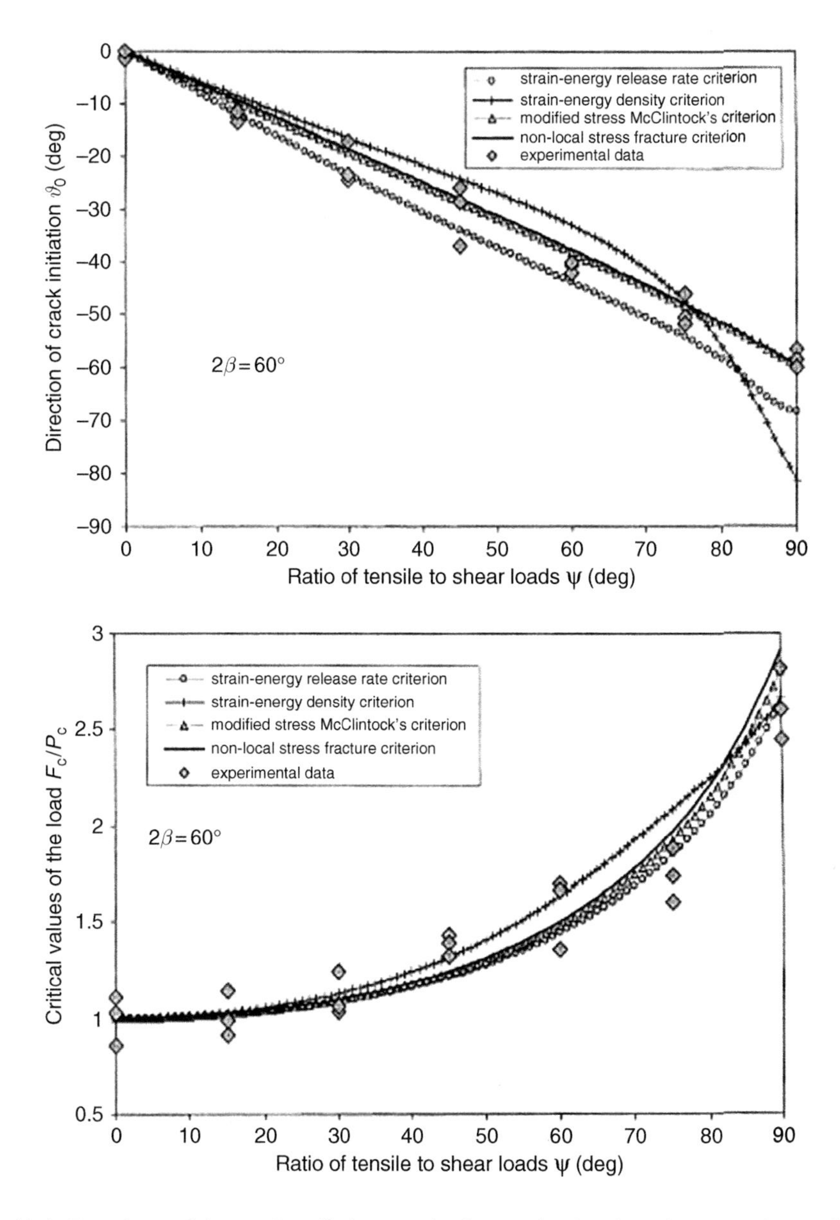

Fig. 11.4. Experimental data and predictions on the fracture load and crack propagation angle for brittle fracture of sharp V-notches in PMMA (Seweryn and Lukaszewicz, 2002). The tension–shear ratio is defined by the loading angle Ψ, which is 0° for pure tension and 90° for pure shear. The 'strain-energy release rate criterion' is approximately equivalent to our FFM, the 'modified stress McClintock's criterion' to our PM and the 'non-local stress fracture criterion' to our LM.

in which data for a V-shaped notch of opening angle 60° was analysed (Seweryn and Lukaszewicz, 2002). Notched specimens were mounted in a testing machine which allowed axial tension to be applied at any chosen angle, thus varying the ratio of tensile to shear forces. Four different theories were used, which correspond more or less to the PM, LM, FFM and strain-energy density approaches, though there were some differences evident in the manner in which the critical distance and critical stress were chosen. Good predictions were achieved for all tension–shear ratios: other notch angles were also tested, and whilst there were some larger errors evident, on the whole the analysis was successful.

11.7 Multiaxial Fatigue

This section reports on some work carried out by Luca Susmel and myself to investigate the application of the TCD to the prediction of the fatigue limit under multiaxial loading conditions. To our knowledge this is the only investigation of this kind to have been conducted to date, though Pluvinage and co-workers have applied their own critical distance method to multiaxial problems (Quilafku et al., 2001).

In our initial study (Susmel and Taylor, 2003), we conducted experiments using V-shaped notches with sharp radii, loaded in tension at various angles of inclination. Figure 11.5 shows an example of the type of test specimen: six different angles were used. Three methods were used to analyse the data:

(i) The PM, applied in exactly the same way as used previously to predict fatigue limits in uniaxial tension. Thus, the stress parameter used was the first principal stress, its critical value being the plain specimen fatigue limit in tension.

(ii) The LM, again applied in the same way, and using the same critical stress, as previously.

(iii) The PM, using the Susmel–Lazzarin critical plane criterion: This will be referred to as the PM/S—L approach. This involved calculating the stresses on a critical plane centred at a point $L/2$ from the notch root.

The direction in which to draw the line for the analysis is not obvious. For example, in using the PM the 'correct' point could lie anywhere on an arc of radius $L/2$ (see Fig. 11.5). Our procedure was to make predictions at all points on the arc, choosing the prediction which gave the lowest fatigue limit. We found that both the LM and the PM/S–L approaches gave good predictions – within 20% of the experimental fatigue limits in all cases. The conventional PM approach was slightly less accurate, with some predictions falling outside the 20% error band. Examination of the specimens showed clear evidence of the two classic stages of crack growth (Fig. 11.6). In general, the initial Stage I growth occurred on the plane of maximum shear stress, whilst the Stage 2 growth coincided with the direction of maximum normal stress. The transition point, at which the crack turned from Stage 1 to Stage 2, occurred at a distance of approximately $L/2$, thus giving a physical interpretation to the critical distance.

Subsequent work (Taylor and Susmel, 2004; Susmel, 2004) involved applying the same approaches to data from the literature. This included a variety of material types, having τ_{oa}/σ_{oa} ratios from 0.6 to 0.9, and a wide range of notches with K_t values from 1.5 to 18. The majority of results came from in-phase tension/shear or tension/torsion

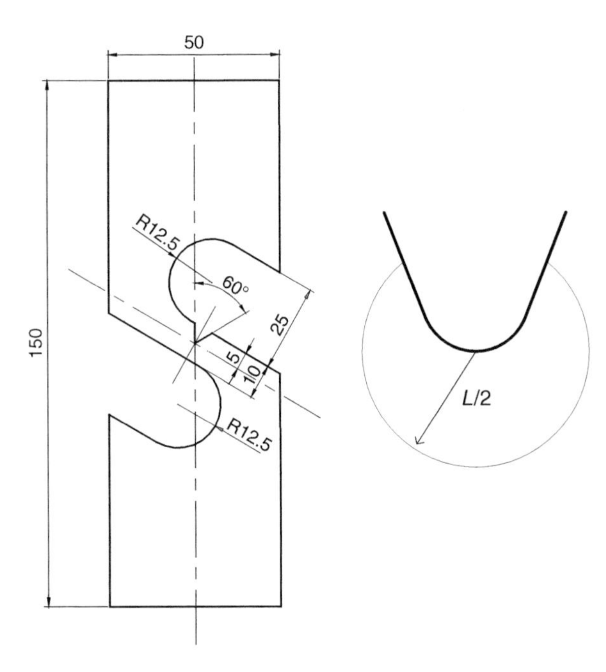

Fig. 11.5. Example of an inclined-notch specimen used for mixed mode I/II testing, and a schematic showing that, for mixed mode loading, the critical point for the PM might lie anywhere on an arc of radius $L/2$.

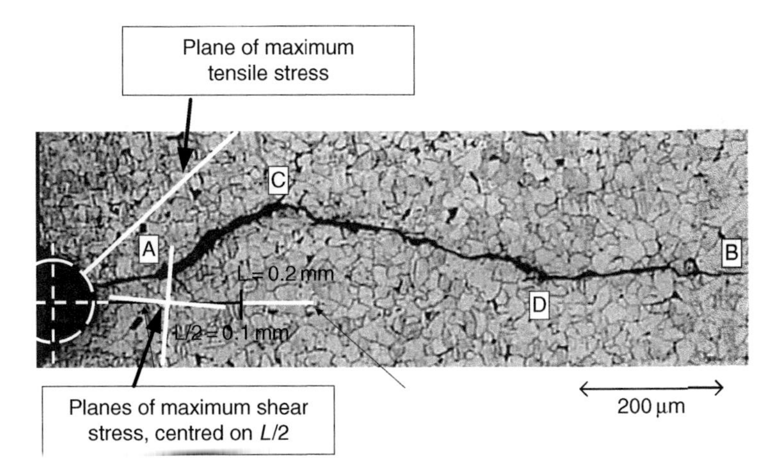

Fig. 11.6. Stage 1 growth (up to $L/2$) occurs on a plane of maximum shear, Stage 2 growth occurs on a plane of maximum tension (from Susmel and Taylor, 2003).

loading, though there were some out-of-phase loadings also. Best results were achieved when the critical plane theory was employed, that is the PM/S–L approach; Fig. 11.7 shows a summary of the predictions. Considering the number of variables involved,

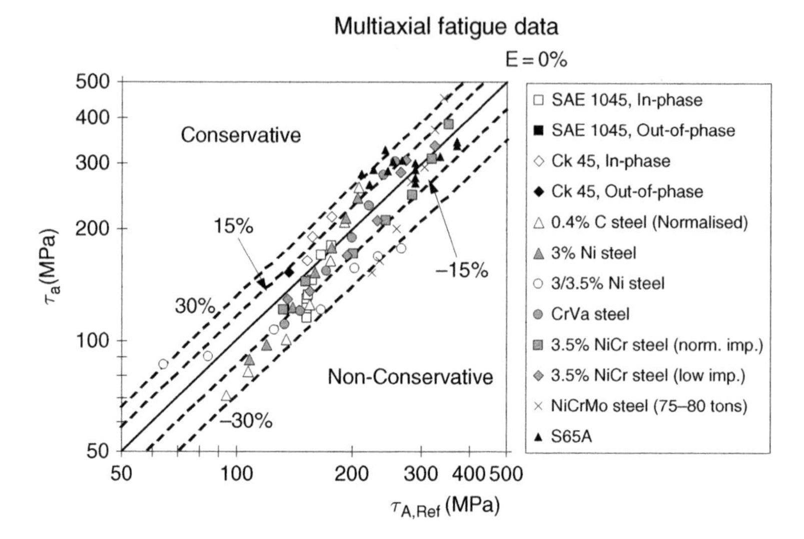

Fig. 11.7. Predicted and experimental fatigue limits for notched specimens under multiaxial loading, using the PM/S–L approach (Susmel, 2004).

and the large amount of scatter in much of the original data, these findings are very encouraging.

11.8 Size Effects in Multiaxial Failure

We have seen in several other places in this book how the TCD is able to predict size effects: changes in the failure stress, or stress-intensity, with changes in the size of the notch or crack, or even of the specimen itself. Such effects are a powerful argument for the use of a critical distance method: they demonstrate that material behaviour changes significantly when the dimensions of the notch, or of the specimen, become similar to the material's critical distance. Size effects in multiaxial loading are particularly interesting because the experimental data clearly show that the nature of the size effect changes with the type of loading. Here we will consider two examples: fatigue of metals and monotonic fracture of bone.

11.8.1 Fatigue

Figure 11.8 shows some data from Murakami, who tested specimens which had small holes drilled into their surfaces, to simulate defects such as porosity and inclusions (Murakami, 2002). His results for a 0.46% C steel tested in rotating bending (which effectively creates a loading of pure tension at the hole) have already been discussed in Chapter 9 (Section 9.2.2) where it was shown that the TCD could accurately predict the rather strong effect of hole diameter on fatigue limit. These data are again shown in Fig. 11.8, accompanied by some data for the same steel, tested in torsion. The plain fatigue limits in bending and torsion are included as horizontal lines. It is immediately clear that the size effect is much greater in torsion than in tension. Holes with diameters

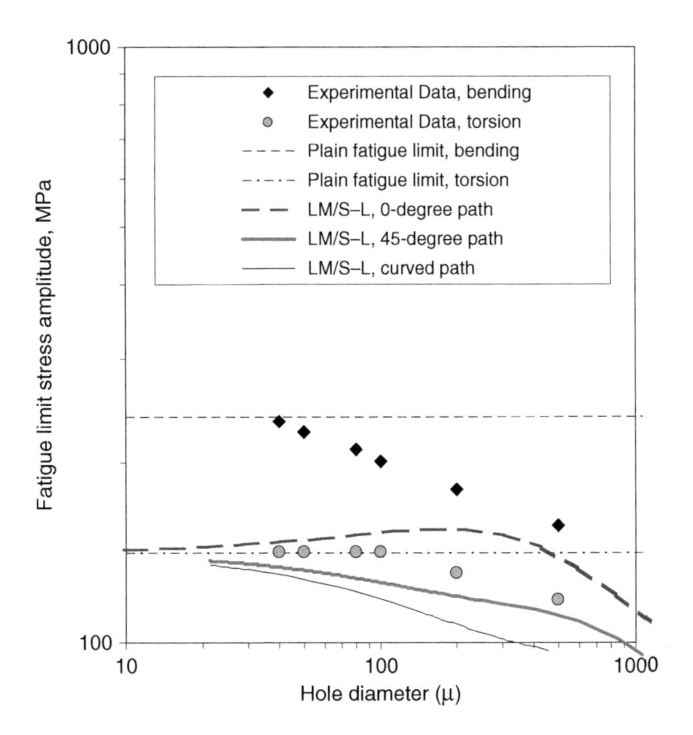

Fig. 11.8. Data from Murakami on holes tested in bending and torsion. Predictions of the torsion data using LM/S–L on three different paths.

up to 100 μm have no effect at all on the fatigue limit – in fact, the fatigue failures occurred not at the holes themselves but elsewhere on the specimens. By contrast, a 100 μm hole reduced the fatigue limit in tension significantly.

Figure 11.9 illustrates the differences that exist between the stress fields around these holes for the tension and torsion cases. Tension creates two points of maximum stress concentration – two 'hot spots' at opposite points across the hole diameter. At these hot spots the tensile stress is concentrated by a factor of 3. In torsion, on the other hand, if the loading is fully reversed (i.e. at $R = -1$), then four hot spots occur, at which the tensile stress is concentrated by a factor of 4 whilst the shear stress is concentrated by a factor of 2.

Since we had previously shown that the LM could predict this tension data very accurately, we decided to investigate its use to predict the torsion data. Obviously an LM prediction using the first principal stress as the characteristic stress (and the tensile fatigue limit as its critical value) will fail because it will tend to the tensile fatigue limit at small hole diameters, rather than the torsion fatigue limit. This approach might possibly be applicable to very large holes in torsion but certainly not to the holes considered here. We therefore used the LM in conjunction with the *S*–L critical plane criterion. This is similar to the analysis described in the previous section, except that, because we are using the LM instead of the PM, we must decide on a suitable focus path: the line

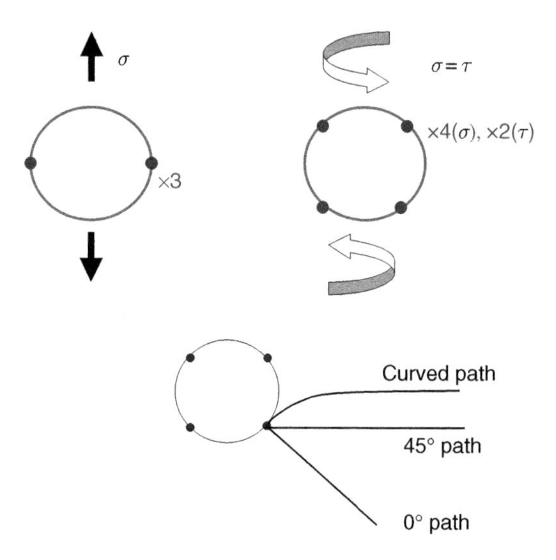

Fig. 11.9. Above – hot spots and stress concentration factors for a hole in tension and torsion; below – the three paths used in the torsion analysis.

on which the stresses will be averaged. Figure 11.9 shows the three paths considered, which were as follows:

(a) A line drawn perpendicular to the hole surface at the hot spot (and therefore also perpendicular to the first principal stress): this will be called the 0° path.

(b) A 45° path, starting at the hot spot. This corresponds to the direction of maximum shear stress and therefore to the critical plane at the hot spot.

(c) A path, starting at the hot spot, which follows the maximum value of the shear stress: this gives a curved path as shown.

Figure 11.8 shows the predictions made using the three paths. Analytical solutions can be obtained using stress fields predicted from the Airy stress functions. For example, the LM/S–L analysis performed on the 0° path (method (a)) yields the following solution for the fatigue limit τ_{oh} for a specimen containing a hole of radius a:

$$\tau_{oh} = \frac{2L}{C}\left[\tau_{oa} - 2\left(\tau_{oa} - \frac{\sigma_{oa}}{2}\right)a^2\frac{D}{C}\right] \tag{11.9}$$

where:

$$C = 2L - a + \frac{2a^2}{(a+2L)} - \frac{a^4}{(a+2L)^3};\ D = \frac{1}{a} - \frac{1}{a+2L}$$

Predictions were also made using FEA, which is more convenient for complex paths such as the curved line.

As figure 11.8 shows, all three paths give reasonable predictions – in fact the maximum error for any of the three predictions is 19%. The curved path of maximum shear stress gives, predictably, the lowest estimates of the fatigue limit. The predictions of the 0° path seem rather counter intuitive, as they increase to a maximum value which is greater than the plain fatigue limit. In fact this behaviour is reflected by the experimental observation that, for holes up to a certain size, failure occurred elsewhere on the specimen. Such holes could be said to be 'stronger' than the rest of the specimen (at least when statistical size effects are taken into account). The lowest errors were obtained using the 45° path, which gave predictions which were slightly conservative and always within 10% of the experimental values.

Of course, for this approach to be generally applicable it must also be able to predict the data from the specimens loaded in tension: as Fig. 11.10 shows, the same approach (LM/S–L) using the 45° and 0° paths is capable of good predictions in this case also. This gives us considerable confidence that the same methodology will be able to predict any intermediate type of loading, that is any mixture of tension and torsion. It is of significance that, in making these predictions, we have not needed to change the value of L, that is the critical distance is a material constant, independent of the type of loading or of the criterion used to predict the fatigue limit. This was also true for the predictions made in the previous section, using the PM/S–L approach.

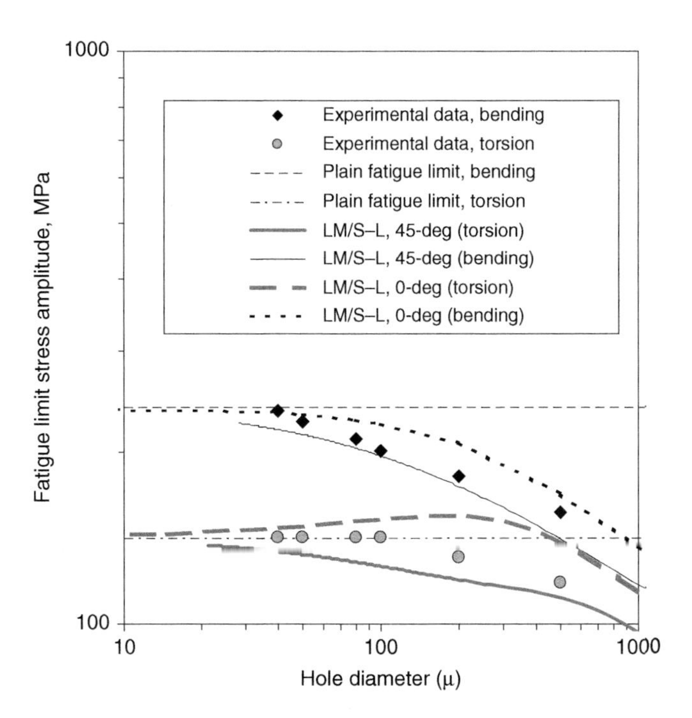

Fig. 11.10. Predictions of both the bending and the torsion data using the 45° and 0° paths. All four give predictions within acceptable limits of accuracy.

These results provide some insight into the question 'Why is the size effect different in torsion and in tension?' The explanation can be found by comparing the stress fields created by these two types of loading, as shown in Fig. 11.11, which displays results obtained from FEA for the case of a hole loaded with a nominal stress of 100 MPa in either tension or torsion.

Consider a hole of diameter $d = L = 150\,\mu\text{m}$: this is a crucial value because holes of this size have almost no effect in torsion but quite a strong effect in tension. When making the LM/S–L prediction, we use the shear and normal stresses averaged over a distance which, for this particular hole, will be $r = 2d$. Over this distance the average value of the stresses in the tension case will be significantly larger than their nominal values of 50 MPa, but in torsion the shear stress, whose value plays a dominant role in the S–L criterion (Eq.11.8), remains almost constant until much smaller distances, of the order of $0.2d$. The rising value of normal stress near the hole does exert some effect but its role in the equation is relatively minor. The consequence of this is that the stress concentration effect of the hole in torsion will not really be felt until we are averaging over distances of less than $0.5d$, that is when the hole diameter itself is of the order of $4L$, which is $600\,\mu\text{m}$.

The success of the 45° path might be attributed to the fact that this corresponds to the classic definition of the critical plane, that is the plane of maximum shear stress at the hot spot. This suggests that the same approach may well be successful for other notches. However, Murakami's observations showed that cracks did not grow in this direction: no Stage 1 growth was evident, rather the crack grew perpendicular to the maximum

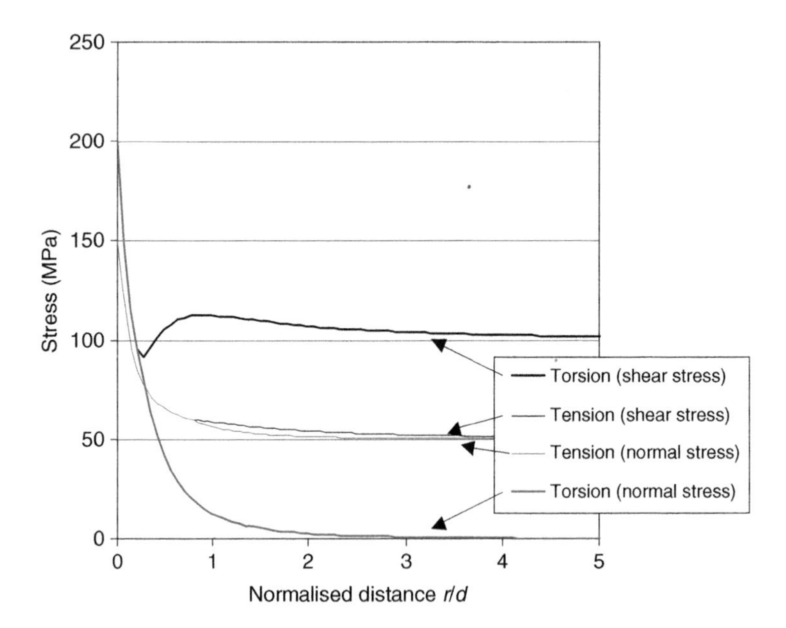

Fig. 11.11. Stress–distance curves for holes in tension and torsion.

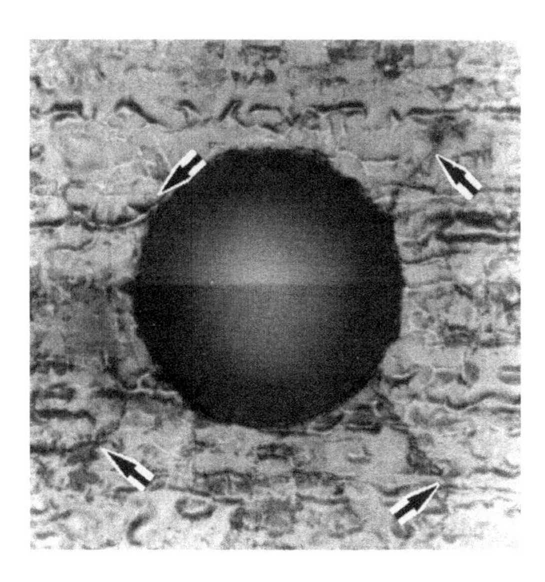

Fig. 11.12. Cracks emanating from a 200 μm diameter hole after testing in cyclic torsion (Murakami, 2002).

tensile stress right from the start (Fig. 11.12). This differs from our own observations of cracks growing from sharp V-notches, which showed definite Stage-1/Stage-2 behaviour (Fig. 11.6). Thus the 0° path may be more physically reasonable. Another possible explanation for the success of the 45° path is that, even if crack growth does not occur in this direction, it is nevertheless the optimum path for dislocation motion. Fatigue crack growth, whether in Stage 1 or Stage 2, invariably occurs by plastic deformation, that is by local dislocation motion near the crack tip. Thus the use of a critical distance approach such as the LM, applied to this path, may reflect the effect of stress gradient on dislocation motion. It is well known that dislocation motion is more difficult if high gradients of strain are present: this phenomenon is known as 'strain gradient plasticity' (Hutchinson, 2000). It has been used to explain other effects, such as the apparent increase in measured hardness of a material with decreasing size of indenter. This effect occurs for indent sizes less than about 100 μm, so it would seem to operate on a similar size scale to the present problem.

11.8.2 Fracture of bone

There is an interesting parallel between the results in the previous section and some data on the fracture of bone. Several researchers have conducted tests on whole bones, drilling holes into them and measuring their strength in bending or torsion (Hopper et al., 1998; Seltzer et al., 1996; Specht et al., 1990;). As Fig. 11.13 shows, there is a clear effect of hole size, which differs with loading mode in just the same way as we saw in Murakami's fatigue results. Quite large holes, of the order of 8 mm diameter, have only a small effect on bending strength, whilst the effect is much greater in torsion. The motivation for this research was to assess the effect of holes and other stress concentration features created during surgery, for example when attaching a metal fracture plate or taking a

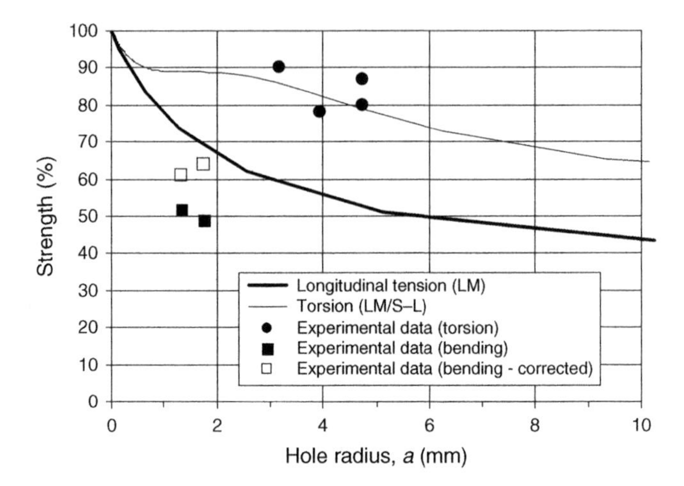

Fig. 11.13. The effect of hole size on the strength of whole bones in bending and torsion: experimental data and predictions. The strength is expressed as a percentage of the strength of intact bones.

biopsy sample. In normal use bones experience both bending and torsion, creating a truly multiaxial situation.

Some other work on bone fracture has already been discussed in Section 8.7, where we presented some of our own test results, showing that bone fracture can be predicted using the TCD. Figure 11.13 shows some predictions using the same multiaxial criterion which was used above to predict metal fatigue: the LM/S–L approach. Note that the bending data had to be corrected because these tests involved clusters of holes, which had a greater effect than single holes. Once this correction was made, however, the predictions were quite accurate. We used the 45° path, and in this material that had an obvious physical meaning, because it coincided with the longitudinal axis of the bone, on which cracks grow preferentially due to the material's strong anisotropy.

11.9 Out-of-Plane Shear

So far in this chapter, we have concentrated primarily on multiaxial loading situations involving mixtures of tension and in-plane shear: Mode I and Mode II in the terminology of fracture mechanics. We now consider the case of Mode III or out-of-plane shear, such as will occur, for example, if we have a specimen of circular cross section containing a circumferential notch, loaded in torsion. Some of the general theories described above are also intended to cover this kind of situation, but it is well known that difficulties arise due to changes in the mechanism of growth. In general, the plane of crack growth will not coincide with the plane of the original crack or notch; in order to grow, the crack front will have to rotate through some angle. This happens, for example, in Stage-2 fatigue crack growth and in the fracture of brittle materials. The crack front divides up into a series of smaller cracks, creating what has been aptly described as a 'factory roof' fracture surface. Continued crack growth is hindered by physical interactions between the crack faces, giving rise to local crack closure and rubbing, all of which can have a significant effect on crack growth rates and thresholds.

We have conducted some investigations to find out whether the TCD can still be used under these circumstances. In fatigue, using data from the literature on torsion (Taylor and Susmel, 2004) and mixed tension/torsion (Susmel, 2004), we found that the PM could be used along with the S–L critical plane theory; reasonable predictions could be made using the normal value of L as calculated from data on simple tension, but the predictions improved significantly if we used a slightly larger critical distance, L_T, derived from estimates of the threshold and fatigue limit in torsion.

A similar result emerged when we tested a brittle polymer, PMMA, using circumferentially notched specimens. In this case the effect was more dramatic, with the critical distance increasing from 0.11 mm in tension (a value similar to that reported previously in Chapter 5) to 0.4 mm in torsion. However, though the critical distance changed significantly, the critical stress (defined as the maximum principal stress) was constant at 113 MPa. Tests on plain specimens showed that this material behaved in a classic brittle fashion, failure occurring at a constant value of the maximum principal stress, equal to 66.6 MPa, in both tension and torsion. In this case the change in L may be linked to a change in the failure mechanism: in tension the material failed suddenly, as soon as a craze formed, but in torsion many small crazes formed at the notch root prior to failure, creating a damage zone (see Fig. 11.14). Thus it may be appropriate to regard the material as classically brittle in tension (with an L value dependant on craze length) but quasi-brittle in torsion (with an L value dependant on the size of the damage zone at failure). We found that accurate predictions could be made for notches in mixed tension/torsion loading, using a critical distance estimated by linear interpolation between the two extreme values, taking the ratio between the minimum and the maximum principal stresses as a measure of the degree of multiaxiality.

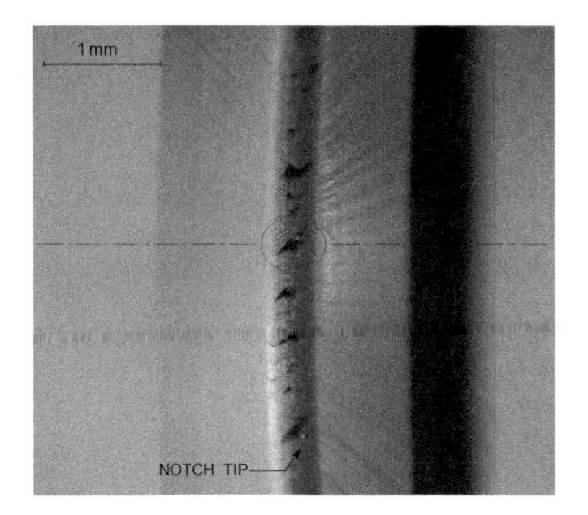

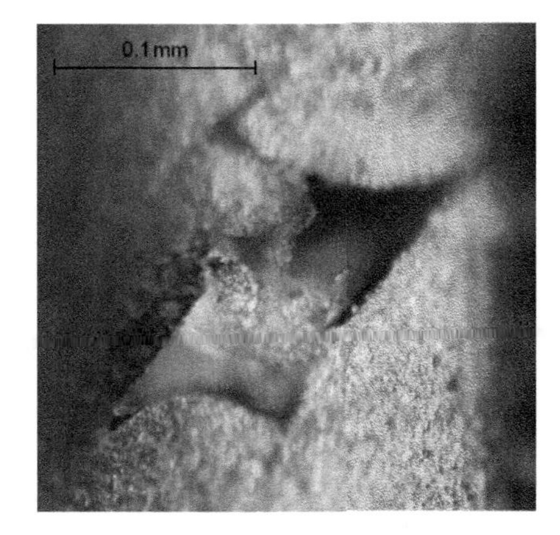

Fig. 11.14. Damage prior to failure at the root of a notch in a specimen of PMMA loaded in torsion (courtesy of F.Pessot).

11.10 Contact Problems

It was mentioned in the previous chapter that stress fields created due to local contact are generally multiaxial in character. A number of workers have used critical plane theories to predict contact related cracking such as fretting fatigue. In one case, a critical plane criterion has been applied in conjunction with the TCD, with considerable success in predicting fretting fatigue limits (see Section 10.4).

11.11 Concluding Remarks

In conclusion, then, the findings reported above suggest that the TCD can certainly be used for multiaxial problems in both fracture and fatigue, though a number of decisions need to be made regarding the failure criterion to be used and the direction of the focus path. Work to date in this area has been limited, but it seems that, at least for in-plane tension/shear problems, existing criteria such as critical plane theories can be successfully adapted for use with the PM or LM. Out-of-plane shear creates further complications. Further work is certainly needed in this area, with the aim of developing a systematic approach which takes account not only of shear stresses but also of the other tensile stresses which arise due to constraint, controlled by specimen thickness, notch orientation and the T-stress.

However, it should be recognised that multiaxial fatigue is a much more complex problem than its uniaxial counterpart, as evidenced by the large number of competing criteria being used in current practice. It is highly likely that there is no one perfect solution, and that different approaches may be optimal for different materials and different types of loading. The important conclusion, for our purposes, is that use of the TCD confers clear advantages when it comes to assessing notches and other stress concentrators subjected to complex loading states.

References

Hopper, S.A., Schneider, R.K., Ratzlaff, M.H., White, K.K., and Johnson, C.H. (1998) Effect of pin hole size and number on the in vitro bone strength in the equine radius loaded in torsion. *American Journal of Veterinary Research* **59**, 201–204.

Hutchinson, J.W. (2000) Plasticity at the micron scale. *International Journal of Solids and Structures* **37**, 225–238.

Maccagno, T.M. and Knott, J.F. (1989) The fracture behaviour of PMMA in mixed modes I and II. *Engineering Fracture Mechanics* **34**, 65–86.

Murakami, Y. (2002) *Metal fatigue: Effects of small defects and nonmetallic inclusions.* Elsevier, Oxford.

Papadopoulos, I.V. (1998) Critical plane approaches in high-cycle fatigue: On the definition of the amplitude and mean value of the shear stress acting on the critical plane. *Fatigue and Fracture of Engineering Materials and Structures* **21**, 269–285.

Quilafku, G., Kadi, N., Dobranski, J., Azari, Z., Gjonaj, M., and Pluvinage, G. (2001) Fatigue specimens subjected to combined loading. Role of hydrostatic pressure. *International Journal of Fatigue* **23**, 689–701.

Seltzer, K.L., Stover, S.M., Taylor, K.T., and Willits, N.H. (1996) The effect of hole diameter on the torsional mechanical properties of the equine third metacarpal bone. *Veterinary Surgery* **25**, 371–375.

Seweryn, A. (1998) A non-local stress and strain energy release rate mixed mode fracture initiation and propagation criteria. *Engineering Fracture Mechanics* **59**, 737–760.

Seweryn, A. and Lukaszewicz, A. (2002) Verification of brittle fracture criteria for elements with V-shaped notches. *Engineering Fracture Mechanics* **69**, 1487–1510.

Smith, D.J., Ayatollahi, M.R., and Pavier, M.J. (2001) The role of T-stress in brittle fracture for linear elastic materials under mixed-mode loading. *Fatigue and Fracture of Engineering Materials and Structures* **24**, 137–150.

Specht, T.E., Miller, G.J., and Colahan, P.T. (1990) Effects of clustered drill holes on the breaking strength of the equine third metacarpal bone. *American Journal of Veterinary Research* **51**, 1242–1246.

Suresh, S. (1998) *Fatigue of materials*. Cambridge University Press, Cambridge UK.

Susmel, L. (2004) A unifying approach to estimate the high-cycle fatigue strength of notched components subjected to both uniaxial and multiaxial cyclic loadings. *Fatigue and Fracture of Engineering Materials and Structures* **27**, 391–411.

Susmel, L. and Lazzarin, P. (2002) A biparametric wohler curve for high cycle multiaxial fatigue assessment. *Fatigue and Fracture of Engineering Materials and Structures* **25**, 63–78.

Susmel, L. and Taylor, D. (2003) Two methods for predicting the multiaxial fatigue limits of sharp notches. *Fatigue and Fracture of Engineering Materials and Structures* **26**, 821–833.

Susmel, L. and Taylor, D. (2006) On the use of the conventional high-cycle multiaxial fatigue criteria reinterpreted in terms of the theory of critical distances. In *Proceedings of Fatigue 2006* Atlanta USA.

Taylor, D. and Susmel, L. (2004) La teoria delle distanze critiche per la stima del limite di fatica a torsione di componenti intagliati. In *Proc. XIV ADM-XXXIII AIAS* (Edited by Demelio, G.) pp. 235–236. AIAS, Bari.

CHAPTER 12

Case Studies and Practical Aspects

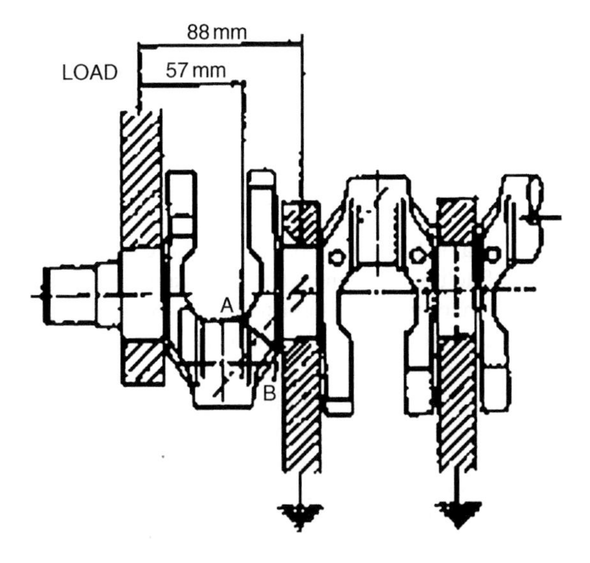

Fig. 12.1. The complexity of engineering components is typified by this automotive crankshaft detail, shown here in a fixture designed to test its fatigue properties.

12.1 Introduction

In previous chapters, the TCD has been explained, and its various applications discussed, largely with reference to simple test specimens containing notches. But, of course, the ultimate test of any theory of fracture is its ability to make predictions in real engineering situations, where stress concentrations arise as a result of the complex shapes and loading modes experienced by real components and structures. In this chapter, we will examine the issues surrounding the application of the TCD in practical situations, beginning with case studies which illustrate its use for some actual components, and component features such as joints. This will hopefully be of value to readers who intend to make use of the technique in design or failure analysis. After some discussion of size effects and 3D features, the chapter concludes by considering a practical problem: the need to make simplified models in the stress analysis of components, and how these models can be analysed using the TCD.

12.2 An Automotive Crankshaft

We will begin with a component which has in fact already been mentioned earlier in this book, in Section 2.3. This crankshaft, which was studied as part of a collaborative project with Rover Ltd, UK, is a good example of the kind of complex geometry and loading that can arise on a real component. In this case, the geometrical complexity arises from the need to accommodate the various rotating bearings: sharp corners arise at the edges of the bearing surfaces and these tend to be the sites of fatigue failure in practice. Loading includes both bending as a result of forces coming up from the pistons and torsion due to rotation of the entire crankshaft. In this project, we decided to test these two loading modes separately. Figure 12.1 shows the set-up in which a bending load was applied to one of the bearings; in a different rig we applied axial torsion, in both cases cyclic loads with a mean of zero (i.e. an R ratio of -1) were used. Failures invariably occurred at the bearing corners: Colour Plate 6 shows an example of a failed shaft after torsion loading.

Standard tests on specimens of the material – which was a spheroidal graphite cast iron, established its fatigue limit in tension as $\Delta\sigma_o = 590\,\text{MPa}$ and its threshold as $23.5\,\text{MPa(m)}^{1/2}$, giving a value for L of 0.505 mm. In this case, we obtained the threshold value not from standard fracture mechanics tests, but rather by finding the fatigue limit for specimens containing sharp notches. In principle the two methods will give the same results, provided the notches are sufficiently sharp, as demonstrated by Smith and Miller (see Section 9.2.1), which in practice means that the notch root radius should be less than the critical distance, L. The notch used should also be sufficiently long to avoid short-crack problems, which means that it should be at least $10L$ in length. In most materials these requirements are not difficult to achieve, and the testing is much simpler and less error-prone than that required to establish ΔK_{th} by standard fracture-mechanics methods. Some threshold values have already been tabulated for various materials (e.g. Taylor, 1985; Taylor and Li, 1993) but care should be taken in using these results to make sure that the material concerned is exactly the same, since variations in composition and thermomechanical treatment can considerably change fatigue properties.

In Chapter 2 we already showed the analysis of the crankshaft in bending, so this will not be repeated here. One point, however, which is worth further discussion is the choice of the focus path: the line from which the stress-distance curve is obtained. As Fig. 2.3 shows, this line was drawn from the hot-spot (the point of local maximum stress at the surface) in a direction perpendicular to the surface at that point. Since the hot spot occurs at a free surface, this line will also be perpendicular to the direction of principal stress at the hot spot, and in this particular case the maximum principal stress was considerably larger than the other two principal stresses, so the situation approximated that of a simple notched tensile specimen. No multiaxial analysis was needed, and in practice the plane of crack growth was seen to coincide with this focus path.

The analysis of the same component loaded in torsion, however, obviously necessitated a multiaxial approach. At the bearing corner, the maximum and minimum principal stresses were approximately equal and opposite, indicating a state close to that of pure shear. Cracks, at least when viewed macroscopically, tended to grow perpendicular to the maximum principal stress, which is to be expected in this relatively brittle material. A multiaxial critical-plane approach was used – the Susmel Lazzarin (S–L) criterion – which has already been described in Section 11.3.3. This involves determining tensile and shear stresses on the expected plane of crack growth, which in this case was the plane of maximum principal stress. The issues involved in the choice of critical plane have been discussed in Section 11.8: there are merits in choosing either the maximum principal stress plane or the maximum shear stress plane, but in cases where the plane of crack growth is already known from practical experience, this will be the obvious choice.

The other parameter required for this multiaxial analysis is the fatigue limit of the material in torsion, $\Delta\tau_o$. This was not available, so it was necessary to estimate it. Using data from similar materials, it was decided that a suitable value for the ratio $\Delta\tau_o/\Delta\sigma_o$ would be 0.85: this reflects the relatively brittle nature of this cast iron. Table 12.1 summarises our results, which show very reasonable accuracy in the prediction of the fatigue limit load ranges both in bending and torsion. For the case of bending, two different FEAs were used: one which had an element size typical of that which would be used in the industry for routine analysis of this component and one which had a more refined mesh. Element sizes in the critical region were 2.8 and 0.8 mm respectively. The finer mesh was better but even the coarse mesh gave sufficient accuracy: the use of coarse meshes is discussed further in Section 12.10.

In this case study, we have not tested the component under its actual in-service loading, but rather under simplified test conditions. In practice it is difficult to determine the actual loadings, but it is known that they consist of a mixture of these two simple cases of bending and torsion. The fact that we have been able to make accurate predictions in both cases suggests that it will also be possible to predict the fatigue limit in any intermediate case of mixed tension/torsion at the critical location. This component was the subject of a paper written a number of years ago (Taylor et al., 1997), at which time we used a different technique known as the CMM, which has been described in Section 3.5. This approach is expected to be useful in cases where the feature is sufficiently sharp and sufficiently large that it can be modelled as a long crack, which proved to be the case here. The advantage of CMM over the TCD is that it can be implemented with coarser mesh, though this was not a particular issue in this case.

Table 12.1. Summary of the Crankshaft analysis

Loading	FE mesh	Fatigue limit (experimental)	Fatigue limit (predicted)
Bending	Coarse	12.0 kN	14.5 kN
Bending	Fine	12.0 kN	12.4 kN
Torsion	Fine	1.55 kNm	1.58 kNm

12.3 A Vehicle Suspension Arm

This second case study came about as a result of a project with the Fiat Research Centre (CRF) in Italy; the component studied was a suspension arm – the so-called 'wishbone' – from a Fiat Punto. The analysis has been published previously (Taylor et al., 2000). This case study illustrates two new aspects: the prediction of finite fatigue life and the prediction of failure location on a component which has more than one stress concentration feature.

Figure 12.2 shows a FE model of the component. In laboratory tests, it was subjected to blocks of variable amplitude loading cycles to simulate repeated braking and acceleration. Failure occurred after an average of 656 blocks, corresponding to 78,720 cycles.

The component has many geometrical features which will give rise to stress concentrations, but two in particular were of interest: (1) a fillet with a very sharp root radius and (2) a region of curvature in the area of a bushing (see Fig. 12.2). The FEA revealed that the maximum stress occurred at the fillet, and commercial software purchased by CRF for fatigue analysis predicted that failure would occur at this feature. In practice, however, failure occurred not at the fillet but at the bushing feature, despite the fact that the maximum stress at this feature was lower and, in addition, the surface in this region was protected by the application of shot blasting, which increased the fatigue life of the material.

Figure 12.3 summarises the analysis of this component using the TCD. The available test data consisted of stress–life curves for plain specimens tested in bending and two

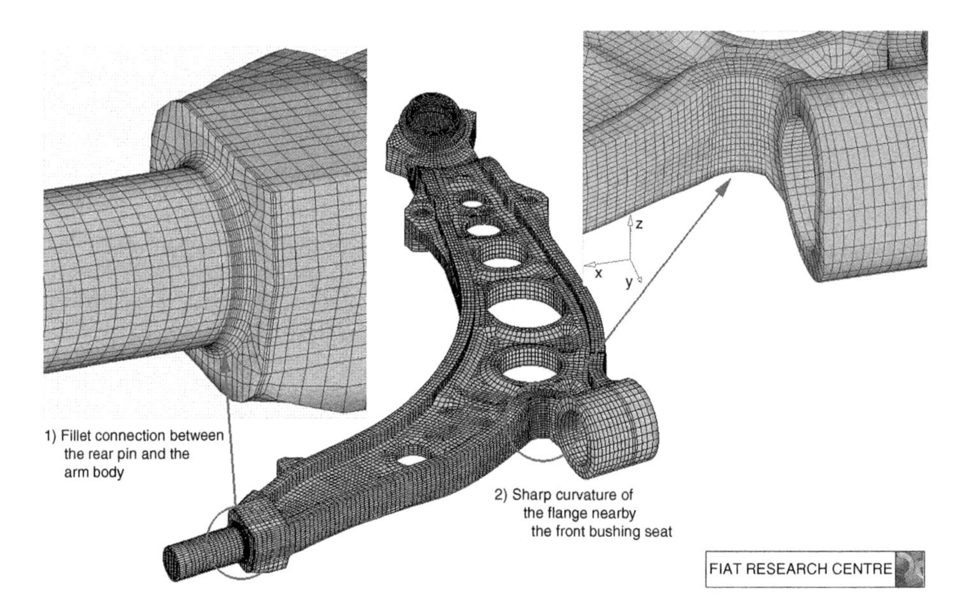

Fig. 12.2. Finite element analysis of the Fiat suspension component, showing two areas of stress concentration: (1) fillet and (2) bushing area.

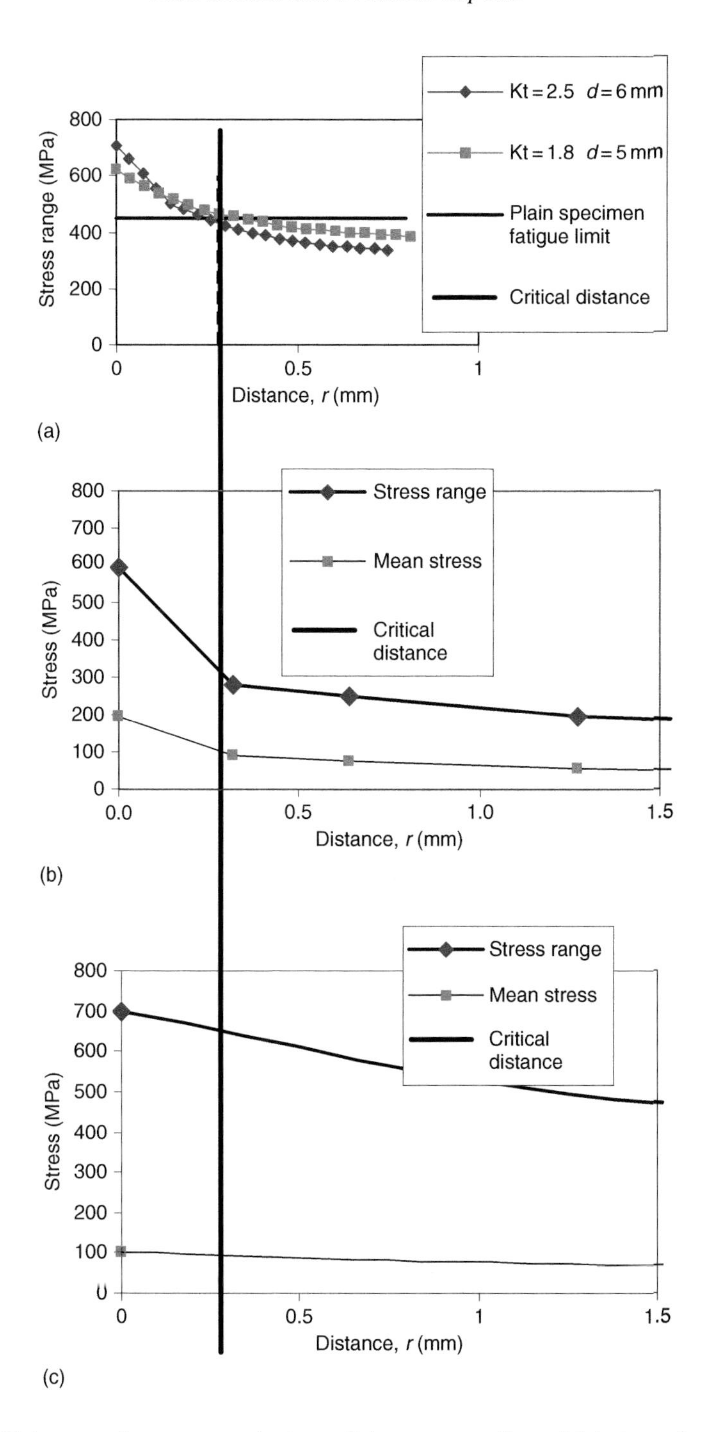

Fig. 12.3. Fiat suspension component: stress–distance curves from: (a) test specimens at their fatigue limit; (b) the fillet feature and (c) the bushing feature. The thick vertical line shows $L/2$ for all three graphs.

different types of notched specimen. Unfortunately, both of the notches were quite blunt, having radii of 1.25 and 0.63 mm and K_t factors of 1.8 and 2.5. Since we did not have data from cracks or sharp notches, a different method was used to estimate the critical distance. This is illustrated in Fig. 12.3a which shows stress–distance curves for the three specimen types when loaded at their respective fatigue limits. If the PM is valid, then all three curves should intersect at a single point, giving us the values of $L/2$ and $\Delta\sigma_o$. In fact there is some scatter but, with reasonable accuracy, we can estimate values of $L = 0.56$ mm and $\Delta\sigma_o = 450$ MPa. This illustrates the fact that we can obtain the necessary parameters for the TCD provided we have data from two different specimen types, such as a plain and a notched specimen or two different notches. However, as one can appreciate from Fig. 12.3a, better accuracy will be obtained if we have data from more than two types of notches, and it is better to have sharper notches, which will have steeper stress–distance curves.

The other two graphs in Fig. 12.3 show stress–distance curves taken from the two component features. The three graphs have been aligned to show a common value for the critical distance $L/2$. The fillet has a very high stress gradient, so the stress at the critical point is much lower than the hot-spot stress: this explains why this feature, though selected by the commercial software as the critical one, is in fact less dangerous than the corner feature, which has a lower hot-spot stress but a much shallower stress gradient.

An extra factor to be taken into account in predicting the fatigue life was the effect of the shot blasting: plain-specimen fatigue data on blasted material was available which could be used to find the increased fatigue strength; the value of L was assumed to be unchanged. The complex loading pattern was analysed using a linear damage accumulation law (Miner's law). The fatigue life was predicted to be 1210 blocks (145,200 cycles), larger than the experimental result by a factor of 1.85, which is very reasonable considering the complexity of the analysis.

This case study has illustrated a very important fact about stress concentration features: the feature which causes the highest stress is not necessarily the one at which failure will occur, if this high stress is also accompanied by a high stress gradient, and therefore a small stressed volume. The TCD, unlike most methods of fatigue prediction currently in use, is able to anticipate this result. This analysis has also illustrated that we can predict fatigue life in the medium- to high-cycle range.

12.4 Failure Analysis of a Marine Component

This case study demonstrates the use of the TCD as a tool in failure analysis. A few years ago, I was approached by a company which was experiencing failures of a certain component in the field. For reasons of confidentiality I cannot give details of this component, suffice it to say that it was a large cast-iron structure with marine applications, which was suffering fatigue cracking from a sharp, right-angle corner. Two modifications were introduced to try to solve the problem. In the first – which I will call Modification 1 – the root radius of the corner was increased from 0.3 to 3.2 mm. Normally this would be a good idea, but in this case it did not work: fatigue failures continued. In the second modification – Modification 2 – the original sharp corner was retained, but the maximum load on the component was reduced by 14%. This was successful in preventing further failures.

Figure 12.4 shows the test data obtained for the material: values of $\Delta\sigma_o$ from plain specimens and ΔK_{th} from standard fracture mechanics tests, as a function of R ratio. Also shown is the calculated value of L, which is almost constant but does decrease slightly with increasing R. Note the particularly large values of the threshold and of L in this material, which was a low strength, grey cast iron. The R ratio of the loading in the original design and in Modification 1 was 0.65; in Modification 2 the reduction in maximum stress lowered the R ratio slightly to 0.56. The value of L changes only slightly in this range so we used a mean value of 3.8 mm. Figure 12.5 shows the stress-distance curves, calculated from FEA, for the three different designs, loaded as they would be in service. It is clear that the sharp corner in the original design causes a very high degree of stress concentration, which is greatly reduced when the radius is increased in Modification 1.

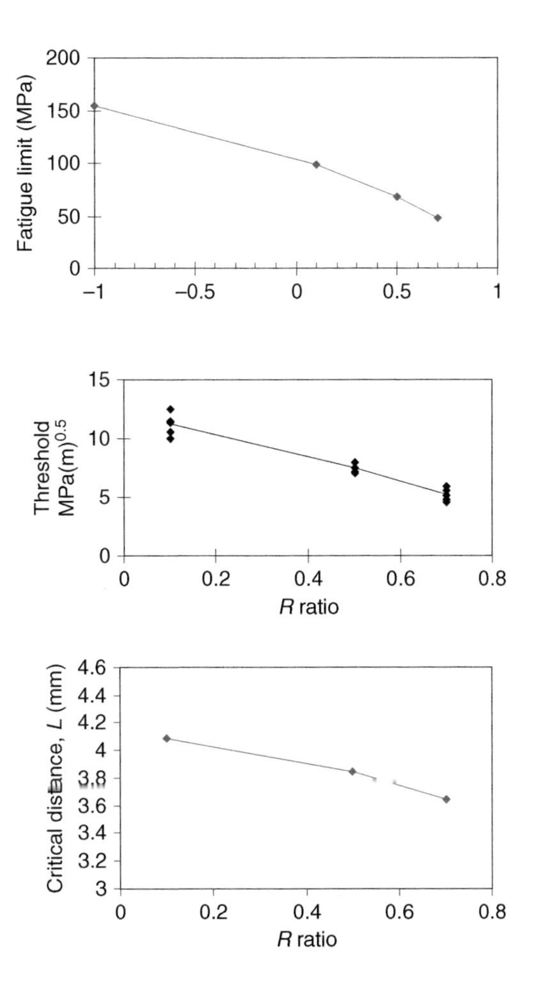

Fig. 12.4. Test data for the cast iron used in the marine component.

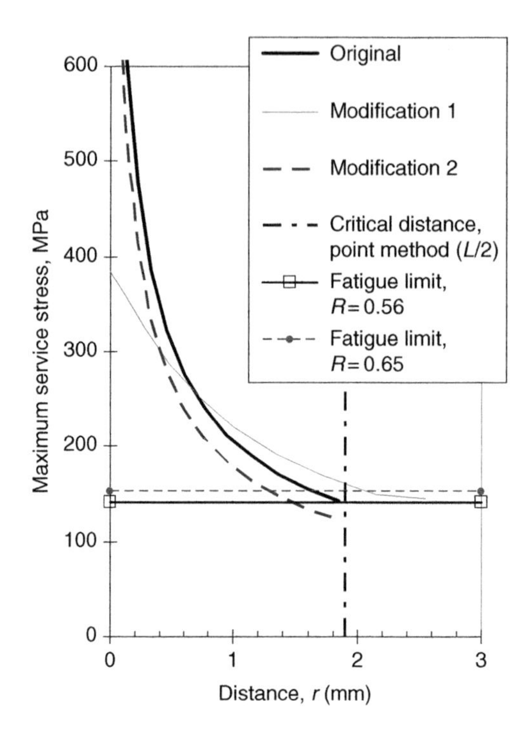

Fig. 12.5. Stress–distance curves for the marine component, also showing the critical distance for the PM and the fatigue limit at $R = 0.56$ and 0.65.

By contrast the load reduction of Modification 2 causes only a modest drop in the hot-spot stress. However, when we look at the stresses at the critical point we see a very different story. The stresses are now much more similar, but if anything Modification 1 has a slightly higher stress than the original design. The fatigue limits at the two R ratios of interest are also shown. In all three cases the stress at the critical point is very close to the respective fatigue limit value, so it would be difficult to be sure whether fatigue failures will occur or not, given a realistic error margin of 20% for this analysis.

If we take the results at face value we would predict that Modification 1 will fail, that the original design will be in danger of failing (being loaded at exactly its fatigue limit), whilst Modification 2 will survive (being below its fatigue limit by 25%). In fact this is exactly what happened in practice.

It is worth noting that the story could have been very different if a different material had been used. This cast iron has a particularly large L value; if we were to replace it with a low-carbon steel, for which L would be less than 0.5 mm, then the critical point will move much closer to the surface and the relative position of the three curves will change: now the increased radius of Modification 2 will have a real benefit.

This case study has illustrated the value of the TCD in the analysis of component failures after the fact, and in the consideration of various design options. A full description of this project has been published elsewhere (Taylor, 2005).

12.5 A Component Feature: Angled Holes

This work was conducted to investigate an issue raised by Goodrich, a company which makes components for aircraft. Some of their components consist of blocks of aluminium alloy in which are drilled a complex series of intersecting holes and cylindrical cavities. At points where these features intersect, FEA revealed some very high, but highly localised stresses. A method was required for conducting a fatigue limit analysis under these conditions: the TCD seemed to offer a potential solution.

Rather than analysing an entire component – which would have been very complex and difficult to test experimentally – we decided to design a specimen which would mimic the essential features of the highly stressed regions already identified. Figure 12.6 shows the result: a rectangular tensile specimen containing a circular hole, the hole being drilled at an angle of 45° to the surface. This creates a stress concentration near a very thin metal edge – which we called 'the knife-edge'. In fact the hot spot occurs on the hole surface but slightly away from the specimen surface. The K_t value for this kind of hole is a function of the angle: in the present case it had a value of 4.8. Similar specimens had previously been used in brittle fracture tests on PMMA (see Section 6.2.2).

A cast aluminium alloy L51 was used; tests were carried out at $R = -1$, for which the material had a plain-specimen fatigue limit (at 2 million cycles) of 121 MPa. The threshold and critical distance were found by testing sharply notched specimens. One problem which arose here was that, due to the small sizes of specimens available, the

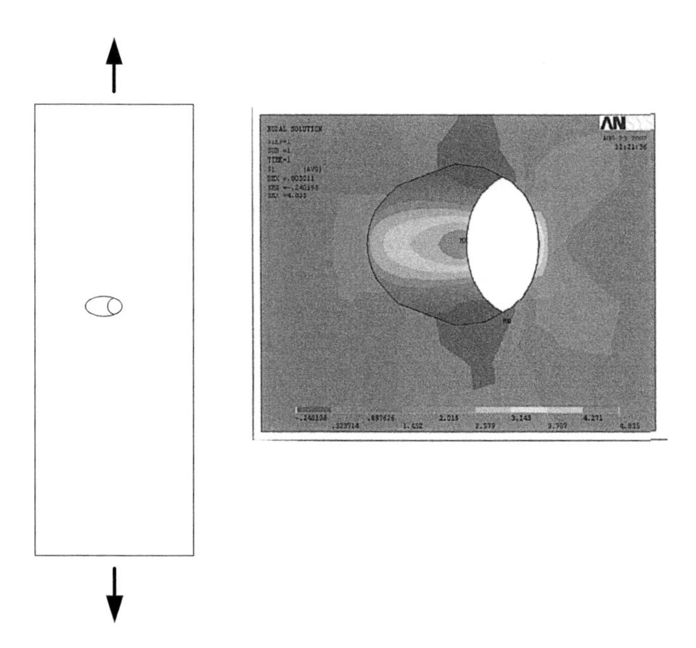

Fig. 12.6. Angled hole 'knife-edge' specimen, showing contours of maximum principal stress obtained from FEA. The hole radius was 4.5 mm.

depth of the notch was rather small, necessitating a correction for notch size effects. There are various ways that this could be done – in this case we assumed that the El Haddad approach (see Section 9.2.3), which was developed for short crack, would also be valid for this small, sharp notch. As noted previously (Section 3.3.2), the El Haddad approach, which is an ICM, is in fact identical to the LM when applied to cracks. This gave a ΔK_{th} value of $6.05\,MPa(m)^{1/2}$. The predicted limit at 2 million cycles for the angled-hole specimens was a nominal stress range of 44.7 MPa, which compared favourably with the experimental value of 54.4 MPa. This means that, whilst the K_t of this hole was 4.8, its K_f (fatigue strength reduction factor) was only 2.2, a good example of the notch insensitivity of low-strength materials. This kind of analysis, as well as giving the company a justification for leaving rather high stresses in their components, allows a more rational choice to be made between different materials – a choice in which the best material may not be the one with the highest plain fatigue limit.

12.6 Welded Joints

Welded joints provide several challenges: they are geometrically complex, creating stress concentrations; the details of the local geometry are highly variable, such as the weld bead size and shape and the degree of penetration, and the material in and around the weld has different properties from the base metal. Welds are very often the weak points in structures, especially when it comes to fatigue failure.

A lot of effort has gone into developing procedures for the analysis of welded joints, but these procedures still have many shortcomings. For industrial designers, the normal way to predict the fatigue behaviour of welds is to use a national standard such as BS7608:1993 *Fatigue Design and Assessment of Steel Structures* or the recently published Eurocode 3. The basic procedure for analysing a weld is to place it within a certain class, based on the geometry of the joint and type of loading and, to a certain extent, the quality of the welding – for example, the degree of penetration of weld metal. Once a class is established for the weld, the fatigue life and fatigue limit are expressed in terms of the nominal stress applied to the joint. A lot of thought has gone into these standards and, in cases where a weld class and nominal stress can be clearly identified, they work very well, at least for steels of low and medium strength for which plenty of experimental data exist. This is partly because, for these steels, fatigue behaviour is more or less the same, independent of that of the base metal. Unfortunately the choice of the appropriate class is often not obvious, and an incorrect choice can make a very big difference.

A number of procedures have been suggested to allow welded joints to be assessed more rationally, especially making use of FEA. Radaj considered various ways to improve the definition of the nominal stress (Radaj, 1990), and, along with Sonsino, applied a number of critical distance methods, including Neuber's imaginary radius method (Section 3.3.1) and the ICM (see Section 3.3.2) (Sonsino et al., 1999). Atzori, Lazzarin and co-workers used both the ICM and the NSIF (Section 4.5) (Atzori et al., 1999; Lazzarin et al., 2003) in conjunction with a volume-averaged strain energy (Livieri and Lazzarin, 2005).

A different approach, developed specifically for welded joints, is the extrapolation or 'hot-spot stress' approach. The maximum stress value predicted by FEA and similar methods is unreliable because it depends critically on the local geometry near the weld, which is variable. If the edge of the weld toe is modelled as having zero radius, then a singularity exists at which the local elastic stress is theoretically infinite. In practice, this will mean that the maximum stress depends on the density of the FE mesh, always increasing with mesh refinement. The extrapolation method avoids this by defining the stress at the weld by extrapolation of the stress gradient distant from the weld. The idea is that the joint possesses a certain stress-concentration factor which can be determined, independently of the local geometry. This approach has been highly developed by some workers (Niemi, 1995); personally I am not fond of it because I am not convinced of the soundness of the basic idea, and I find that in practice it is very difficult to specify how to carry out the extrapolation. Nevertheless it is fair to say that it is being used quite widely, and is included in some national standards in addition to the classification methods.

Other workers have developed detailed models similar to those used for fatigue studies elsewhere, in which the various stages of crack initiation, short-crack growth and long-crack growth are explicitly modelled (e.g. Toyosada et al., 2004).

12.6.1 Application of the TCD to fatigue in welded joints

We have explored the use of the TCD to predict fatigue in welded joints in steels (Taylor et al., 2002) and in aluminium alloys (Crupi et al., 2005), with considerable success. In order to find the necessary mechanical properties, we tested plain and notched butt welds as shown in Fig. 12.7. In a simple butt weld which is ground flat to make a plain specimen, failure occurs in the heat affected zone (HAZ) close to the edge of the weld metal.

By testing specimens which had a sharp notch machined at this point we were able to measure ΔK_{th} and L using the same approach as described above. For the low-carbon steel studied, at a fatigue life of 5×10^6 cycles, the results were $\Delta\sigma_o = 153\,\text{MPa}$, $\Delta K_{th} = 5.62\,\text{MPa(m)}^{1/2}$, $L = 0.43\,\text{mm}$; these values are probably applicable to welds made from most low- and medium-strength steels, which all tend to have similar fatigue behaviour.

These results were used to make a number of predictions. First we considered a T-shaped weld made from the same material (Fig. 12.8): the fatigue limit was predicted within 10% error using the PM, LM or the CMM, with reasonable predictions possible even with quite coarse FE meshes (see Table 12.2). This is important because the mesh density is limited by practical concerns in many cases, especially in large structures with many joints such as vehicle chassis components. The mesh shown in Fig. 12.8, though clearly quite coarse, was sufficiently fine to ensure good predictions from all three methods. Another important practical aspect is the accuracy with which the weld area can be modelled. We used a simple model of the weld bead as a triangular prism with zero radius at its edges, as shown in Fig. 12.8; the resulting singularity was of no concern because we were not examining stresses at the singular point, so it was possible to arrive at a converged solution. This issue is discussed again in Section 12.10.

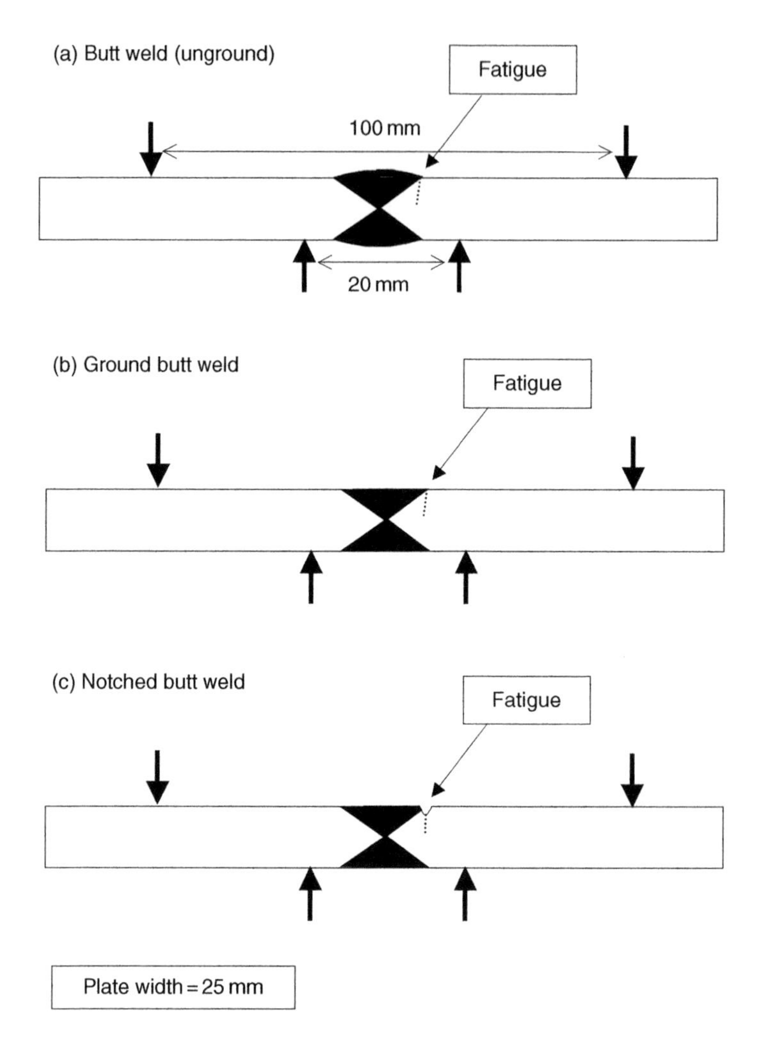

Fig. 12.7. Test specimens of butt welds: unground, ground and notched. The plain fatigue limit was obtained from the ground specimens, and the threshold value from the notched specimens.

Two other well-known problems in welded joints were analysed using this approach. Figure 12.9 shows the effect of bead-to-base-metal angle for unground butt welds (Gurney, 1979): our PM prediction describes the experimental data very nicely. Figure 12.9 also shows data and predictions for cruciform welds (non-load-carrying) which display a strong size effect: increased plate thickness results in lower fatigue strength (Lazzarin and Livieri, 2001). Various theories have been advanced to explain this phenomenon, but the use of the TCD shows that it is just the same effect which occurs in many other stress concentration features (see Section 12.9) which the TCD predicts very easily.

The above examples were characterised by two features which simplified the analysis: they were essentially 2D in character, and stresses at the weld toe were dominated by

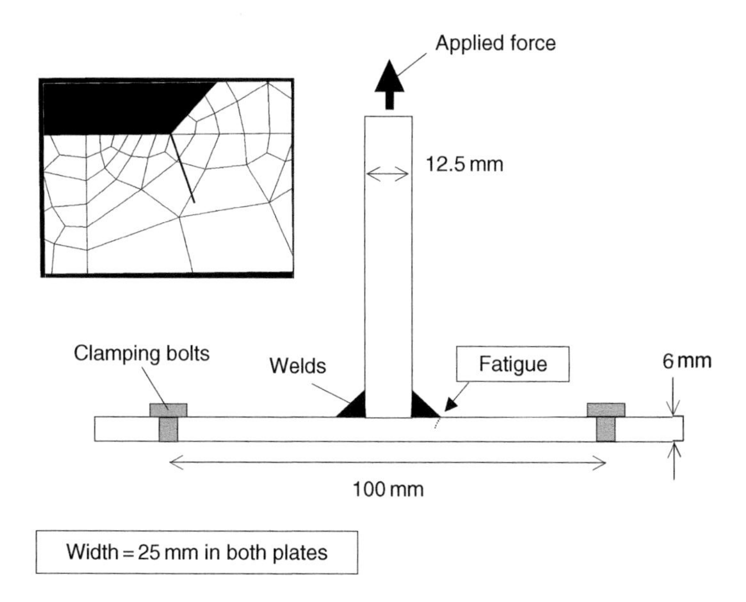

Fig. 12.8. T-shaped weld specimen: fatigue failure occurred from the corner of the weld bead as indicated. Also shown is the FE model at the failure location.

Table 12.2. Experimental and predicted fatigue limits for the T-shaped weld and unground butt weld, using different methods and mesh sizes

Weld and FE model number	Experimental fatigue limit	FE mesh size (mm)	Crack modelling method (CMM)	Point method (PM)	Line method (LM)
T-shape 1	2.5 kN	0.05	2.53 kN	2.24 kN	2.46 kN
T-shape 2	2.5 kN	0.19	2.38 kN	2.32 kN	2.56 kN
T-shape 3	2.5 kN	0.19	2.61 kN	2.38 kN	2.77 kN
T-shape 4	2.5 kN	1.5	2.60 kN	2.15 kN	2.66 kN
T-shape 5	2.5 kN	3.0	3.04 kN	3.27 kN	3.59 kN
Butt 1	140 MPa	0.22	102 MPa	131 MPa	138 MPa
Butt 2	140 MPa	1.5	96 MPa	125 MPa	139 MPa

a single tensile principal stress. In general, welded joints can have complex 3D shapes (see Section12.8) and display multiaxial stress states (Susmel and Tovo, 2006).

12.7 Other Joints

There are many other methods of joining parts in a structure, all of which lead to stress concentration and in many cases, points of structural weakness. These include adhesive joints and mechanical joints using pins, rivets, bolts etc. The TCD has been applied quite extensively to the failure of mechanical joints in fibre composite materials, as described in Section 8.5. Here it was shown that the TCD, combined with a suitable multiaxial

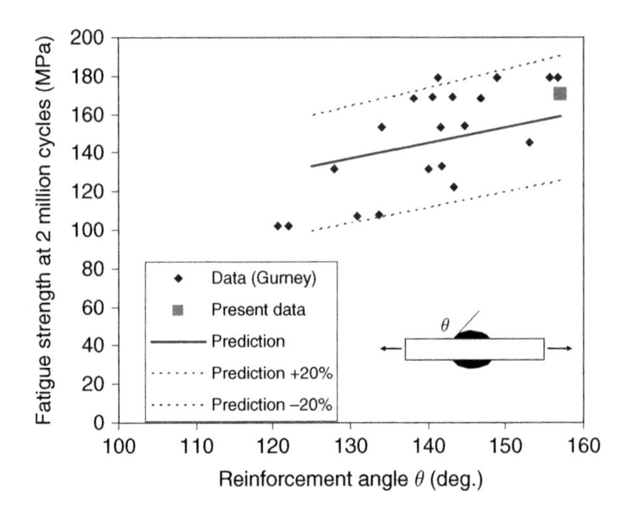

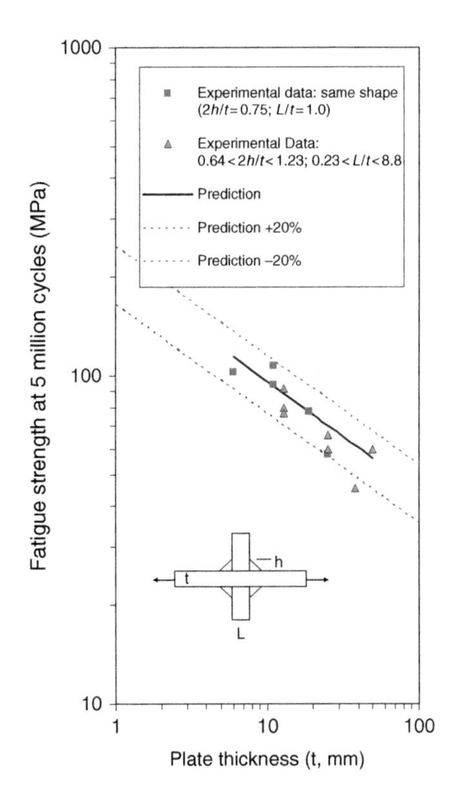

Fig. 12.9. Data and PM predictions for (above) butt welds, showing the effect of angle θ and (below) cruciform welds (showing the size effect).

criterion, was able to predict both compressive and tensile failures around features such as pin-loaded holes and repair patches.

The case of adhesive joints is an interesting one, involving as it does two different materials and two possible types of failure: adhesive failure at the interface or cohesive failure within the glue itself. Many adhesive joints are dominated by shear loading: Fig. 12.10 shows an example of the distribution of shear stress in double lap joint. Note that, unlike a typical notch, the maximum stress occurs not at the geometrical discontinuity itself but some distance from it. Nevertheless, it seems that the TCD can still make valid predictions: Fig. 12.10 shows some data generated in our laboratories

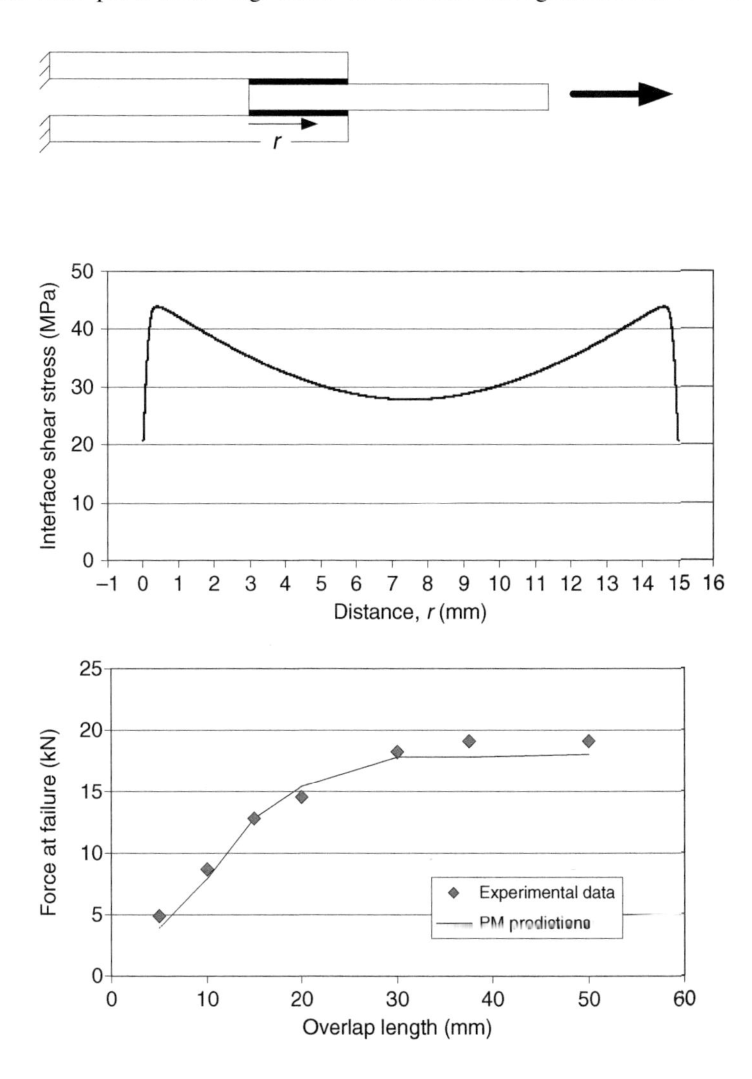

Fig. 12.10. An adhesive joint (shear, double-overlap) showing (above) the interface shear stress as a function of distance from one end (in this case for an overlap of 15 mm) and (below) experimental and predicted joint strength.

by A.Martinez which displays the classic behaviour of lap joints – the failure load initially increases with overlap length but eventually levels off, indicating a maximum useful length for the joint. This behaviour occurs because in long overlaps there is very little stress carried in the central portion. Predictions using the PM are very successful, as shown.

It is clear that, in the general case, a multiaxial stress parameter will be needed to capture the effects of shear and normal stresses across the joint (see Chapter 11), but in this particular case the shear stress was very much the dominant term so a prediction based on this stress alone was successful. Interestingly the value of L here was very large, equal to 3.9 mm. Some under-prediction is occurring with the smallest overlap, of 5 mm, as one would expect, since now the critical distance occurs at more that halfway along the overlap distance. It is likely that a modified method such as the FFM/LM (see Section 3.3.6) would be needed for smaller overlaps, and in fact such combined stress/energy methods have been used for predicting the initiation of failure in adhesive lap joints (Braccini et al., 2005) and sealing joints between ceramic components (Muller et al., 2005). Ribeiro-Ayeh and Hallstrom used the PM to predict the strength of bi-material joints in which a polymer foam was joined to aluminium or PMMA (Ribeiro-Ayeh and Hallestrom, 2003). They used the maximum principal stress, whilst noting that it would not be appropriate for all types of joints.

12.8 Three-Dimensional Stress Concentrations

Many researchers studying stress-concentration and stress-gradient effects have considered only 2D problems, such as through-thickness notches in flat plates. Whilst this is an obvious place to start, it is clearly not sufficient for a complete theory of fracture, because many situations involve 3D features. Features in real components often have stress gradients in three dimensions, and even in simple notches a 3D element arises in multiaxial loading states such as tension/torsion. As we saw in Chapter 11, the TCD can still be used in such cases by combining it with a critical plane theory, which implies the identification of the appropriate plane, after which we return to an essentially 2D approach applied on that plane.

We have seen in the preceding sections of this chapter that the TCD is capable of making accurate predictions for 3D features, such as crankshaft corners and angled holes. But when developing a theory, one should always try to find problems that the theory cannot solve, and to this end we discovered some types of 3D features for which the PM and the LM gave poor predictions (Bellett et al., 2005). Two examples are shown in Figs 12.11 and 12.12. We first came across the problem when analysing welds, and indeed the geometry of Fig. 12.11 often arises when two overlapping sheets are welded together. But we subsequently found that the same type of problem can be created even in specimens made from solid material: the specimen in Fig. 12.12 was made specially to illustrate this. When we used the TCD we found that the predictions were underestimating the actual fatigue strength of the specimens; the typical error was of the order of a factor of 2, but varied with specimen design, being as high as 4 in some cases. The predictions were always conservative, which is some consolation for engineering designers, but nevertheless the errors were unacceptably high.

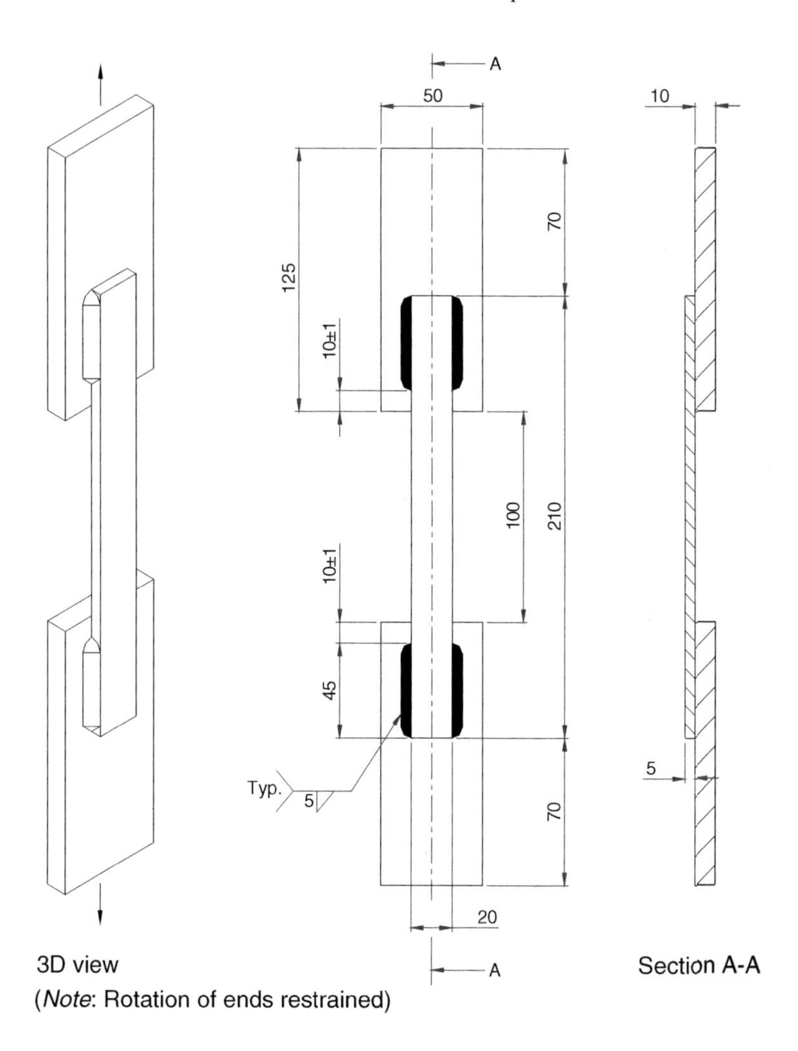

Fig. 12.11. A test specimen containing welded joints. Four weld beads run longitudinally to connect steel plates of different widths. The specimen was loaded in axial tension with restraints to prevent rotation. Fatigue failures occurred from the ends of the weld beads: the measured fatigue limit was considerably higher than predicted using the TCD.

It became clear that the essential element in these features, which is lacking in other 3D features, is that there is a high gradient of stress in orthogonal directions. Features such as the crankshaft corner (Section 12.2) are 3D in the sense that the hot spot occurs at a single point, rather than along a line as would occur in a 2D notch, so there is a finite stress gradient in all directions. However, for features like the crankshaft corner, there is one dominant direction on which the stress gradient is high – in this case the direction going into the material, normal to the surface – whilst in other orthogonal directions (i.e. moving along the surface) the stress gradient is much smaller. Figure 12.13 shows another example of such a feature, this is a normal notched specimen which has been rotated through 90° before testing in bend; the hot spot is on the top surface but the

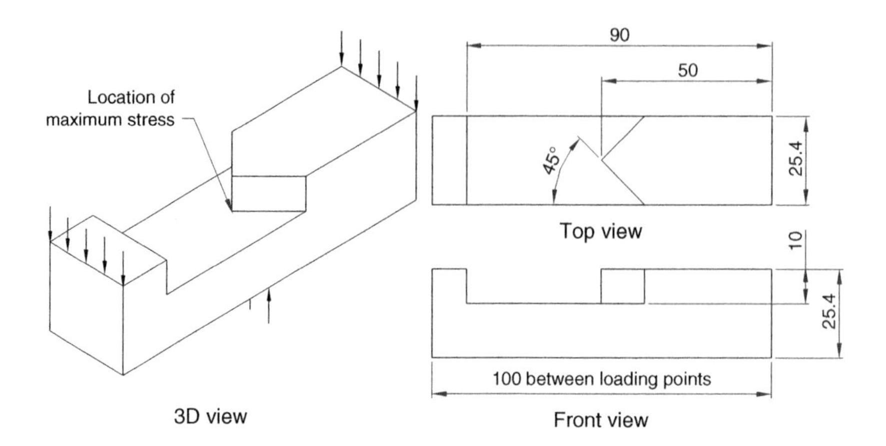

Fig. 12.12. A specimen machined from solid steel, which contains a three-dimensional stress concentration feature (arrowed). Tested in three-point bending, this specimen had a fatigue limit which was higher than predicted using the TCD.

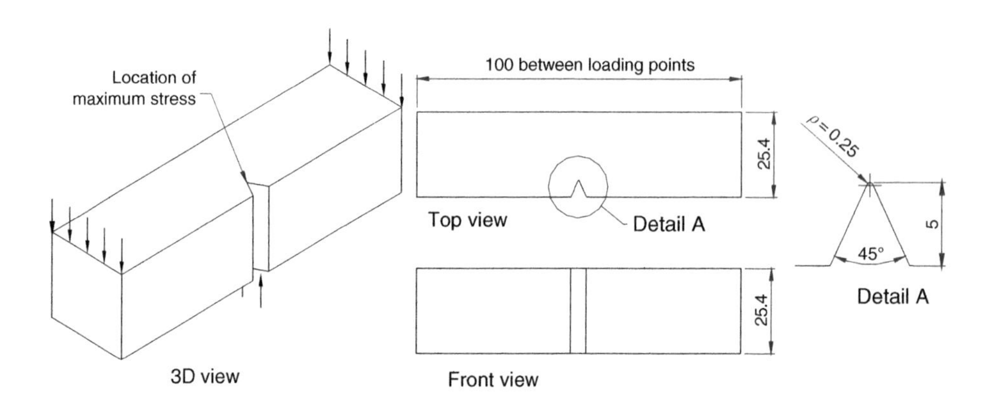

Fig. 12.13. A conventional single-edge-notched specimen tested in three-point bend with the loading direction parallel to the notch root. This creates a stress gradient along the notch root; however, TCD predictions of this specimen's fatigue limit were accurate.

stress gradient found when moving along the notch root is relatively mild. Figure 12.14 compares the stress gradients for these two specimens.

We found that this is not just a problem for the TCD, other theories such as the local strain approach (based on maximum plastic strain range) and Pluvinage's volumetric method (a critical-distance approach using elastic/plastic analysis) also gave highly conservative predictions: details of these analyses can be found in the published paper (Bellett et al., 2005). It has been shown elsewhere that the NSIF method also gives conservative errors when applied to 3D features (Tanaka et al., 2002). To date we have not come up with a satisfactory explanation for this problem, which certainly merits further study. Some possible reasons for the discrepancy include changes in the shape of the initiating cracks

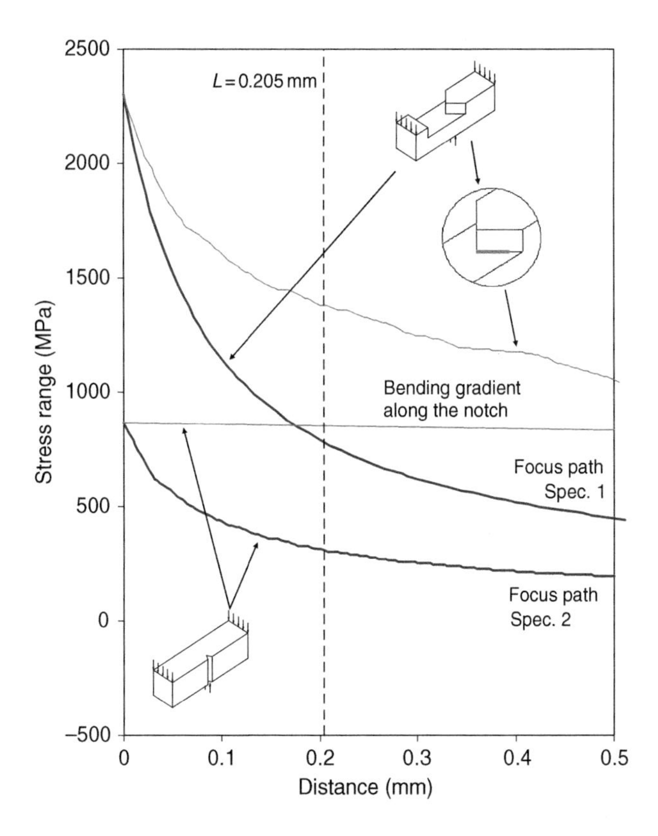

Fig. 12.14. Stress–distance curves taken from the two specimens illustrated in figs 12.12 and 12.13. The thick lines represent a path normal to the specimen surface, the thin lines a path along the surface at the notch.

(this idea was developed in a recent publication (Bellett and Taylor, 2006) but found to give only a partial explanation), changes in the stressed volume (suggesting a statistical size effect) and differences in the level of constraint, which appears to be lower in these features.

12.9 Size Effects and Microscopic Components

Size effects have already been mentioned at several points in this book, though mostly when considering the size of geometric features. It has been shown that, for a given shape, the strength of a notched specimen decreases as the notch size increases. A common example is the effect of hole radius, as shown for example in Section 8.2 regarding the monotonic strength of composites, and again in Section 9.2.2 for the fatigue strength of metals. If one increases the size of a hole in a specimen of fixed width, then one will of course also decrease the remaining ligament, with a consequent rise in the stress concentration factor of the notch. However, it is clear that this is not the main reason for the size effect, which persists even when the notch size is much smaller than the specimen size.

Another, related, effect is the change in strength which occurs when we change the size of the entire specimen, including the notch. Changes in size which occur in this way, at constant shape, are often referred to as *scaling effects*. It is obvious that, if we scale up the entire specimen and adjust the applied loads so that the nominal stress is still the same, then the stress field in the larger specimen is identical to that in the smaller one, except for the change of scale. In classical mechanics there is no reason why this should lead to a change in strength, since the maximum stress is unchanged. In fracture mechanics, however, a reduction in strength will be predicted because we have changed the length of the crack; so the expected scaling law in the case of a cracked body (assuming it obeys LEFM) is that strength is proportional to the square root of size. For notched bodies having stress gradients which are lower than those of cracks we can expect a weaker scaling law. Size effects can also be expected in any other cases where a stress gradient arises, from the relatively low stress gradients in plain beams loaded in bending (Section 5.3) to the severe gradients found in contact problems (Chapter 10).

These size effects, which occur as a result of the geometry of the body, and in particular as a result of the stress gradient which this geometry causes, are referred to as *geometrical size effects*. We have seen in several places in this book that the TCD can be very successful in predicting geometrical size effects. There is, however, another type of size effect, which is the *statistical size effect*. Larger specimens tend to be weaker, even when loaded in simple tension; now it is not the stress gradient which is causing the effect but the absolute size of the specimen. This effect is normally explained on statistical grounds, using approaches such as the Weibull method (see Sections 4.4 and 5.2.4). The essential assumption here is that failure follows the 'weakest-link' rule, that is failure will occur from the worst place in the specimen. For example, if failure occurs by brittle fracture from pre-existing defects (as is often the case in ceramic materials), then the worst place will correspond to the largest defect; alternatively the worst place may be the weakest grain in the material, from which a fatigue crack might initiate most easily. In practice, it is often difficult to define exactly what we mean by the worst place, but this doesn't really matter provided we assume that it can be described by a statistical distribution, since this means that the strength of the body will also follow a statistical distribution. The most commonly used form is the Weibull distribution:

$$P_f = 1 - \exp\left[-\left(\frac{\sigma}{\sigma^*}\right)^b\right] \tag{12.1}$$

Here P_f is the cumulative probability of failure for an applied stress of σ, with b and σ^* being material constants. There are other forms of the Weibull equation, and also other distributions which can be used: the essential feature is the weakest-link assumption, since this naturally leads to the size effect. For a larger specimen, there will be a greater probability of finding a defect of a given size, and therefore a greater probability of failure at any given applied stress.

In situations where stress gradients exist, we have two possible sources for the size and scaling effects: increasing the scale will reduce the stress gradient, but it will also increase the volume of material experiencing high stresses. It is difficult to separate geometric size effects from statistical size effects, and at the present time this is a problem which is largely unresolved. In some fields, especially that of building materials such as concrete, statistical size effects have traditionally been used. The main issue here is the fact that these materials are used to construct very large structures, such as buildings and dams, which can be expected to have lower strength than the test specimens that are used to measure material properties. Weibull-type methods can give useful predictions, but some modifications are needed, especially regarding the extremely low probability end of the distribution. Carpinteri and co-workers have developed an ingeneous theory based on the concept of fractals, which is also capable of predicting size effects in these materials (Carpinteri and Cornetti, 2002; Carpinteri et al., 2003). Process zone models have also been successfully used in this field (Bazant, 2004; Carpinteri et al., 2002). Though it is not obvious at first sight, it turns out that process zone models contain an inherent scaling law. If the remaining ligament is much larger than the process zone, then failure tends to occur at constant zone size, but if the ligament size is reduced, the process zone at failure also reduces. It turns out that this occurs at sizes of the same order of magnitude as our critical length parameter L, suggesting a link between process zone models and the TCD which will be explored in the next chapter (Section 13.4.2).

We have also been able to predict size effects in concrete using the TCD, even for the case of plain, unnotched specimens in which the stress gradient arises due to bending. This was mentioned in Section 5.3 where it was shown that a TCD-like method, the FFM, was able to give reasonable predictions. However, there is a crisis in the predictions which will obviously arise when the width of the remaining ligament becomes less than the critical distance. In using the PM and LM we will now be considering stresses in material which is outside the specimen! Equally absurd is the case of the energy-based methods (i.e. the ICM and the FFM) because now the crack that occurs will be large enough to completely break the specimen. This problem is discussed in Section 3.3.6, where it is shown that this crisis can be avoided by combining the stress-based and energy-based methods in a double criterion. Interestingly, the result is rather similar to that of the process zone models: a critical distance which reduces in the case of small ligaments, remaining always less than the ligament width.

Weibull type methods have been used successfully in other fields, for example to predict the fatigue strength of bones in animals of different sizes (Taylor, 2000) and to predict brittle cleavage fracture from cracks and notches in steel (Beremin, 1983), though it is worth mentioning that most realisations of the Beremin method also use a constant critical distance parameter as well. At the present time, then, it is not really possible to unravel these two size effects, since many problems can be successfully solved using either one or the other, or a combination of both. This is not a very satisfactory situation, because the underlying mechanisms assumed are very different in the two approaches, so this is a very useful area for future research.

An appreciation of size and scaling effects is important not only for the design of large structures such as concrete dams but also for the analysis of very small components such as micro-scale and nano-scale electro-mechanical systems, the so-called MEMS and NEMS devices, and nano-materials such as carbon nanotubes. Some analysis of the materials was mentioned in Section 5.5. At an intermediate level we find components which we may call 'microscopic' in the sense that the size of the load-bearing section is of the same order of magnitude as microstructural features. This would include, for example, thin wires and components used in electronic devices and small biomedical components such as cardiovascular stents, which are used to reinforce arteries after angioplasty operations. We investigated these stents, showing that stress concentration features in them could be analysed using the TCD (Wiersma and Taylor, 2005; Wiersma et al., 2006). The value of L required was smaller than that for macroscopic specimens of the same material, which was 316L stainless steel. At first this appeared to be an effect similar to that described above for the analysis of concrete, but it emerged that the lower L value was occurring due to a reduced value of the threshold ΔK_{th} which was probably due to the very low thickness of the material tested, since it continued to occur even when the notch size and remaining ligament were quite large.

12.10 Simplified Models

The FEA and other numerical modelling methods have greatly expanded our ability to estimate the stresses in components and structures. However, there are still many cases where the models that we make are not as accurate as we would like them to be. We have to settle for simplified models. In particular, there are two common simplifications: low mesh density and defeaturing (Chaves, 2002).

12.10.1 Mesh density

If the structure to be modelled is large, or if the features of interest are relatively small, sharp or detailed, then it may be difficult to achieve the necessary mesh density to obtain a precise stress analysis. If the stress gradient is low, so that there is relatively little change in stress over a distance L from the notch root, then a low mesh density will naturally be sufficient to provide an accurate picture. But in most cases of interest the stress gradients will be higher than this: now, in order to use the TCD, it is not necessary to describe the stress field perfectly, provided we can have a reasonable approximation of how it changes over distances of the order of L. This is possible using linear mesh elements, if the element size near the notch is less than or equal to L. At present, this can be a practical limitation on the use of the TCD in some situations, especially large, complex components or those made from high-strength materials having very small L values. However, computing power is increasing at a great rate and advances are being made in the development of improved methods for handling numerical problems, so as time goes on we can expect that more and more practical problems will become accessible.

12.10.2 Defeaturing

Defeaturing is the name given to the making of a model in which some features of the original component are missing or simplified. Figure 12.15 shows a simple example of defeaturing: a corner with a finite root radius is represented without the radius. Such procedures save time and may be essential shortcuts in making a large complex model. We investigated the effect that defeaturing would have on the accuracy of failure predictions. A useful starting point here is the observation made in several places in this book with respect to the data on notches (e.g. Fig. 5.12, Fig. 9.2): notches with root radii less than some critical value have the same behaviour as cracks. We showed that the TCD could predict this phenomenon, which is also the basis of two other methods of analysis: the CMM (Section 4.5) and the method of Smith and Miller (Section 9.2.1),

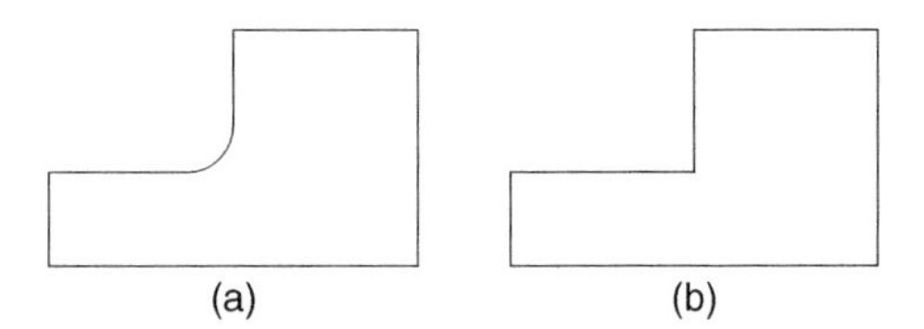

Fig. 12.15. A simple example of defeaturing: the original shape (a) is modelled (b) without the fillet radius.

both of which rely on modelling the notch as a crack. The implication is that if the root radius of a feature is small enough, then it can be changed to zero without altering the outcome, either in the prediction or in reality.

In practice the critical root radius depends on other features of the notch size and shape, but for long, thin notches (notch length being much greater than root radius) a typical value is $2L$; defeaturing can be carried out on notches with radii smaller than $2L$ without significant loss of accuracy. For some types of notches, this limit can be increased to as much as $5L$. The defeatured notch will be a stress singularity in the FE model, but this is not a problem for the TCD as we do not rely on the stress value at the notch root when making the predictions: this is a considerable practical advantage. Care must be taken with defeaturing, however, as it may alter local stresses for other reasons, for example by changing the net load-bearing area.

12.11 Concluding Remarks

This chapter has demonstrated that the TCD is not only capable of predicting the behaviour of simple notched specimens, but can be applied to the analysis of real components and features such as welds and joints, allowing its use as part of the design process and as an investigative tool in failure analysis. The use of a critical length scale allows geometrical size effects to be predicted, though the behaviour of some microscopic components still provides challenges. Practical advantages which the TCD has over other methods are its relative simplicity and the fact that it requires only a few mechanical properties, which can be determined from simple tests or available data on the material. It is insensitive to stress singularities, allowing simplified, defeatured models to be used in some circumstances. Significant problems arise with the assessment of 3D features in which the stress gradient is high in two orthogonal directions, but this appears to be a problem for all current methods of failure prediction. An important disadvantage is the need for a fine FE mesh, with elements of the order of L in size, but this limitation is gradually losing relevance thanks to increases in computing power.

References

Atzori, B., Lazzarin, P., and Tovo, R. (1999) From a local stress approach to fracture mechanics: A comprehensive evaluation of the fatigue strength of welded joints. *Fatigue and Fracture of Engineering Materials and Structures* **22**, 369–381.

Bazant, Z.P. (2004) Quasibrittle fracture scaling and size effect. *Materials and Structures* **37**, 1–25.

Bellett, D. and Taylor, D. (2006) The effect of crack shape on the fatigue limit of three-dimensional stress concentrations. *International Journal of Fatigue* **28**, 114–123.
Bellett, D., Taylor, D., Marco, S., Mazzeo, E., and Pircher, T. (2005) The fatigue behaviour of three-dimensional stress concentrations. *International Journal of Fatigue* **27**, 207–221.
Beremin, F.M. (1983) A local criterion for cleavage fracture of a nuclear pressure vessel steel. *Metallurgical Transactions A* **14A**, 2277–2287.
Braccini, M., Dupeux, M., and Leguillon, D. (2005) Initiation of failure in a single lap joint. In *Proceedings of the 11th International Conference on Fracture* (Edited by Carpinteri, A.) p. 86. ESIS, Turin, Italy.
Carpinteri, A., Chiaia, B., and Cornetti, P. (2002) A scale-invarianty cohesive crack model for quasi-brittle materials. *Engineering Fracture Mechanics* **69**, 207–217.
Carpinteri, A., Chiaia, B., and Cornetti, P. (2003) On the mechanics of quasi-brittle materials with a fractal microstructure. *Engineering Fracture Mechanics* **70**, 2321–2349.
Carpinteri, A. and Cornetti, P. (2002) Size effects on concrete tensile fracture properties: An interpretation of the fractal approach based on the aggregate grading. *Journal of the Mechanical Behaviour of Materials* **13**, 233–246.
Chaves, V. (2002) Use of simplified models in fatigue prediction of components with stress concentrations. MSc Trinity College Dublin.
Crupi, V., Crupi, G., Guglielmino, E., and Taylor, D. (2005) Fatigue assessment of welded joints using critical distance and other methods. *Engineering Failure Analysis* **12**, 129–142.
Gurney, T.R. (1979) *Fatigue of welded structures*. Cambridge University Press, Cambridge UK.
Lazzarin, P., Lassen, T., and Livieri, P. (2003) A notch stress intensity approach applied to fatigue life predictions of welded joints with different local toe geometry. *Fatigue and Fracture of Engineering Materials and Structures* **26**, 49–58.
Lazzarin, P. and Livieri, P. (2001) Notch stress intensity factors and fatigue strength of aluminium and steel welded joints. *International Journal of Fatigue* **23**, 225–232.
Livieri, P. and Lazzarin, P. (2005) Fatigue strength of steel and aluminium welded joints based on generalised stress intensity factors and local strain energy values. *International Journal of Fracture* **133**, 247–276.
Muller, A., Hohe, J., and Beckler, W. (2005) On the evaluation of interfacial crack initiation by means of finite fracture mechanics. In *Proceedings of the 11th International Conference on Fracture* (Edited by Carpinteri, A.) p. 407. ESIS, Turin, Italy.
Niemi, E. (1995) *Stress determination for fatigue analysis of welded components*. Abington Publishing, Cambridge, UK.
Radaj, D. (1990) *Design and analysis of fatigue resistant welded structures*. Abington Publishing, Cambridge UK.
Ribeiro-Ayeh, S. and Hallestrom, S. (2003) Strength prediction of beams with bi-material butt-joints. *Engineering Fracture Mechanics* **70**, 1491–1507.
Sonsino, C.M., Radaj, D., Brandt, U., and Lehrke, H.P. (1999) Fatigue assessment of welded joints in AlMg 4.5Mn aluminium alloy (AA 5083) by local approaches. *International Journal of Fatigue* **21**, 985–999.
Susmel, L. and Tovo, R. (2006) Local and structural multiaxilal stress states in welded joints under fatigue loading. *International Journal of Fatigue* **28**, 564–575.
Tanaka, K., Okajima, H., and Koibuchi, K. (2002) Fatigue strength CAE system for three-dimensional welded structures. *Fatigue and Fracture of Engineering Materials and Structures* **25**, 275–282.
Taylor, D. (1985) *A compendium of fatigue threshold and growth rates*. EMAS, Warley, UK.
Taylor, D. (2000) Scaling effects in the fatigue strength of bones from different animals. *Journal of Theoretical Biology* **206**, 299–307.
Taylor, D. (2005) Analysis of fatigue failures in components using the theory of critical distances. *Engineering Failure Analysis* **12**, 906–914.
Taylor, D., Barrett, N., and Lucano, G. (2002) Some new methods for predicting fatigue in welded joints. *International Journal of Fatigue* **24**, 509–518.

Taylor, D., Bologna, P., and Bel Knani, K. (2000) Prediction of fatigue failure location on a component using a critical distance method. *International Journal of Fatigue* **22**, 735–742.

Taylor, D., Ciepalowicz, A.J., Rogers, P., and Devlukia, J. (1997) Prediction of fatigue failure in a crankshaft using the technique of crack modelling. *Fatigue and Fracture of Engineering Materials and Structures* **20**, 13–21.

Taylor, D. and Li, J. (1993) *Sourcebook on fatigue crack propagation: Thresholds and crack closure*. EMAS, Warley, UK.

Toyosada, M., Gotoh, K., and Niwa, T. (2004) Fatigue life assessment for welded structures without initial defects: An algorithm for predicting fatigue crack growth from a sound site. *International Journal of Fatigue* **26**, 993–1002.

Wiersma, S., Dolan, F., and Taylor, D. (2006) Fatigue and fracture of materials used for micro-scale biomedical components. *Bio-medical Materials and Engineering* **16**, 137–146.

Wiersma, S. and Taylor, D. (2005) Fatigue of material used in microscopic components. *Fatigue and Fracture of Engineering Materials and Structures* **28**, 1153–1160.

CHAPTER 13

Theoretical Aspects

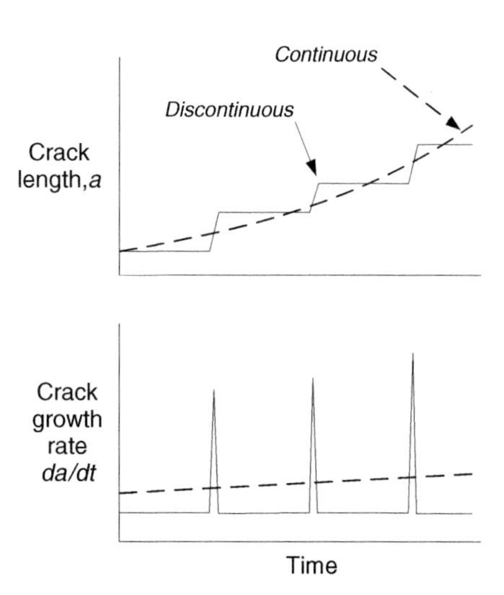

Fig. 13.1. Crack growth is often discontinuous, leading to periodic spikes on the plot of growth rate versus time. This type of growth is captured by the TCD in the form of finite fracture mechanics (FFM).

13.1 Introduction

In this final chapter we focus on the theoretical arguments surrounding the TCD. Many of the previous chapters have concentrated on practical aspects, showing how the TCD can be used in various forms and how it can be applied to predict different failure modes in different materials and structures. I hope that the reader will have been convinced by the wealth of examples showing how the theory can predict experimental data: this is, in my view, the most important test of any scientific theory and should give confidence that the approach is valid for the solution of many industrial problems.

However, there are some other questions which need to be considered in order to complete the picture. Fundamental questions such as 'Why does the TCD work?', 'What is the theoretical basis of the TCD?' and 'How does the TCD compare with other theories of fracture?'. It is natural to seek answers to these questions and to be curious about how the TCD – an extremely simple theory to use – can be so effective in practice. This chapter will consider these questions, bringing together information and ideas presented in earlier chapters. In my view, we are still not in a position to give complete answers to these fundamental questions: the TCD, despite its 50-year history, is still a work in progress. However, some important advances have been made recently which allow us to see what kind of theory the TCD is and to begin to understand why it is so successful in practice.

13.2 What Is the TCD?

In trying to think about what kind of theory the TCD is, it is useful first to recall some points made in Chapter 1 about the various types of theories that are used to predict fracture processes. On one side there are the continuum mechanics theories, which assume that the material is a homogeneous continuum with certain mechanical properties such as a Young's modulus and a fracture toughness. LEFM is one such theory. Continuum approaches allow us to predict failure in complex engineering structures, and they also allow us to define mechanical properties such as K_c to enable comparisons to be made between materials. But such approaches will never explain why one material has a better toughness than another. For insight into this question we turn to micro-mechanistic approaches – theories which model the physical mechanisms that occur during fracture. These models are necessarily simplifications of reality because the mechanical behaviour of a material at the microstructural scale is highly complex.

Between these two extremes lie methods which have elements of both continuum mechanics and mechanistic models. Some of these were discussed in Chapter 4, including process zone models and local approaches. The TCD is one such model, lying closer to the continuum mechanics end of the spectrum. There has been very little work done with the aim of placing the TCD on a firm theoretical foundation within continuum mechanics (but see Adib and Pluvinage, 2003; Mikhailov, 1995). In my view, the TCD is essentially a modification of LEFM in which the existence of mechanisms at the microstructural level is represented by the introduction of a single length scale parameter, L. It is important to take this view of the TCD because it avoids us getting too preoccupied with questions such as 'Why does L take a particular value in a particular material?'. If we apply the same question to K_c in the context of LEFM, we realise that it is a question which LEFM cannot answer, so there is no reason to suppose that the TCD will provide insight into this question either. On the positive side, the TCD shares with LEFM the advantage of continuum mechanics theories that it is not confined to one particular mechanism of failure: LEFM can predict failure (within certain limitations) if the failure occurs by crack propagation, irrespective of the mechanism of crack propagation involved.

13.3 Why Does the TCD Work?

Over the years, a number of explanations have been offered to explain the success of the various methods which make up the TCD. These will be explored below as the TCD is compared to other theories. Briefly, the PM can be compared directly to some mechanistic theories such as RKR and void-growth models (see Chapter 4) in which the stress at a particular distance from the notch root is the operative parameter. Likewise the LM can be considered in the context of mechanisms that involve a line of a certain length, such as a crack or a simplified process zone. More realistic plastic zones and damage zones can be associated with the Area and Volume methods. These analogies are useful but we should not expect them to be exact, for the reasons noted above: micro-mechanisms are complex so it is unlikely that we could ever represent them using a theory with so few parameters.

In my view, the key to understanding the TCD lies in Finite Fracture Mechanics (FFM). This approach, which was described in Section 3.3.5, forms one of the four methods which I classify under the general heading of the TCD. In this approach, crack propagation is assumed to be a discontinuous process, occurring in steps of length $2L$ rather than continuously and smoothly as in the traditional LEFM theory. Figure 13.1 represents this process schematically; the outcome in mathematical terms is to replace the differential form of the Griffith's energy balance:

$$\frac{dW}{da} = G_c \tag{13.1}$$

with an integral form in which the strain energy is summed over the interval of crack growth:

$$\int_{a}^{a+2L} dW = G_c \Delta a \tag{13.2}$$

We showed earlier that this approach was capable of making predictions which were very similar to those of the stress-based TCD methods, especially the LM. Indeed for some simple situations the LM and FFM can be shown to be mathematically identical. Further demonstrations of this similarity can be found in a recent publication (Taylor et al., 2005).

The importance of FFM, in my opinion, is that whilst being completely valid from the point of view of continuum mechanics, it is also representing an important feature of the physical mechanism of cracking processes. When we examine crack growth in real materials, we often find that it is indeed a discontinuous process. Figures 13.2 and 13.3 show two examples, the first from work on fatigue crack growth in an aluminium alloy (Blom et al., 1986) and the second from slow cracking at constant load in bone (Hazenberg et al., 2006). In both cases the crack growth rate rises and falls, the minima being associated with microstructural features: grain boundaries in the case of fatigue, Volkman's canals in the case of bone. We can postulate that such microstructural barriers to crack growth occur in virtually all materials; often they act as a vital toughening mechanism. It is normally assumed that such barriers operate to make crack growth discontinuous even when it occurs too rapidly for us to observe it, in cleavage fracture

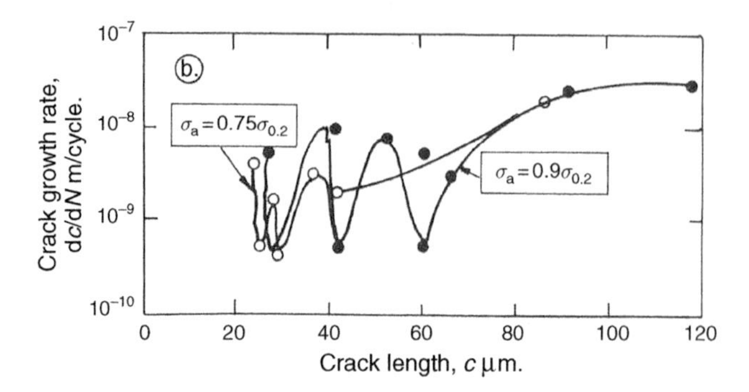

Fig. 13.2. Discontinuous crack growth in an aluminium alloy. Periods of slow growth were seen to coincide with grain boundaries (Blom et al., 1986).

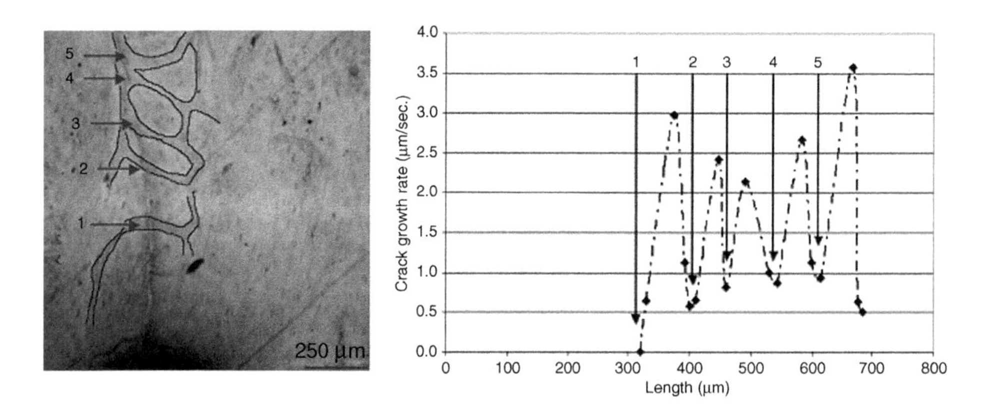

Fig. 13.3. Discontinuous crack growth in bone at constant applied stress. Periods of slow growth (numbered 1–5 on the graph) coincide with Volkmann's canals (enhanced by sketching onto the micrograph).

of steels for example (Qiao and Argon, 2003). In some materials there may be no microstructural barriers, but discontinuous growth occurs nevertheless: this is seen for example in amorphous polymers such as epoxy resin (Kinloch et al., 1983) and in rubbers (Gdoutos in ECF). In this so-called 'stick-slip' growth the operative mechanism is a periodic blunting and sharpening of the crack tip.

The theoretical argument can therefore be stated as follows. The FFM is valid at the continuum-mechanics level, unlike the PM and LM which make simplifying assumptions about the stress distribution, or the ICM with its simplified criterion for crack growth. The FFM also represents, albeit in simplified form, the real nature of discontinuous crack growth processes. Therefore the predictions of the FFM are both valid and realistic. In this argument, the PM and LM are assumed to work simply because they are approximations to the FFM. These approximations work because of the link between the form of the elastic stress field and the value of the stress intensity. In certain cases this link breaks down, notably when the size of the specimen becomes small as discussed in Section 3.3.6. In these cases the predictions of the PM and LM deviate from those of the FFM. However, in most practical cases this deviation does not occur and so the PM and LM can be used, which is very convenient because they are easier to implement for bodies of complex shape than the energy-based FFM.

We have seen that in some cases the critical stress parameter, σ_o in monotonic fracture or $\Delta\sigma_o$ in fatigue, is identical to the strength of a plain, unnotched specimen of the material. However, in quite a lot of situations this is not the case: we saw that σ_o takes a higher value for brittle fracture in polymers and metals, as does $\Delta\sigma_o$ for fatigue in polymers. This discrepancy can be explained by the fact that the PM and LM do not use the actual elastic/plastic stress distribution but rather the elastic one, so the stresses do not always correspond to actual stresses found in the material. To put this another way, we can say that K_c and L are the two fundamental parameters which define the TCD, whilst σ_o is a parameter whose value can be calculated from K_c and L, which enables us to use the stress-based approaches but which does not necessarily have any physical meaning.

13.4 The TCD and Other Fracture Theories

In this section we will examine the relationship between the TCD and other methods for predicting fracture. Many of these issues have already been mentioned elsewhere in this book and so will not be repeated in detail here. The aim of the present section is rather to gather these various observations together in order to be able to take a comprehensive view of the TCD in the context of other theories.

13.4.1 Continuum mechanics theories

It has already been pointed out that the most appropriate description for the TCD is as a continuum mechanics theory which has been modified by the addition of a length scale. Traditional continuum mechanics theories, which lack this length scale, can be considered as either stress based (i.e. a simple approach which defines failure in terms of the maximum stress in the body) or energy based (i.e. fracture mechanics). In many practical cases we see a transition from a regime of behaviour which is described by a stress-based argument to one which is described by an energy argument. For example, consider the effect of crack length on fracture stress in brittle ceramic materials (e.g. Fig. 13.4, taken from Chapter 5). For very small crack lengths, failure occurs at a constant stress, for large crack lengths it occurs at constant K. In between these two regimes is one in which neither theory is applicable. The TCD is capable of predicting behaviour throughout the whole range, coinciding with the stress and energy criteria in cases where L is large or small, respectively, compared to crack length.

We have defined the TCD to include both stress-based (PM, LM) and energy-based (FFM, ICM) methods, emphasising this transition. It is interesting to note that other problems in mechanics have both stress and energy solutions, such as plasticity, for which we use either the Tresca stress-based argument or the Von Mises energy-based approach.

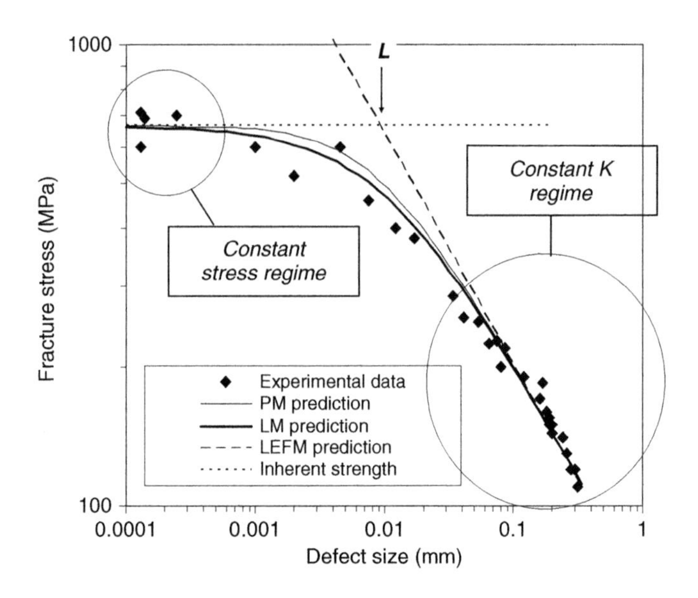

Fig. 13.4. Graph reproduced from Fig. 5.4, indicating regimes of constant stress and constant K.

13.4.2 Process zone models

These models were described in Section 4.6; based in the early work of Dugdale and Barenblatt, they are enjoying increased popularity in recent years thanks to developments in numerical modelling and computer power. Like the TCD, these methods are essentially continuum mechanics approaches, modified by creating a zone ahead of the crack or notch in which material is assigned special properties. This zone is a simple simulation of the process zone: the essential simplification being to reduce it to a line, rather than an area (the approach is generally confined to 2D problems at present). All the mechanisms of non-linear deformation and failure – mechanisms such as void growth or micro-cracking – are assumed to be represented by the stress–displacement curve (Fig. 4.4), which describes how the stress $p(u)$ varies with displacement u.

This stress/displacement curve can take a variety of forms, leading to different predictions. However, one can show that the approach coincides with that of the TCD in a particular case, as follows. Consider a crack under load, having a process zone ahead of it. The length of this process zone, λ, is not assumed to take any particular value in the model, it simply develops from the previous assumptions. The distribution of stress $p(x)$ with distance x from the process-zone tip will be as sketched in Fig. 13.5. We can find the stress intensity, K, associated with this process zone by using a crack-line loading argument (Lawn, 1993); the result is

$$K = -\left(\frac{2}{\pi}\right)^{1/2} \int_0^{\lambda} \frac{p(x)}{x^{1/2}} dx \tag{13.3}$$

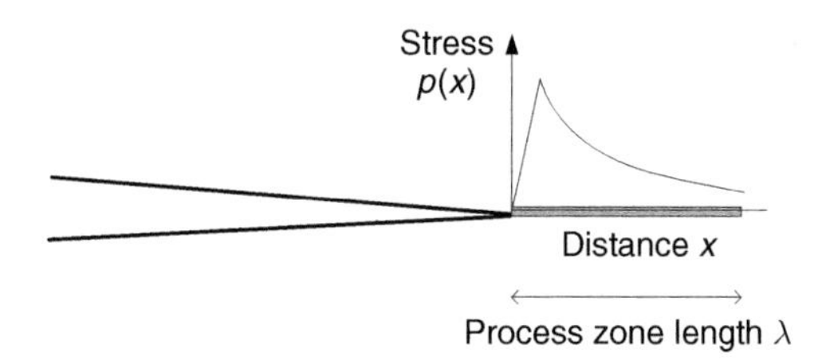

Fig. 13.5. Stress as a function of distance x in the process zone near the crack tip.

This will be numerically equal to the toughness K_c because it represents the stress intensity which must be overcome by external forces in order to propagate the crack. The solution to this equation depends on the form of the function $p(x)$. We consider a simple case in which this function is a constant, equal to the average value of $p(u)$. This is equivalent to modelling the material as rigid and perfectly brittle: considering an uncracked specimen we see that this average value of $p(u)$ must be equal to the tensile strength of the material, σ_u. Solving Eq. (13.3) gives us a value for the size of the process zone in this case:

$$\lambda = \frac{\pi}{8}\left(\frac{K_c}{\sigma_u}\right)^2 \tag{13.4}$$

Clearly this equation is identical to the one that we normally use to define L, with the exception of the constant $\pi/8$ (which is equal to 0.393) rather than $1/\pi$ (which is 0.318). This shows that predictions obtained using this form of the process zone model, in which a constant value is assumed for $p(u)$, will be almost identical to predictions using the PM, taking the critical distance at $L/2$. For other functions $p(u)$ we can show that the result for λ will, in general, always take the form of a function of (K_c/σ_u), though not necessarily that of Eq. (13.4).

In fact, equations such as 13.4, with or without the initial constant in π, are frequently used in process zone theory to indicate the general order of magnitude of the size of the process zone at failure (Bazant, 2004). It was noted above that the process zone is not a fixed size at failure in these models, but in fact its size is pretty much constant when located inside a large body, but changes when the dimensions of the body become small. This behaviour is similar to that of our combined stress-energy models as described in Section 3.3.6, such as the FFM/LM model. Such models are probably the closest parallel between the TCD and process zone theories.

13.4.3 Mechanistic models

A direct comparison can be made between the TCD and some mechanistic models of fracture processes. Perhaps the most obvious is the RKR model of cleavage in steels, which was discussed in Section 4.3. This model envisages failure to be controlled by the behaviour of a cracked carbide particle: since the particles are small and located in the grain boundaries the condition for failure ends up looking very much like the PM – a certain stress must occur at a certain fixed distance from the crack tip. Though superficially similar, these models have some important differences. The RKR model takes as its distance a real, microstructural parameter - the grain size d (actually it was found for various reasons that $2d$ was more appropriate) – and a real stress, the stress needed to propagate the carbide crack. When we applied the TCD to data in this area we found that the critical distance $L/2$ was generally somewhat smaller, and the critical stress considerably larger, than the values used by Ritchie et al. The reason for this is that we are using the elastic stress field, not taking account of the effects of plasticity. Of course, the RKR model must use the real elastic/plastic stress distribution, whilst the TCD should retain the assumption of linear elasticity, for the same reasons that this assumption is used in LEFM.

Other micro-mechanistic models also use length parameters which correspond to microstructural distances – for example, the critical length feature in void growth models such as those of Rice and Tracey (see Section 7.3.2) is the spacing between inclusions which act as sites for void nucleation. The size at which crazes form in PMMA is critical in modelling brittle fracture in that material (see Section 6.6). Again parallels can be drawn between these models and the TCD which are useful in general terms in gaining insight into the operative modes of fracture, but one should avoid trying to push these analogies too far.

13.4.4 Weibull models of cleavage fracture

In Section 7.3.2 we described how the PM-like RKR model of cleavage later evolved by first introducing probabilistic parameters to describe the distribution of carbide sites, in some cases losing the link with the original micro-mechanism as it developed into the Weibull-based Beremin model (also called the 'Local Approach'). The Beremin model is a continuum mechanics model in which the probability of failure is calculated by combining the probabilities of failure for each small volume element of the material within the plastic zone.

A Weibull model, in its simplest form, uses two material constants, σ^* and b, to calculate the probability of failure:

$$P_f = 1 - \exp\left[-\left(\frac{\sigma}{\sigma^*}\right)^b\right] \tag{13.5}$$

The constant σ^*, being a normalising parameter for the applied stress, is essentially a measure of the strength of the material, though it is also a function of the volume of material under stress. The exponent b dictates the degree to which the material's strength is stochastic, as opposed to deterministic: a material having a large value of b is essentially deterministic in its behaviour; it will fail if the stress exceeds a fixed value. If we combine this concept with the idea of a varying stress field – at a notch for instance – then we see that a high b value implies that failure will be controlled simply by the maximum stress value, thus the critical distance would be essentially zero and failure will always occur at the hot spot. For lower values of b there will exist the possibility that failure could occur from other locations away from the hot spot, in regions of slightly lower stress, and this effect will be augmented by the fact that an increasing volume of material will be involved.

This argument suggests a loose link between b and our critical distance L, but it is not an argument which can be pushed too far, because in the case of the Weibull analysis the shape of the stress distribution will also have an effect on the size of the potential failure zone. In Chapter 12 a distinction was made when talking about size effects between statistical size effects (controlled by changes in the stressed volume) and geometric size effects (controlled by stress gradient effects). These two effects may both operate simultaneously, but they should be recognised as different effects, requiring different methods of analysis.

Before we leave this discussion it is useful to recall an observation made in Section 7.3.2: in order to obtain accurate predications from the Beremin model, the size of volume elements is usually fixed at a specific value, V_0. The value used for V_0 is typically of the same order of magnitude as the grain size but it is essentially an empirical parameter chosen to allow the model give accurate predictions of experimental data. It is not clear what the reason for this procedure is, but certainly this volume averaging process is very reminiscent of the TCD, implying that a critical distance philosophy is being used to smooth out stress-gradient effects, in conjunction with the statistical analysis. One can find the use of critical distance ideas such as this in almost every modern theory of fracture.

13.4.5 Models of fatigue crack initiation and growth

Modelling of fatigue processes was considered previously in Chapter 9. Predictions of fatigue behaviour normally consider the two processes of crack initiation and crack growth separately. The crack growth stage is further subdivided into a long-crack regime where fracture mechanics can be used, and a short-crack regime where modifications are needed to fracture mechanics. Currently the principal modification is that made to allow for crack closure, which is assumed to alter the effective value of ΔK and therefore the driving force for crack growth.

There are clear links to be found between the behaviour of short cracks and the TCD: the crack length at which short crack behaviour is most noticeable is the same as the value of L, as shown in Fig. 9.11 for instance. This value is typically somewhat larger than the grain size of the material, usually by a factor in the range 3–10. However, it is rather curious that a TCD method such as the LM or PM is capable of predicting the effect of crack length, as in Fig. 9.11, despite the fact that closure is not included in the TCD. A possible explanation, which has been suggested previously (DuQuesnay et al., 1986), is that L may correspond to the distance over which a crack must grow in order for closure to develop. It is worth noting that exactly the same kind of short-crack effect can be found in the data on brittle fracture (see for example Fig. 5.4) despite the fact that the mechanisms involved are completely different.

A phenomenon commonly observed when fatigue cracks grow from notches is that crack growth can be initially quite fast, slowing down as the crack moves away from the notch. It is generally accepted that this is an effect of short-crack growth: near the fatigue limit this crack may stop growing, becoming a 'non-propagating crack'. The lengths of such cracks are invariably within the short-crack regime, and I have argued in a previous paper that a link can be made to the LM, assuming that the length of the non-propagating crack is $2L$ and that the average stress over this length is a measure of the driving force for crack propagation. The relevant theory is the crack-line-loading approach, in which stress intensity can be calculated based on the stresses which exist across the crack faces. This argument was developed in detail elsewhere (Taylor, 2001). There are, however, some deficiencies in this approach: the crack-line loading method is only approximate for the case of sharp notches, becoming increasingly inaccurate as the notch root radius decreases. So whilst predictions of the same general form could be achieved in this way, they were not as accurate as the predictions of the TCD itself. And in any case this is not an argument which would apply to very blunt notches, for which non-propagating cracks of this type do not occur, though smaller, grain-sized cracks do arise.

13.5 Values of L

We have seen that L can take different values in different materials and different failure processes. Commonly encountered values range from microns to millimetres, and there are reasons to believe that in some cases L may be as small as the atomic separation (Pugno and Ruoff, 2004) or as large as several metres (Dempsey et al., 1999). It is interesting to consider why a particular value of L occurs, as this may give us insight into the operative mechanisms of failure.

Small values of L are associated with microstructural features such as the grain size: we saw this for the case of brittle fracture both in engineering ceramics such as silicon nitride (Section 5.2.1) and in steel (Section 7.2.3). When we find that L is equal to the grain size, or a small multiple thereof, this implies that the grain boundary is acting as an effective barrier to crack propagation: the discontinuity in crack growth occurs at the level of the grains. Amorphous polymers such as PMMA have no microstructure: here L values of the order of 100 μm may correspond to the size at which crazes form.

Larger values of L can be associated with two different observed mechanisms. The first is the growth of a crack which becomes non-propagating. This phenomenon is common in fatigue and in fretting; it is normally explained using a resistance curve (R-curve) concept (illustrated in Fig. 9.4) whereby the threshold stress intensity for crack growth increases with increasing crack length. This creates the situation where the crack may initially be able to grow, but then stops as its stress intensity drops below the threshold. The lengths to which such cracks grow will depend on the mechanism which is creating the shape of the R-curve, that is the mechanism of toughening in the material. In metal fatigue this is generally attributed to crack closure; closure itself can be created by several different means, some of which are strongly related to microstructure (e.g. crack-face roughening due to deviations at the grain boundaries) whilst others are determined by continuum properties such as the development of a plastic wake, though even this is probably affected by the yielding behaviour of individual grains at this scale of operation. This implies that the full, long-crack threshold value will be established over lengths which are about an order of magnitude larger than the grain size, since the crack will have to grow through several grains to experience the benefits of the toughening mechanism. This is exactly what we find in practice. Because this phenomenon is also responsible for short-crack behaviour it is not surprising that the length of the curved portion of the R-curve is also a small number of multiples of L. This argument explains why in metal fatigue and in fretting, L values of the order of 3–10 times the grain size are common. A similar mechanism may also be operating during the monotonic fracture of some relatively brittle materials which make use of toughening mechanisms, such as bone. We found L values of the order of 1 mm (Section 8.7), which is consistent with the increase in toughness over crack lengths of several millimetres seen in R-curves (Nalla et al., 2005).

A second phenomenon associated with large L is the creation of a damage zone; this can be seen in composite materials (Chapter 8), in building materials (Section 5.3) and in some polymers such as polystyrene which display multiple crazing. The largest values of L, of the order of millimetres and centimetres, are associated with this phenomenon. This can be viewed in the same light as the R-curve, since it represents mechanisms in the material by which toughening is achieved, for example the dissipation of energy through micro-cracking and the establishment of supporting bridges behind crack faces (Nalla et al., 2003). The difference is only that we see a diffuse area of damage rather than a single crack. Again the size of this zone tends to be at least an order of magnitude larger than the size of microstructural features, because many such features are involved in its creation.

It may be true – though I have not seen this argument demonstrated anywhere – that a principle operates here similar to that of the R-curve, by which the toughening effect of the damage zone is exhausted if it grows above a certain size. Certainly it can be demonstrated that the size of the damage zone at failure is approximately constant, at least in cases where the specimen size is much larger than that of the damage zone, and this emerges in theoretical models such as process zone models and combined stress-energy models such as our FFM/LM, both of which were discussed above.

13.6 The Value of σ_o/σ_u

It was noted earlier that the magnitude of the critical stress σ_o is unlikely to have any physical significance, since it relates to the value of the elastic stress in a region where the actual physical stress will be modified by plasticity, damage and other non-linear effects: in a word, σ_o does not actually exist at the relevant location. However, it is interesting to look at values of the ratio between this stress and the actual failure stress of the material, σ_u (or the fatigue limit in the case of fatigue failure). The ratio σ_o/σ_u takes a value of unity for monotonic fracture in ceramics and in composites, and also in metal fatigue. It takes a higher value for monotonic fracture in polymers and metals: values in the range 2–4 are common, though values outside this range have also been measured, giving a continuous spectrum of values from 1 up to almost 10.

Before commenting on the reason for these values we should consider more carefully what is actually meant by the plain specimen strength σ_u. The assumption is that this is the strength of material containing no defects or stress concentration features of any kind. In practice this may be impossible to achieve, since some features such as porosity and inclusions are inevitable in certain processing operations and so should perhaps be considered as an integral part of the material itself. To be more precise, we can define

σ_u as the strength of material containing defects which are all much smaller than L and so will not be expected to exert an effect: we saw that this was possible to achieve even when L is very small, such as in engineering ceramics (Section 5.2.1). However, materials made by normal processing routes may contain larger defects and therefore have tensile strengths which are less than this ideal σ_u value, and this may be one reason for an apparent difference between σ_u and σ_o in some cases. If this happens, then the relationship between defect size and strength would be expected to show a cut-off rather than a smooth transition, as illustrated in Fig. 5.11 for the case of ceramics.

However, there are many cases where σ_o is certainly larger than σ_u and this creates some interesting phenomena, notably the existence of non-damaging notches and cracks. Why is this occurring and is it linked in some way to the operative failure mechanisms? Though a completely clear picture has not yet emerged, a common feature of cases where $\sigma_0/\sigma_u > 1$ is the existence of plasticity. We found the largest values of this ratio in metals, where plasticity is the main toughening mechanism, and values of unity in the most brittle materials – ceramics – where plasticity plays no role. Composites, at least those made using long continuous fibres of brittle materials such as carbon, also make no significant use of plasticity as a toughening mechanism, so this explains the value of σ_0/σ_u in their case. Polymers such as polycarbonate ($\sigma_0/\sigma_u = 2$), though they fail in a brittle manner, do develop plasticity before failure, whilst in polystyrene which has a σ_0/σ_u value only slightly greater than unity the main mechanism is multiple crazing. The PMMA fits rather less well into this picture, having $\sigma_0/\sigma_u = 2$ and showing little plasticity at room temperature; however, the stress–strain curves for this material do show some non-linearity, due either to plastic deformation or non-linear elastic behaviour, and we found that at an elevated temperature of 60°C, where plasticity is clearly displayed, the ratio σ_0/σ_u rose to 2.9. In HCF of metals this ratio turns out to be 1.0, despite the existence of plastic deformation, but in this case the extent of plasticity is very limited, the plastic zone size being much smaller than L.

Thus, whilst there are some cases which need more careful attention, a general picture emerges in which high ratios of σ_0/σ_u are associated with materials which use plasticity to achieve toughness. This can be explained by noting that, in these materials, plain-specimen failure occurs in a different way from the failure of notched specimens. Instead of failing by cracking, ductile materials fail by extensive plastic strain, involving mechanisms such as shear deformation and necking. Since the mechanism has changed, it is not surprising to find that the value of σ_u is different from that of σ_o. This is true even for a material such as steel at low temperature, where, though failure occurs by brittle cleavage, nevertheless in plain specimens this cannot happen until after the yield strength has been exceeded, since plastic deformation is necessary to form the initial cracks.

13.7 The Range and Limitations of the TCD

This final section considers the applicability of the TCD; under what circumstances can it be said to be a valid approach for failure prediction? This question can be answered in two ways: on theoretical grounds and with reference to the experimental data.

In defining the theoretical validity of the TCD it is useful to start by looking at the theoretical basis for fracture mechanics, which has been very well outlined by Broberg in the Introduction to his book *Cracks and Fracture* (Broberg, 1999). A concept which defines the limitations of LEFM is that of 'local control' versus 'global control'. A fracture process – in this case the propagation of a crack – is said to be under local control if it can be predicted based only on conditions in the material close to the crack tip. By contrast, global control exists if conditions remote from the crack tip play a role, for example at the specimen boundaries. Classic, unstable brittle fracture conforms to local control, but failure which involves a period of stable crack growth may create global control due, for example, to relaxations in stress at the specimen boundary. Local control can be expected to break down if the plastic zone or process zone associated with the crack is no longer small compared to the specimen size, since conditions inside the zone will be affected by the proximity of the specimen boundaries; thus the concept of local control leads on to the well-known condition of small-scale yielding which is a crucial requirement for LEFM validity.

Another situation in which local control breaks down is fatigue in the finite-cycle regime. Local control may be said to apply to the fatigue limit, since it is assumed that, once a crack begins to grow from the notch, it will continue growing and eventually cause failure. However, the number of cycles to failure cannot be predicted only from the initial stress conditions at the notch: this is clear from the case where the notch is a sharp crack, for which the initial crack length is now a factor in determining N_f. The size of the specimen also plays a role, though a less important one.

Local control is implied when we use the stress-based TCD methods – the PM and LM – since we only look at stresses in a region near the hot spot defined by the magnitude of L. Thus we can expect that these methods may not work when applied in cases where the size of the plastic zone is a significant proportion of the specimen width, such as the failure of relatively tough, ductile metals or LCF. The same can be said for the size of the zone of damage in brittle materials, implying that the TCD may not be applicable to very small specimens of these materials. The energy-based TCD methods, that is the ICM and, especially, the FFM, use LEFM as their basis and so can also be expected to encounter difficulties when local control is lost. In this respect, there is no theoretical reason why FFM cannot be extended to include global control, using the methods of elastic-plastic fracture mechanics instead of LEFM; to date, however, this exercise has not been attempted.

When we examine the experimental data, we find ample evidence to show that the TCD can be used successfully when the small-scale yielding criterion is fulfilled. In these cases, the TCD provides an excellent extension to LEFM: where LEFM can accurately predict the behaviour of a long crack, TCD can also predict the behaviour of a small crack, notch or other geometric feature. Likewise HCF is a valid area for the TCD and one in which it performs excellently. We can also find clear cases in brittle materials where the TCD becomes invalid as a result of the process zone size becoming too large with respect to the specimen size. This is discussed in Section 5.3 in respect of building materials, for which even specimens of moderate size encounter this problem. The situation is easy to diagnose because the value of L will become large compared to specimen width, perhaps even exceeding it, which makes methods such as the LM impossible to implement. In these cases we found that a modified form of the TCD using two conditions, such as the FFM and LM, was able to provide good predictions; this is promising, though the theoretical basis for this approach needs more careful analysis.

Perhaps more interestingly, the experimental data reveals several instances where the TCD gives good predictions despite the fact that its theoretical validity has been violated. Thus, Susmel shows that the TCD can predict medium-cycle fatigue in steels despite the theoretical objections mentioned above (Susmel and Taylor, 2005), and some workers have used critical-distance approaches rather like the TCD to predict ductile fracture despite the existence of large-scale plasticity (Schluter et al., 1996). It is not uncommon to find that a method can be used outside the strict limits of its validity; at the end of the day the most important test of any theory is its ability to predict the data. However, one should certainly be cautious in using the theory in these areas, at least until its success is better understood. In any case, from an engineering point of view, it is unwise to make use of any theory of fracture until it has been demonstrated to work when applied to test specimens of similar material and geometry to that of the component under consideration.

13.8 Concluding Remarks

This chapter has presented my own thoughts on the theoretical development of the TCD and its relationship to other theories of fracture. I hope that it will stimulate others to make contributions in this area. History clearly shows that scientific theories first become adopted because they are shown to be useful, and only later is a full understanding developed of their theoretical basis. This was certainly the case for LEFM and I am sure that the same will be true for the TCD.

References

Adib, H. and Pluvinage, G. (2003) Theoretical and numerical aspects of the volumetric approach for fatigue life prediction in notched components. *International Journal of Fatigue* **25**, 67–76.

Bazant, Z.P. (2004) Quasibrittle fracture scaling and size effect. *Materials and Structures* **37**, 1–25.

Blom, A., Hedlund, A., Zhao, W., Fathulla, Weiss, B., and Stickler, R. (1986) Short fatigue crack growth behaviour in Al2024 and Al7475. In *The Behaviour of Short Fatigue Cracks (EGF1)* (Edited by Miller, K.J. and delosRios, E.R.) pp. 37–66. MEP, London.

Broberg, K.B. (1999) *Cracks and fracture.* Academic Press, San Diego, California, USA.

Dempsey, J.P., Adamson, R.M., and Mulmule, S.V. (1999) Scale effect on the in-situ tensile strength and failure of first-year sea ice at Resolute, NWR. *International Journal of Fracture, special issue on fracture scaling* 9–19.

DuQuesnay, D.L., Topper, T.H., and Yu, M.T. (1986) The effect of notch radius on the fatigue notch factor and the propagation of short cracks. In *The Behaviour of Short Fatigue Cracks (EGF1)* (Edited by Miller, K.J. and delosRios, E.R.) pp. 323–335. MEP, London.

Hazenberg, J.G., Taylor, D., and Lee, T.C. (2006) Mechanisms of short crack growth at constant stress in bone. *Biomaterials* **27**, 2114–2122.

Kinloch, A.J., Shaw, S.J., Tod, D.A., and Hunston, D.L. (1983) Deformation and fracture behaviour of a rubber-toughened epoxy: 1. Microstructure and fracture studies. *Polymer* **24**, 1341–1354.

Lawn, B. (1993) *Fracture of brittle solids.* Cambridge University Press, Cambridge.

Mikhailov, S.E. (1995) A functional approach to non-local strength conditions and fracture criteria. *Engineering Fracture Mechanics* **52**, 731–754.

Nalla, R.K., Kinney, J.H., and Ritchie, R.O. (2003) Mechanistic fracture criteria for the failure of human cortical bone. *Nature Materials* **2**, 164–168.

Nalla, R.K., Kruzic, J.J., Kinney, J.H., and Ritchie, R.O. (2005) Mechanistic aspects of fracture and R-curve behaviour in human cortical bone. *Biomaterials* **26**, 217–231.

Pugno, N. and Ruoff, R. (2004) Quantized fracture mechanics. *Philosophical Magazine* **84**, 2829–2845.

Qiao, Y. and Argon, A.S. (2003) Cleavage cracking resistance of high angle grain boundaries in Fe-3%Si alloy. *Mechanics of Materials* **35**, 313–331.

Schluter, N., Grimpe, F., Bleck, W., and Dahl, W. (1996) Modelling of the damage in ductile steels. *Computational Materials Science* **7**, 27–33.

Susmel, L. and Taylor, D. (2005) The theory of critical distances to predict fatigue lifetime of notched components. In *Advances in Fracture and Damage Mechanics IV* (Edited by Aliabadi, M.H.) pp. 411–416. EC, Eastleigh UK.

Taylor, D. (2001) A mechanistic approach to critical-distance methods in notch fatigue. *Fatigue and Fracture of Engineering Materials and Structures* **24**, 215–224.

Taylor, D., Cornetti, P., and Pugno, N. (2005) The fracture mechanics of finite crack extension. *Engineering Fracture Mechanics* **72**, 1021–1038.

Author Index

Subject Index

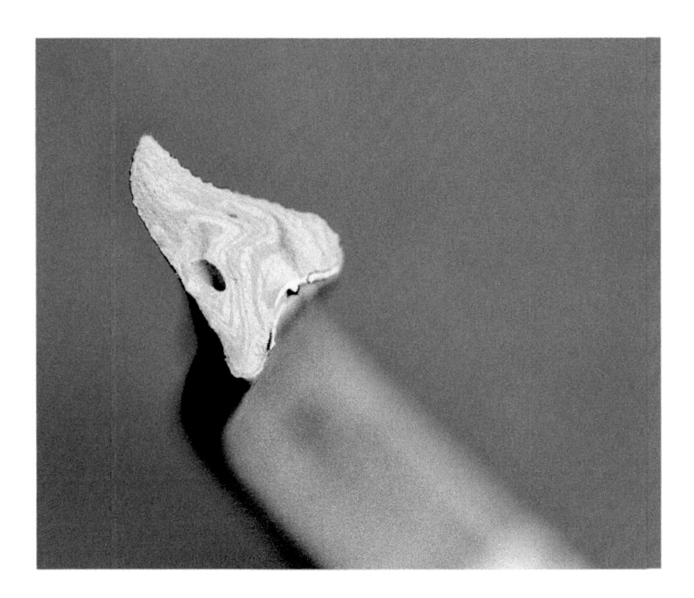

Colour Plate 1. Brittle fracture in the handle of a jug.

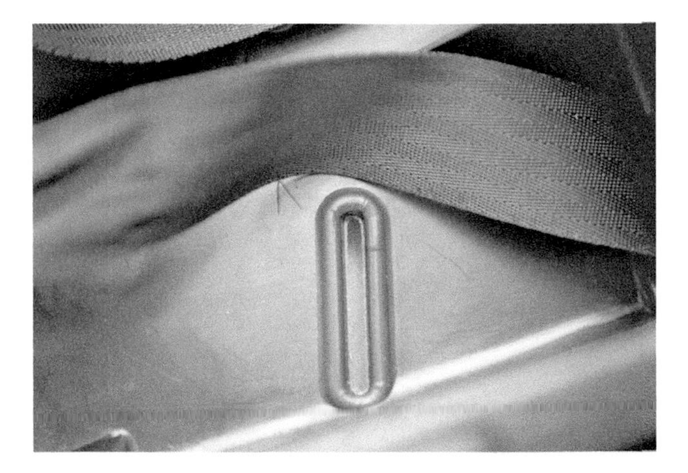

Colour Plate 2. A slot in the back of a child's car seat; the seat belt, shown above, normally passes through the slot. In an accident, the stress-concentrating effect of the ends of the slot caused a brittle fracture.

Colour Plate 3. Brittle fracture in an aluminium ladder, initiated at a stress concentration where the rung meets the stile.

Colour Plate 4. Fatigue failure at a sharp corner in the landing gear of a 737 aircraft.

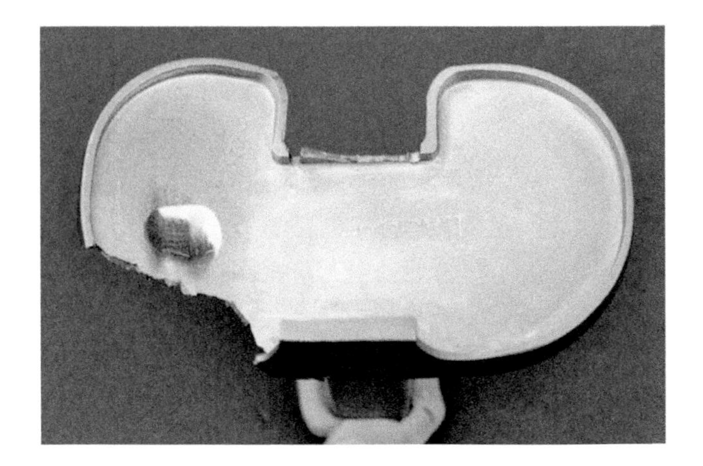

Colour Plate 5. Fatigue failure in the metal component of an aritifical knee joint, originating at the edge of a contact patch.

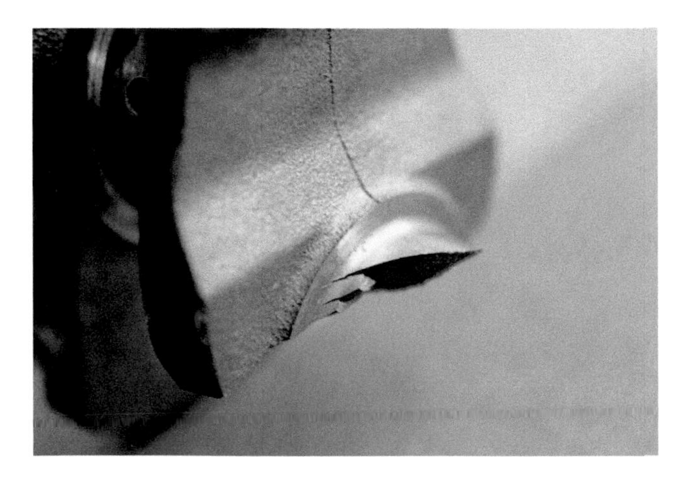

Colour Plate 6. In this automotive crankshaft loaded in cyclic torsion, cracks developed from the bearing corner and grew at approx.45° to the shaft axis.

9780080444789